Physico-Chemical Principles of Color Chemistry

Advances in Color Chemistry Series—Volume 4

Physico-Chemical Principles of Color Chemistry

Edited by

A. T. PETERS
Chemistry & Chemical Technology,
University of Bradford,
Bradford
UK

and

H. S. FREEMAN
Dept. of Textile Engineering, Chemistry & Science,
North Carolina State University,
Raleigh
USA

BLACKIE ACADEMIC & PROFESSIONAL
An Imprint of Chapman & Hall
London · Glasgow · Weinheim · New York · Tokyo · Melbourne · Madras

Published by
Blackie Academic & Professional, an imprint of Chapman & Hall,
Wester Cleddens Road, Bishopbriggs, Glasgow G64 2NZ

Chapman & Hall, 2–6 Boundary Row, London SE1 8HN, UK

Blackie Academic & Professional, Wester Cleddens Road, Bishopbriggs, Glasgow G64 2NZ, UK

Chapman & Hall GmbH, Pappelallee 3, 69469 Weinheim, Germany

Chapman & Hall USA, 115 Fifth Avenue, Fourth Floor, New York, NY 10003, USA

Chapman & Hall Japan, ITP-Japan, Kyowa Building, 3F, 2-2-1 Hirakawacho, Chiyoda-ku, Tokyo 102, Japan

DA Book (Aust.) Pty Ltd, 648 Whitehorse Road, Mitcham 3132, Victoria, Australia

Chapman & Hall India, R. Seshadri, 32 Second Main Road, CIT East, Madras 600 035, India

First edition 1996

© 1996 Chapman & Hall

Typeset in 10/12pt Times by Keytec Typesetting Ltd, Bridport, Dorset

Printed in Great Britain by The University Press, Cambridge

ISBN 0 7514 0210 9

Apart from any fair dealing for the purposes of research or private study, or criticism or review, as permitted under the UK Copyright Designs and Patents Act, 1988, this publication may not be reproduced, stored, or transmitted, in any form or by any means, without the prior permission in writing of the publishers, or in the case of reprographic reproduction only in accordance with the terms of the licences issued by the Copyright Licensing Agency in the UK, or in accordance with the terms of licences issued by the appropriate Reproduction Rights Organization outside the UK. Enquiries concerning reproduction outside the terms stated here should be sent to the publishers at the Glasgow address printed on this page.

The publisher makes no representation, express or implied, with regard to the accuracy of the information contained in this book and cannot accept any legal responsibility or liability for any errors or omissions that may be made.

A catalogue record for this book is available from the British Library

Library of Congress Catalog Card Number: 95-79037

∞ Printed on acid-free text paper, manufactured in accordance with ANSI/NISO Z39.48-1992 (Permanence of Paper)

Preface

At the beginning of this series of volumes on Color Chemistry, the editors pointed to a number of events that have served as stimuli for technological advances in the field, thus preventing dyestuff manufacturing from becoming what might otherwise be viewed by now as a 'sunset industry'. The volumes which followed have provided ample evidence for our belief that the field of colour chemistry is very much alive, though arguably in need of further stimulus. For instance, a viable approach to the design of new chromophores and to the design of metal-free acid, direct, and reactive dyes having fastness properties comparable to their metallized counterparts represent the kind of breakthroughs that would help ensure the continued success of this important field.

While it must be acknowledged that serendipity 'smiled' on our discipline at its inception and has repeated the favor from time to time since then, few would argue against the proposition that most of the significant advances in the technology associated with any scientific discipline result from research designed to enhance our understanding of the fundamental causes for experimental observations, many of which are pursued because they are unexpected, intriguing and intellectually stimulating. Little reflection is required for one who knows the history of the dyestuff industry to realize that this is certainly true in the colour chemistry arena, as it was basic research that led to fiber-reactive dyes, dyes for high technology, and modern synthetic organic pigments. In fact, the final chapter in the previous volume provided a clear indication of the potential utility of the fundamental principles of quantum chemistry in addressing problems of interest to dyestuff chemists. With these thoughts in mind, we have chosen to make the fourth volume in the series a book that focuses on some fundamental considerations important to the design and development of technically useful new organic dyes and pigments.

Although there are many commercial azo dyes and pigments currently being manufactured in very high reaction yields, there exists a significant number of azo dyes derived from heteroaromatic amines that are manufactured in yields below 80%. While this is a well-known problem, a viable general solution has not been developed. As a result, the synthesis of disperse dyes from diazonium salts of heteroaromatic amines continues to be an important research problem. In addition to providing an update on basic research on this specific problem, recent developments pertaining to the conduct and monitoring of diazotization and coupling reactions are covered in the first chapter. In the chapter which follows, the kinetics associated with dyeing with disperse dyes is reviewed. The review covers a number of mathematical models that have been formulated to define dye diffusion in a heterogeneous system, dye dissolution processes, and dye transport phenomena. Chapter two also contains a summary of the heterogeneous kinetics associated with dyeing cellulose with fiber reactive dyes, and includes fundamental considerations such as fiber porosity and dye fixation versus hydrolysis.

It is well-known that dye aggregation plays an important role in the fastness properties of synthetic dyes on textile substrates. The third chapter covers factors responsible for dye aggregation and the stability of the formed aggregates. Difficulties associated with measuring aggregation of anionic (acid, direct, and fiber reactive) azo dyes in an aqueous medium and the relationships between dye aggregation and the development of dyes for a liquid crystalline medium are emphasized. Similarly, in a chapter dealing with the contribution of crystal form/habit to the properties of organic colourants, the authors illustrate how the physical characteristics of organic pigments affect their technical properties and how those characteristics can be controlled to give predictable performance properties. Among the pigment characteristics assessed are crystal size and aggregation.

The present volume is concluded with review chapters covering fundamental principles pertaining to the solubility, photostability, and genotoxicity of synthetic dyes. These three chapters also provide a summary of the details associated with currently employed experimental methods, a summary of available experimental data, structure-property-relationships, and implications for future dye design and development.

It is anticipated that this volume will provide fertile soil for the germination of creative ideas and new research programs that afford colour chemistry the opportunity to expand and flourish in the 21st century. The editors wish to express their appreciation to the group of researchers who captured our vision of the need for a book of this type and took time to author a chapter.

A. T. P.
H. S. F.

Contributors

S. Banerjee	Institute of Paper Science and Technology, Atlanta, Georgia 30318, USA
G. Baughman	US Environmental Protection Agency, Environmental Research Laboratory, Athens, Georgia 30605, USA
C. M. Brennan	Zeneca Specialities, Hexagon House, Blackley, Manchester M9 8ZS, UK
J. F. Bullock	Zeneca Specialities, Hexagon House, Blackley, Manchester M9 8ZS, UK
D. J. Edwards	Zeneca Specialities, PO Box 42, Hexagon House, Manchester M9 8ZS, UK
J. Esancy	Sandoz Chemicals Corporation, Box 669245, Charlotte, NC 28266, USA
H. S. Freeman	North Carolina State University, Department of Textile Engineering, Chemistry and Science, College of Textiles, Box 8301, Raleigh, North Carolina 27695, USA
D. Hinks	North Carolina State University, Department of Textile Engineering, Chemistry and Science, College of Textiles, Box 8301, Raleigh, North Carolina 27695, USA
A. F. M. Iqbal	Ciba Geigy Limited, Pigments Division, Research Center, Marly CH-1723, Switzerland
A. A. Jaber	Physics Department, Keele University, Keele, Staffs ST5 5BG, UK
N. Kuramoto	Division of Organic Materials, Osaka Prefectural Industrial Technology Research Institute, 2-1-53 Enokojima, Nishi-ku, Osaka 550, Japan
A. Mahendrasingham	Physics Department, Keele University, Keele, Staffs ST5 5BG, UK
R. B. McKay	Ciba Pigments, Hawkhead Road, Paisley PA2 7BG, UK

B. Medinger	Ciba Geigy Limited, Pigments Division, Research Center, Marly CH-1723, Switzerland
A. P. Ormerod	Chemical Sciences Institute, Science Research Institute, Salford University, Salford M5 4WT, UK
T. A. Perenich	University of Georgia, Department of Textiles and Merchandising and Interiors, Athens, Georgia 30602, USA
A. T. Peters	Reader in Colour Chemistry, Chemistry & Chemical Technology, University of Bradford, Bradford, West Yorkshire BD7 1DP, UK
P. Rys	Technisch-Chemische's Laboratorium, Eidgenossische Technische Hochshule, Zurich CH-8092, Switzerland
G. J. T. Tiddy	Chemical Sciences Institute, Science Research Institute, Salford University, Salford M5 4WT, UK

Contents

1	**Diazotization of weakly basic aromatic amines: kinetics and mechanism** P. RYS	**1**
1.1	Prologue	1
1.2	Dependence of diazotization rates on acidity	7
	1.2.1 Carbocyclic aromatic amines	8
	1.2.2 Heterocyclic aromatic amines	17
	1.2.3 Diazotization of cyano-substituted aromatic amines	21
1.3	Diazotization under industrial reaction conditions	23
	1.3.1 Analytical methods and practical units	24
	1.3.2 2,6-Dichloro-4-nitroaniline	25
	1.3.3 2-Bromo-4,6-dinitroaniline	26
	1.3.4 2-Bromo-6-cyano-4-nitroaniline	27
	1.3.5 2-Aminothiazole	29
	1.3.6 2-Amino-4-chloro-5-formylthiazole	31
	1.3.7 2-Amino-4-chloro-5-formyl-3-thiophenecarbonitrile	34
	1.3.8 Evaluation of optimum reaction and process conditions	34
1.4	Conclusions	41
	Acknowledgements	41
	References	42

2	**The heterogeneous kinetics of reactive and disperse dyeing** C. BRENNAN and J. BULLOCK	**44**
2.1	Introduction	44
2.2	The heterogeneous kinetics of dyeing polyester	44
	2.2.1 Introduction	44
	2.2.2 Description of the dissolution process	47
	2.2.3 Practical factors affecting the rate of dissolution	49
	2.2.4 Finite or infinite kinetics?	50
	2.2.5 Transport to the fibre surface	52
	2.2.6 Practical factors which affect transport to the fibre surface	56
	2.2.7 Description of the process of diffusion within the fibre	56
	2.2.8 Practical factors affecting the rate of diffusion in the fibre	58
	2.2.9 The heterogeneous processes of levelling (migration)	59
2.3	The heterogeneous kinetics of dyeing cellulose with fibre reactive dyes	61
	2.3.1 Introduction	61
	2.3.2 Uptake of dye from bulk solution to internal fibre	65
	2.3.3 Fibre diffusion	70
	2.3.4 Diffusion with simultaneous chemical reaction	73
2.4	Summary	79
	References	80

3 Aggregation and lyotropic liquid crystal formation of anionic azo dyes for textile fibres 83
D. J. EDWARDS, A. P. ORMEROD, G. J. T. TIDDY, A. A. JABER and A. MAHENDRASINGHAM

3.1	Anionic azo dyes	83
	3.1.1 Acid dyes	83
	3.1.2 Direct dyes	83
	3.1.3 Reactive dyes	84
3.2	Aggregation	84
	3.2.1 Factors governing the formation and stability of aggregates	84
	3.2.2 Measurement techniques	86
	3.2.3 Anionic azo dye aggregates	88
3.3	Lyotropic liquid crystal formation	90
	3.3.1 Liquid crystals	90
	3.3.2 Measurement techniques	91
	3.3.3 Lyotropic liquid crystals	94
	3.3.4 Chromonic liquid crystals	94
3.4	Textile dyeing	103
	References	105

4 Contribution of crystal form/habit to colorant properties 107
A. IQBAL, B. MEDINGER and R. B. McKAY

4.1	Introduction	107
4.2	Molecular and crystal lattice characteristics of organic pigments and their effect on pigmentary performance	108
	4.2.1 Molecular constitution	108
	4.2.2 Crystal lattice properties	112
	4.2.3 Crystal habit	125
4.3	Control of crystal form/habit	136
	4.3.1 In-synthesis control	137
	4.3.2 Post-synthesis control	137
4.4	Concluding remarks	141
	References	143

5 Dye solubility 145
G. L. BAUGHMAN, S. BANERJEE and T. A. PERENICH

5.1	Introduction	145
	5.1.1 Scope of the review	145
5.2	Solubility measurement	146
	5.2.1 Difficulties in solubility measurement	146
	5.2.2 Measurement methods	149
5.3	Measured solubilities	151
	5.3.1 Solubility in water	151
	5.3.2 Solubility in organic solvents	180
5.4	Estimation methods	184
	5.4.1 Computational	184
	5.4.2 Solubility parameters	187
	5.4.3 Regression analysis	190
	5.4.4 Empirical/estimation	192
	Acknowledgment	193
	References	193

6	**The photodegradation of synthetic colorants**	**196**
	N. KURAMOTO	

    6.1   Introduction   196
    6.2   Basic photochemical principles   197
          6.2.1   Excited electronic states and photochemistry   197
          6.2.2   General photochemical reactions of dyes   199
          6.2.3   Photosensitized oxidation   201
    6.3   Physical factors affecting dye lightfastness   202
    6.4   Mechanisms of dye photodegradation   204
          6.4.1   Azo dyes   204
          6.4.2   Anthraquinone dyes   212
          6.4.3   Triphenylmethane dyes   217
          6.4.4   Indigoid dyes   220
          6.4.5   Quinophthalone dyes   221
          6.4.6   Fluorescent dyes   223
          6.4.7   Miscellaneous dyes   224
    6.5   Catalytic fading   225
    6.6   Photodecomposition and stabilization of dye–polymer matrices   228
          6.6.1   Natural fibers   228
          6.6.2   Polyamide fibers   232
          6.6.3   Polyester fibers   235
          6.6.4   Polypropylene fibers   235
          6.6.5   Sensitized photodegradation of dye–polymer systems   236
    6.7   Photochemistry and photostabilization of functional dyes   237
    6.8   Conclusions   244
    Acknowledgment   244
    References   244

7	**Genotoxicity of azo dyes: bases and implications**	**254**
	H. S. FREEMAN, D. HINKS and J. ESANCY	

    7.1   Introduction   254
    7.2   Test methods   254
          7.2.1   The *Salmonella* mutagenicity assay   255
          7.2.2   The Prival modification   257
          7.2.3   Rat bacterial reduction system   258
          7.2.4   Other mutagenicity test methods for azo dyes   258
    7.3   Aromatic amines   260
          7.3.1   Benzidines and related amines   260
          7.3.2   Anilines and related amines   264
    7.4   Mechanisms and metabolism   265
          7.4.1   Monoazo dyes   267
          7.4.2   Disazo dyes   273
    7.5   Structure–activity relationships   275
          7.5.1   Empirical   275
          7.5.2   Computer-assisted   280
    7.6   Environmental   283
          7.6.1   General   283
          7.6.2   Metallized dyes   284
    7.7   Conclusions   289
    References   290

Index   **293**

1 Diazotization of weakly basic aromatic amines: kinetics and mechanism

P. RYS

1.1 Prologue

The diazotization of weakly basic carbocyclic and heterocyclic aromatic amines is a key aspect of the industrial production of a whole series of disperse and cationic monoazo dyes. Until the early 1950s the colour gamut of the existing disperse monoazo dyes was rather limited and covered only the spectrum range from yellow to red. At that time, no blue monoazo disperse dyes were commercially available. Instead, almost all blue disperse dyes were based on anthraquinone. Subsequently, however, these relatively expensive anthraquinone-based dyes have been gradually replaced by blue, greenish-blue and brilliant bluish-red azo compounds. According to Zollinger,[1] the proportion of monoazo disperse dyes has since increased from about 50% to 70%, while the percentage of anthraquinone disperse dyes has decreased from 25% to 15%. This success of the much cheaper azo dyes can be attributed to various factors, such as to the simplicity of their synthesis, to the more extensive possibilities for structural variations and to the often very high molar extinction of azo compounds.[2]

The search for bluish-red, blue and greenish-blue monoazo disperse dyes was initiated by the observation that such dyes can basically be formed by a combination of very acidic diazo components with highly basic coupling components (Figure 1.1). Typical examples of such dyes are of the types illustrated by the structures **1**, **2** and **3**.

However, due to shortcomings in the diazotization and coupling reaction not every desired combination has proved to be technically feasible. Summaries and reviews on disperse dyes have been published by Schwander,[2] Weaver and Shuttleworth,[3] Dawson,[4,5] Hallas[6] and Annen et al.[7]

The very fact that highly acidic diazonium ions must be combined with highly basic coupling components (Figure 1.1) poses great challenges to dye chemists and chemical engineers. Some of the major problems which they face are outlined in Figures 1.2 and 1.3.

Since weakly basic amines are poor nucleophiles, they can only be diazotized using strong nitrosating electrophiles, such as the nitrosonium ion. The latter is formed virtually quantitatively from nitrous acid at

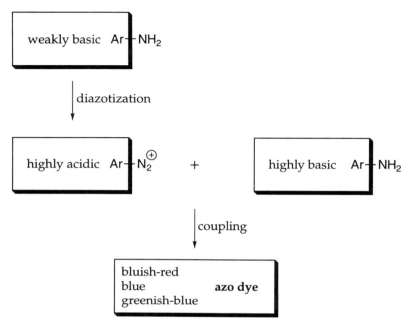

Figure 1.1 Schematic representation of the synthetic route for blue disperse monoazo dyes (Ar = carbocyclic or heterocyclic aromatic residue).

acidities greater than ~ 65% perchloric acid or 65% sulphuric acid.[8-10,13] Another reason why weakly basic aromatic amines have to be diazotized in concentrated acids, such as nitrosyl sulphuric acid, is that many of the highly acidic diazonium components formed are stable only in sulphuric acid concentrations greater than 85%.

However, this in turn brings about a critical situation for the subsequent coupling reaction with highly basic aromatic amines (Figure 1.3). On the one hand, the azo coupling must be performed at pH values at which there is a sufficiently high concentration of the free base of the coupling component present to ensure a fast enough coupling reaction. On the other hand, the transition of the very acidic diazo component solution to the moderately acidic or neutral coupling reaction mixture is associated with very rapid decomposition of the diazonium ions at acidities lower than ~ 60% sulphuric acid.[14] In this context, it is interesting to draw a parallel between the decomposition of diazotized weakly basic aromatic amines and that of nitrosyl sulphuric acid. The highest instability of nitrosyl sulphuric acid has been found to exist within the acidity range of 40% to 60% sulphuric acid. At these acidities both nitrous acid and the nitrosonium ion coexist in appreciable concentrations forming dinitrogen trioxide, the anhydride of the nitrous acid. The

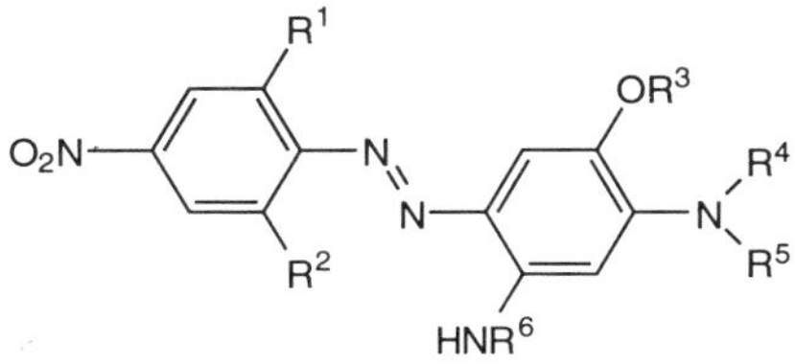

1

R¹ = H, CN, Cl, NO₂, Br
R² = H, CN, Cl, NO₂, Br
R³ = Alkyl
R⁴ = H, alkyl, C₂H₄OH, C₂H₄CN, C₂H₄–OCO–CH₃
R⁵ = H, alkyl, C₂H₄OH, C₂H₄CN, C₂H₄–OCO–CH₃
R⁶ = H, CO–CH₃, SO₂CH₃

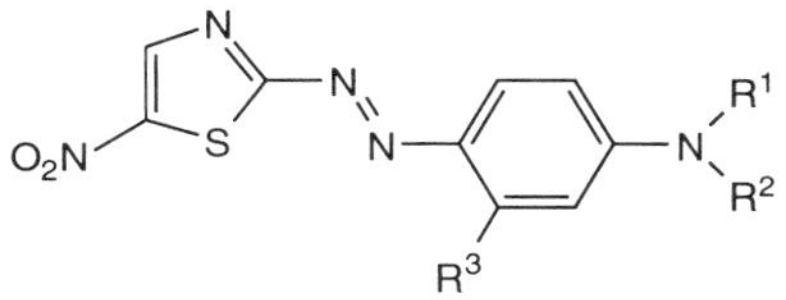

2

R¹ = Alkyl
R² = Alkyl, C₂H₄OH, CH₂–CHOH–CH₂OH, CH₂–CHOH–CH₂Cl
R³ = H, alkyl

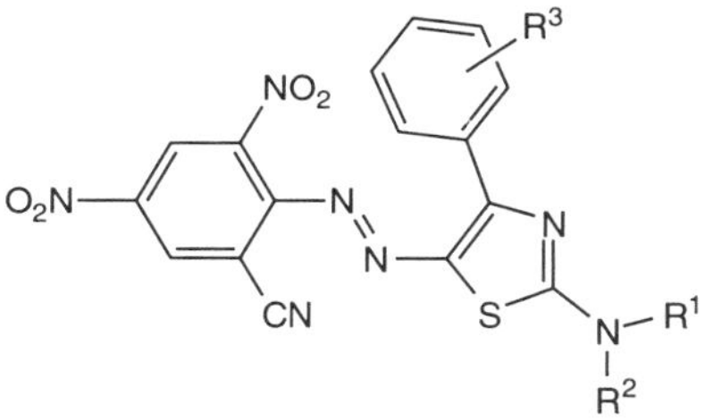

3

R¹ = Alkyl, alkenyl, cycloalkyl, aralkyl
R² = Alkyl, alkenyl, cycloalkyl, aralkyl
R³ = H, alkyl, alkoxy

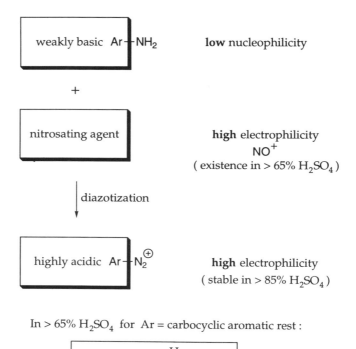

Figure 1.2 Schematic representation of the diazotization of weakly basic aromatic amines (Ar = carbocyclic or heterocyclic aromatic residue, H_0 is the Hammett acidity function).

homolytic dissociation of this into nitric oxide and nitrogen dioxide is thought to play an important role in the decomposition of nitrosyl sulphuric acid (Figure 1.4).[10-12,15] It must therefore be considered that the decomposition of diazotized weakly basic aromatic amines also occurs via the diazoanhydrides (Figure 1.5).

The maximum equilibrium concentration of such anhydrides exists at acidities equal to the pK_m values. If the acidity of the diazo component is increased by introducing acidifying 'electron-withdrawing' substituents, the pK_m values decrease,[19-22] thus shifting the maximum equilibrium concentration of the corresponding diazoanhydrides into a range of moderate acidities.

Possible alternative decomposition routes of diazotized weakly basic amines are depicted in Figure 1.6. These routes involve a reduction of the strongly electrophilic diazonium ion by electron transfer from a nucleophilic reagent.

This reagent can be either the strongly basic coupling component required for the formation of a blue azo disperse dye, or compounds

DIAZOTIZATION OF WEAKLY BASIC AROMATIC AMINES 5

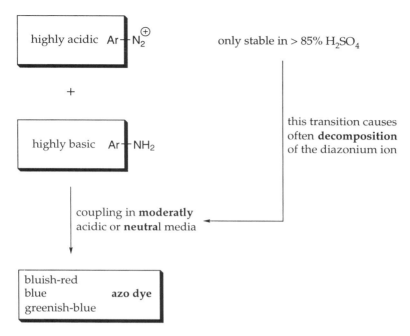

Figure 1.3 Schematic representation of the azo coupling reaction of highly acidic aromatic diazo components with highly basic coupling components (Ar = carbocyclic or heterocyclic aromatic residue).

$$H^{\oplus} + HONO \xrightleftharpoons{K_R} NO^{\oplus} + H_2O$$

$$NO^{\oplus} + HONO \rightleftharpoons N_2O_3 + H^{\oplus}$$

$$\downarrow$$

$$NO + NO_2$$

$$K_R = \frac{h'_R [HNO_2]}{[NO^{\oplus}]} \quad ; \quad pK_R = -7.86 = H'_R - \log\frac{[HNO_2]}{[NO^{\oplus}]}$$

Figure 1.4 Decomposition of nitrosyl sulphuric acid within the acidity range of 40% to 60% sulphuric acid.[10–12,15] (H'_R is the acidity function for protonation–dehydration processes.[16,17,47,51] $H'_R = -\log_{10} h'_R$.)

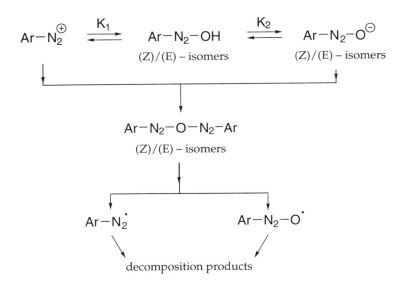

Figure 1.5 Proposed decomposition route of diazotized weakly basic aromatic amines in moderately acidic reaction mixtures. (H_- is the acidity function characterizing the acid–base equilibrium of a monoanionic base.[18,47,51] $H_- = -\log_{10} h_-$. H'_R is defined in Figure 1.4.
(a) Dilute alkaline, neutral or moderately acidic solutions. (b) Strong acidic solutions.

originating from the decomposition of the nitrosyl sulphuric acid. This decomposition inevitably takes place in the course of the mixing of the liquors containing the diazo and the coupling component, respectively. The rate of this mixing process determines the time that elapses as the reaction mixture passes across the dangerous acidity range favourable for the decomposition of the diazo component. Thus, the yield of the azo dyes can be improved by optimizing the efficiency of the mixing process.

In spite of the established industrial importance of diazotizing weakly basic aromatic amines it was not until the disastrous explosion[23] at Ciba in Basle in December 1969 that chemists started to investigate the kinetics

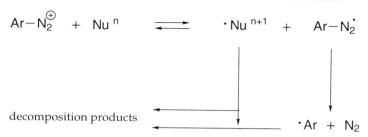

Figure 1.6 Proposed decomposition routes of diazotized weakly basic aromatic amines initiated by an electron transfer reaction. Nu^n is the highly basic coupling component or compound generated during the decomposition of the nitrosyl sulphuric acid.

and the mechanism of the diazotization at very high acidities. The two main studies[24,25] appeared in 1971 and 1978. Although some kinetic measurements were reported, both investigations were carried out at very low concentrations (10^{-3} mol l^{-1} to 10^{-4} mol l^{-1}) in order to allow the kinetics to be followed directly by UV/Vis spectroscopy. However, if the diazotization is performed at concentrations significant for industrial processes, we have found that the experimentally measured rate constants are approximately 200 times larger! This surprising result will be explained later in this chapter.

1.2 Dependence of diazotization rates on acidity

The diazotization of aromatic amines is one of the most widely studied and best known N-nitrosation reaction. The thorough investigation of its kinetics under various experimental conditions played an important part in identifying the specific nitrosating agents involved and gave a valuable insight into the reaction mechanism itself.[26-30] These studies led to the establishment of a mechanistic framework put forward and summarized in review articles by Ridd.[31,32] In spite of the vast amount of new experimental data that have been acquired since, the general principles of this framework still remain valid.[13,33]

For many years, the diazotization has been recognized as a multistage reaction (Figure 1.7) and it is also known that the dependence of its rate on acidity is rather complex.

The aim of this review is to discuss our new experimental results on the diazotization of weakly basic aromatic amines in nitrosyl sulphuric acid in the light of the mechanism in Figure 1.7. However, this is preceded by a short summary of the present knowledge on the kinetics of diazotization in aqueous perchloric and sulphuric acids. In this regard, use has been made of the comprehensive review of Williams.[13]

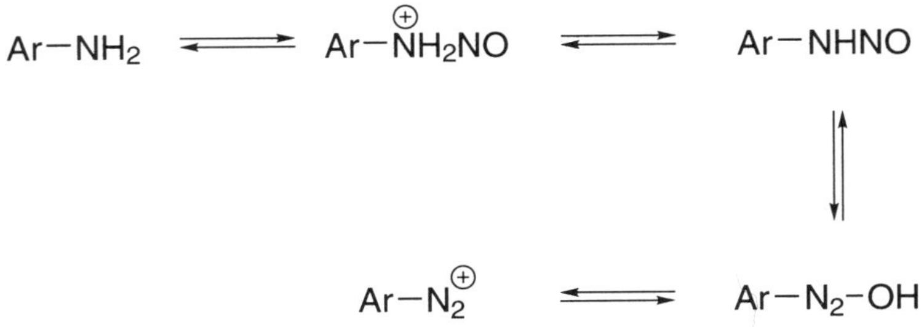

PHYSICO-CHEMICAL PRINCIPLES OF COLOR CHEMISTRY

$$Ar-NH_2 \rightleftharpoons Ar-\overset{\oplus}{N}H_2NO \rightleftharpoons Ar-NHNO$$

$$Ar-\overset{\oplus}{N_2} \rightleftharpoons Ar-N_2-OH$$

Figure 1.7 Simple representation of the complex multistage diazotization reaction.

1.2.1 Carbocyclic aromatic amines

In addressing this subject, it is more convenient to discuss the rate profiles for three broad acidity ranges separately. In each case the transition between these ranges is characterized by a remarkable change in the dependence of the rate of diazotization upon the acidity of the medium. The position of these transitions with reference to the acidity scale also depends on the nature of the amine.

1.2.1.1 Low acidities. This acidity level (up to ~ 0.1 M perchloric or sulphuric acid) is one of the three regions that has been most thoroughly investigated and is now thought to be well understood.[13,32] The kinetic behaviour is characterized by an inverse dependence of the experimentally measured rate constant k_{exp} on the acidity. The latter is defined[18] by the Hammett acidity function ($H_0 = -\log_{10} h_0$) which can be considered as an extension of the concept of pH to strongly acidic media, equation (1.1).

$$k_{exp} = kh_0^{-1}; \log k_{exp} = H_0 + \text{const.} \tag{1.1}$$

The rate of diazotization at a given acidity within this low acidity range obeys one of the two possible equations (1.2) or (1.3).

$$\text{Rate} = k_{exp}[ArNH_2]_T[HNO_2]_T^2 \tag{1.2}$$

$$\text{Rate} = k_{exp}[ArNH_2]_T[HNO_2]_T h_0 \tag{1.3}$$

The subscript T denotes the total (stoichiometric) concentration of the corresponding reagent. These concentrations are defined in Figure 1.8. The rate expressions found experimentally can be interpreted in the light of the two slightly different mechanistic pathways depicted in Figures 1.9 and 1.10. Assuming that the transformation of the *N*-nitrosamine to the diazonium ion is not rate determining, the kinetic expressions for

$$K_A = \frac{[ArNH_2]\,h_o}{[Ar\overset{+}{N}H_3]} \quad ; \quad [ArNH_2]_T = [ArNH_2] + [Ar\overset{+}{N}H_3]$$

$$[ArNH_2] = \frac{K_A\,[ArNH_2]_T}{(K_A + h_o)} \quad ; \quad [Ar\overset{+}{N}H_3] = \frac{h_o\,[ArNH_2]_T}{(K_A + h_o)}$$

$$K_N = \frac{[NO_2^-]\,h_o}{[HNO_2]} = 4.5 \times 10^{-4} \quad ; \quad [HNO_2]_T = [NO_2^-] + [HNO_2]$$

$$[NO_2^-] = \frac{K_N\,[HNO_2]_T}{(K_N + h_o)} \quad ; \quad [HNO_2] = \frac{h_o\,[HNO_2]_T}{(K_N + h_o)}$$

Figure 1.8 Equilibrium concentrations relevant for diazotization at low acidities.

$$HNO_2 + H^+ \underset{k_{-1}}{\overset{k_1}{\rightleftarrows}} H_2\overset{+}{N}O_2$$

$$H_2\overset{+}{N}O_2 + NO_2^- \underset{k_{-2}}{\overset{k_2}{\rightleftarrows}} N_2O_3 + H_2O$$

$$N_2O_3 + ArNH_2 \underset{k_{-3}}{\overset{k_3}{\rightleftarrows}} Ar\overset{+}{N}H_2NO + NO_2^-$$

$$Ar\overset{+}{N}H_2NO + S \overset{k_S}{\rightleftarrows} ArNHNO + HS^+$$

$$ArNHNO \longrightarrow ArN_2^+$$

Figure 1.9 Mechanism of diazotization via dinitrogen trioxide, where S is a proton acceptor.

diazotization via dinitrogen trioxide (Figure 1.9) or via nitrous acidium ion (Figure 1.10) are given by the equations (1.4) and (1.5), respectively.

$$\text{Rate} = \frac{k_1 k_2 k_3 k_S h_0 [ArNH_2][HNO_3][NO_2^-][S]}{k_3 k_S [ArNH_2][S](k_2[NO_2^-] + k_{-1}) + k_{-1} k_{-2}[H_2O](k_{-3}[NO_2^-] + k_S[S])} \tag{1.4}$$

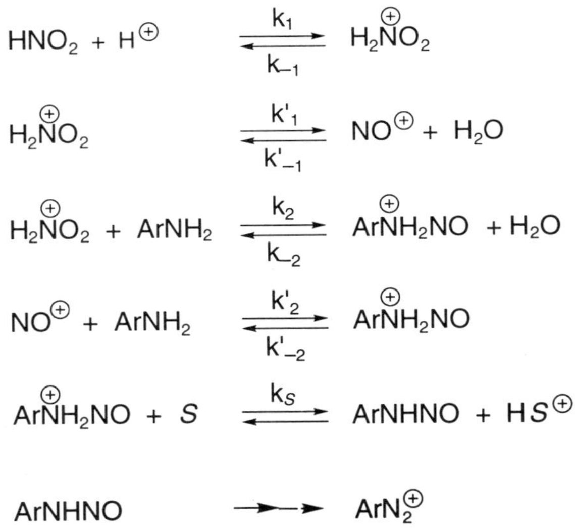

$$HNO_2 + H^{\oplus} \underset{k_{-1}}{\overset{k_1}{\rightleftarrows}} H_2\overset{\oplus}{N}O_2$$

$$H_2\overset{\oplus}{N}O_2 \underset{k'_{-1}}{\overset{k'_1}{\rightleftarrows}} NO^{\oplus} + H_2O$$

$$H_2\overset{\oplus}{N}O_2 + ArNH_2 \underset{k_{-2}}{\overset{k_2}{\rightleftarrows}} Ar\overset{\oplus}{N}H_2NO + H_2O$$

$$NO^{\oplus} + ArNH_2 \underset{k'_{-2}}{\overset{k'_2}{\rightleftarrows}} Ar\overset{\oplus}{N}H_2NO$$

$$Ar\overset{\oplus}{N}H_2NO + S \underset{}{\overset{k_S}{\rightleftarrows}} ArNHNO + HS^{\oplus}$$

$$ArNHNO \longrightarrow ArN_2^{\oplus}$$

Figure 1.10 Mechanism of diazotization via nitrous acidium ion or the free nitrosonium ion. S is a proton acceptor.

$$\text{Rate} = \frac{k_1 k_2 k_S h_0 [ArNH_2][HNO_2][S]}{k_S[S](k_2[ArNH_2] + k_{-1}) + k_{-1}k_{-2}[H_2O]} \quad (1.5)$$

Provided that the N-nitrosation step is rate limiting and that $h_0 \gg K_N$ and $h_0 \gg K_A$, expressions (1.4) and (1.5) reduce to the experimentally observed kinetic equations (1.2) and (1.3), respectively, with k_{exp} representing the expressions shown in equations (1.6) and (1.7).

$$k_{exp} = \frac{k_1 k_2 k_3 K_A K_N}{k_{-1} k_{-2}[H_2O]} \times \frac{1}{h_0} \quad (1.6)$$

$$k_{exp} = \frac{k_1 k_2 K_A}{k_{-1}} \times \frac{1}{h_0} \quad (1.7)$$

It is worthwhile to draw attention to the fact that both mechanistic pathways (Figures 1.9 and 1.10) lead to the same experimental rate equation (1.3), if the rate-limiting step is the base catalyzed deprotonation of the protonated N-nitrosamine, if $h_0 \gg K_N$ and $h_0 \gg K_A$, and if the base S is H_2O. The corresponding k_{exp} expressions for the respective pathways via dinitrogen trioxide and via the nitrous acidium ion are described in equations (1.8) and (1.9) respectively.

$$k_{exp} = \frac{k_1 k_2 k_3 k_S K_A}{k_{-1} k_{-2} k_{-3}} \times \frac{1}{h_0} \quad (1.8)$$

$$k_{\exp} = \frac{k_1 k_2 k_S K_A}{k_{-1} k_{-2}} \times \frac{1}{h_0} \quad (1.9)$$

At high amine and low acid concentrations, the reaction becomes zero order in the amine, equation (1.10).

$$\text{Rate} = k'_{\exp}[\text{HNO}_2]_T^2 \quad (1.10)$$

$$k'_{\exp} = \frac{k_1 k_2 K_N}{k_{-1}} \quad (1.11)$$

The formation of dinitrogen trioxide is then rate limiting and the experimentally measured rate constant $k'_{\exp}$ does not depend on acidity, equation (1.11).

1.2.1.2 Moderate acidities. Over the range of acidity from $H_0 = 1$ to about $H_0 = -6$ (corresponding to 0.1–6.0 M perchloric acid or 1%–~65% perchloric or sulphuric acid) the rate of the diazotization of aniline derivatives ($pK_A > ~3$) at a given acidity obeys equation (1.12).

$$\text{Rate} = k_{\exp}[\text{ArNH}_2]_T[\text{HNO}_2]_T \quad (1.12)$$

The striking feature is that within this moderately acidic range the experimentally measured rate constant increases with acidity, as shown in equation (1.13).

$$k_{\exp} = k' h_0; \; \log k_{\exp} = -H_0 + \text{const.} \quad (1.13)$$

This kinetic behaviour has been explained[32,34] in terms of a mechanism in which the reactive species is now the protonated form of the amine and not the free base form. Although the mechanistic interpretation has been and is still the subject of some controversy, there are no published experimental data that contradict the proposed reaction pathway involving an attack of the nitrosonium ion at the anilinium ion to form a charge-transfer complex.

Comparing the kinetic equations describing the reaction pathways that occur via the anilinium ion or the free base form of the aniline, a criterion can be formulated which permits evaluation of the preferred sequence of the reaction steps (Figure 1.11). It is evident that the higher the acidity (h_0) of the reaction mixture and the basicity of the anilines (small K_A), the more likely it is that the diazotization will proceed via the protonated form of the aniline.

It has been suggested that the reaction sequence involving the direct attack of the protonated aniline by the nitrosonium ion followed by a deprotonation step is rate limiting.[32,34] Based on our recent detailed investigations[35,36] it must be concluded that at moderate acidities it is not the formation of the charge-transfer complex but rather the deprotonation step that limits the rate of diazotization; thus $k_{-2} \gg k_3[S]$. According

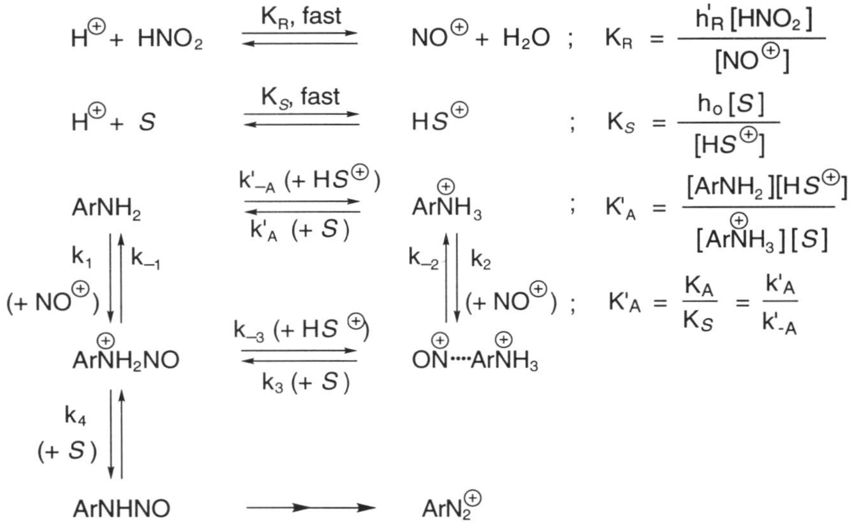

$$H^\oplus + HNO_2 \xrightleftharpoons{K_R,\ \text{fast}} NO^\oplus + H_2O \ ; \quad K_R = \frac{h'_R[HNO_2]}{[NO^\oplus]}$$

$$H^\oplus + S \xrightleftharpoons{K_S,\ \text{fast}} HS^\oplus \ ; \quad K_S = \frac{h_0[S]}{[HS^\oplus]}$$

$$\begin{array}{c} ArNH_2 \\ k_1 \updownarrow k_{-1} \\ (+NO^\oplus) \\ Ar\overset{\oplus}{N}H_2NO \\ k_4 \updownarrow \\ (+S) \\ ArNHNO \end{array} \quad \underset{k'_A\ (+S)}{\overset{k'_{-A}\ (+HS^\oplus)}{\rightleftarrows}} \quad \begin{array}{c} Ar\overset{\oplus}{N}H_3 \\ k_{-2} \updownarrow k_2 \\ (+NO^\oplus) \\ \overset{\oplus}{ON}\cdots Ar\overset{\oplus}{N}H_3 \end{array} \ ; \quad K'_A = \frac{[ArNH_2][HS^\oplus]}{[Ar\overset{\oplus}{N}H_3][S]}$$

$$K'_A = \frac{K_A}{K_S} = \frac{k'_A}{k'_{-A}}$$

$$\underset{k_3\ (+S)}{\overset{k_{-3}\ (+HS^\oplus)}{\rightleftarrows}}$$

$$ArNHNO \longrightarrow Ar\overset{\oplus}{N}_2$$

Criterion for the mechanistic pathway:

$$\frac{k_1 K_A}{h_0} \ \ll \ \frac{k_2 k_3[S]}{k_{-2} + k_3[S]} \quad : \text{via anilinium ions}$$

$$\gg \quad : \text{via free base of the anilines}$$

Figure 1.11 Alternative mechanistic pathways for diazotization under moderately acidic conditions.

to the mechanistic scheme depicted in Figure 1.11 only this type of reaction sequence leads to the expression (1.14) which is similar to the experimental rate equation (1.12). In addition, k_{exp} becomes larger with increasing acidity, giving rise to equation (1.15). Here, h'_R is defined by the acidity function ($H'_R = -\log_{10} h'_R$) for a protonation–dehydration process,[16,17,47,51] such as is visualized by the formation of the nitrosonium ion (Figure 1.10), and a_S is the activity of the proton acceptor S. At moderate acidities, the slopes of the functions h_0 vs. (%H_2SO_4) and h'_R vs. (%H_2SO_4) are fairly similar.[51]

$$\text{Rate} = \frac{k_2 k_3 [HNO_2]_T a_S [ArNH_2]_T h_0 h'_R}{k_{-2}(h'_R + K_R)(h_0 + K_A)} \quad (1.14a)$$

For $h_0 \gg K_A$ and $h'_R \ll K_R$:

$$\text{Rate} = \frac{k_2 k_3 [HNO_2]_T a_S [ArNH_2]_T h'_R}{k_{-2} K_R} \quad (1.14b)$$

Thus

$$k_{\exp} = \frac{k_2 k_3 a_S}{k_{-2} K_R} \times h'_R \qquad (1.15)$$

Also, we were able to collect additional evidence for the formation of the charge-transfer complexes proposed in Figure 1.11. Although the mere existence of such complexes does not prove that they function as reaction intermediates, their presence certainly supports such a mechanism. A nitrosyl sulphuric acid solution having $pK_A > \sim 3$ exhibits an intense colour caused by charge-transfer bands in the UV/Vis spectra. In addition, the nitrosyl sulphuric acid signal in the ^{15}N-NMR spectrum is shifted towards lower fields and is considerably broadened, while the ^{1}H- and ^{13}C-NMR spectra of the aromatic compound remain unaffected.[35,36] It could be checked experimentally that the broadening of the signal is due exclusively to a shortening of the longitudinal relaxation time of the ^{15}N nuclear spins. The strong correlation between the chemical shift and the broadening of the ^{15}N-NMR signal indicates that these effects are interrelated.[36] Following our study of the interaction between nitrosyl sulphuric acid and a series of anilines and heteroaromatic amines such as 4-chloro-, 4-methyl-, 2,4,6-trimethyl-, 4-chloro-N-methylaniline and 2-amino-4-chloro-5-formylthiazole, it can be concluded that both the broadening and the shift of the ^{15}N-NMR signal of nitrosyl sulphuric acid to lower fields (Figure 1.12) are caused by the presence of low concentrations of radicals. Such radicals are most probably formed by a reversible electron transfer within the lifetime of the charge-transfer complex, forming a solvent-caged radical pair. Moreover, as there is only one detectable ^{15}N-NMR signal for the nitrosonium ion, the position of which depends on the presence and concentration of sufficiently basic ($pK_A > \sim 3$) amines, we are inclined to believe that the observed shift is a contact shift or Fermi-contact shift and that there is a fast equilibration between the complexed and the free nitrosonium ion.

The strongest evidence for the proposed mechanism via the direct reaction between an anilinium ion and nitrosonium ion comes from kinetic substituent effects.[32] The significant activating effects of electron-donating substituents can only be explained in terms of their effect on the acidity of the postulated charge-transfer complex.

1.2.1.3 High acidities. As the acidity is increased above that of ca. 65% perchloric or sulphuric acid, the rate of diazotization of anilines passes through a maximum and then decreases very rapidly. The maximum corresponds to an acidity level at which nitrous acid is essentially converted to the nitrosonium ion. The nitrous acid–nitrosonium ion equilibrium has been evaluated by UV spectroscopy.[8,9,12] It can be quantitatively described by the acidity equation shown in Figure

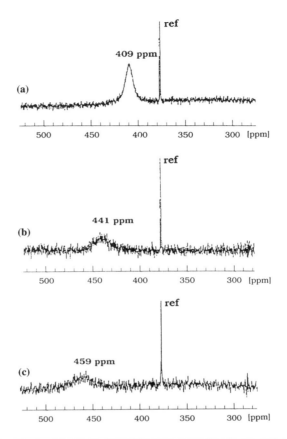

ref = reference signal ($^{15}NO_3^-$).

Figure 1.12 ^{15}N-NMR spectra of nitrosyl sulphuric acid in the presence of 2,4,6-trimethylaniline.

1.4. The results obtained using that equation were recently verified by ^{15}N-NMR spectroscopy.[35,36]

The ^{15}N-NMR signal of nitrosyl sulphuric acid in 10% sulphuric acid at 578 ppm is the signal for the nitrous acid molecule itself. In 96% sulphuric acid the ^{15}N-NMR signal observed at 384.5 ppm is characteristic of the free nitrosonium ion. Within the intermediate acidity range the signal becomes broader due to the coalescence effect as well as to the presence of radical pairs. The latter are formed during the decomposition of nitrosyl sulphuric acid, especially in the acidity range of 40% to 60% sulphuric acid. According to the data shown in Figure 1.13, the concentrations of the nitrosonium ion and the nitrous acid are equal in about 57% sulphuric acid. This corresponds to an H_0 value of -1.5 and an H_R value[16] of -7.86, and is in agreement with the relationship shown in Figure 1.4. However, the dramatic change in the dependence of the rate coefficient (k_{exp}) on the acidity, passing from acid catalysis in equations (1.12) and (1.13) within the moderate acidity range to strong acid inhibition of equations (1.16) and (1.17) at high acidities cannot be ascribed simply to the change in the position of nitrous acid–nitrosonium ion equilibrium.

$$\text{Rate} = k_{exp}[\text{ArNH}_2]_T[\text{HNO}_2]_T \tag{1.16}$$

$$k_{exp} = k''h_0^{-m}; \log k_{exp} = mH_0 + \text{const.} \quad (\text{with } m \geqslant 2) \tag{1.17}$$

Adopting the mechanistic scheme shown in Figure 1.11, the experimentally observed kinetic behaviour is consistent with a reaction sequence involving a rapid and reversible formation of the protonated primary nitrosamine which undergoes a rate-limiting proton transfer to a proton acceptor S followed by several rapid reactions forming the diazonium ion. It is noteworthy that in this case it is kinetically irrelevant whether the protonated nitrosamine is rapidly formed via the nitrosation of the anilinium ion followed by a fast deprotonation or via the nitrosation of the free base of the corresponding aniline. From these

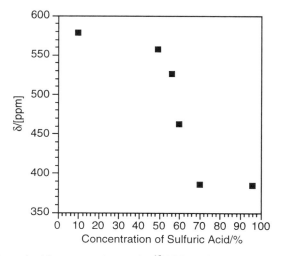

Figure 1.13 Effects of acid concentration on the ^{15}N-NMR signal for the nitrous acid–nitrosonium ion.

considerations and when the proton acceptor S is a neutral species, the rate equation (1.18) for the nitrosation at high acidities can be deduced:

$$\text{Rate} = \frac{ak_4 K_S [S]_T}{(h_0 + K_A)(h_0 + K_S)} [HNO_2]_T [ArNH_2]_T \quad (1.18)$$

Thus, the rate coefficient k_{exp} becomes

$$k_{exp} = \frac{ak_4 K_S [S]_T}{(h_0 + K_A)(h_0 + K_S)},$$

$$\text{with } a = \frac{k_2 k_3}{k_{-2} k_{-3}} \quad K_S = \frac{k'_A k_1}{k'_{-A} k_{-1}} \quad K_S = \frac{k_1}{k_{-1}} K_A$$

Since in the various deprotonation steps more than one Brønsted base can serve as the proton acceptor S, the rate equation (1.18) must be generalized leading to the corresponding coefficient k_{exp} given in equation (1.19):

$$k_{exp} = \frac{a \sum_i (k_{4i} K_{Si} [S_i]_T)}{(h_0 + K_A) \sum_i (h_S + K_{Si})},$$

$$\text{with } a = \frac{k_2}{k_{-2}} \sum_i \left(\frac{k_{3i}}{k_{-3i}} K_{Si} \right) = \frac{k_1}{k_{-1}} \sum_i \left(\frac{k_{Ai}}{k_{-Ai}} K_{Si} \right) = \frac{k_1}{k_{-1}} K_A \quad (1.19)$$

As far as the interpretation of the rate–acidity profile within this high acidity range ($H_0 < -6$) is concerned, the following aspects need to be considered: the acidity dependence of the rate coefficient k_{exp} is determined by the acidity constant of the aniline (K_A), by the charge and the acidity constants of the proton acceptors (K_{Si}) as well as by the acidity dependence of the term $\sum_i (k_{4i} K_{Si} [S_i]_T)$. The charge of base S determines which acidity function $H_S = -\log_{10} h_S$ describes the acidity dependence of the deprotonation step by that particular base. For example, if base S is water, hydrogensulfate or a sulfate anion, the proton transfer is characterized[47,51] by the respective acidity functions H_0, H_- or H_{--}.

If the limiting cases $h_0 \gg K_A$ and $h_S \gg K_{Si}$ (for all i) apply, equation (1.19) can be converted into the relationship shown in equation (1.20).

$$\log k_{exp} - \log A(H_S) = H_0 + H_S + \log a \quad (1.20)$$

with

$$A(H_S) = \sum_i (k_{4i} K_{Si} [Si]_T) = (k_{41} K_{S1} [S_1]_T + k_{42} K_{S2} [S_2]_T + \ldots) \quad (1.21)$$

The dependence of $A(H_S)$ on the acidity function H_S in highly acidic solutions is indeed a complex one.[35] It has been suggested by Challis and

Ridd[37] that both the water as well as the hydrogen sulfate anion present in concentrated sulphuric acid solutions can act as proton acceptors. This was verified by Ernst,[35] who analyzed the observed diazotization rates for 4-nitroaniline, 2,6-dichloro-4-nitroaniline and 2-bromo-4,6-dinitroaniline by means of non-linear regression using a multiparameter kinetic model and detailed information on the composition of sulphuric acid at various acidities.[38,39] He was able to demonstrate that in highly acidic solutions the rate-limiting proton transfer is brought about by water as well as by the hydrogen sulfate and sulfate anions present. Furthermore, from that analysis it is clear that the relative importance of these proton acceptors at a given acidity depends on the nature, and especially the basicity, of the substituted aniline, and on the encounter-controlled deprotonation limit that has to be considered for the more reactive species. In fact, this is to be expected considering Brønsted's relationship for general base catalysis.[40,41] From equation (1.19) it becomes evident why the experimentally evaluated slopes of the $\log k_{exp}$ vs. H_0 plots for various anilines[24,25,35-37] are not an integer of 2.0 as expected from the usual mechanistic consideration, but have values in the range of 2.1 to 2.4. This apparent contradiction is due to the slight H_0 dependence of the term $A(H_S)$ discussed above and the fact that the theoretically expected acidity dependence of the diazotization rate of equation (1.20) is not based on the acidity function H_0, but on H_S, which is a complex combination of the various acidity functions H_0, H_- and H_{--}.

1.2.2 Heterocyclic aromatic amines

Despite the increasing importance of heterocyclic aromatic diazo components in the preparation of azo dyes, very few systematic studies on the kinetics and mechanism of the diazotization of heterocyclic aromatic amines have been published. Up to now, the reason for this paucity of data was probably the lack of efficient analytical tools to investigate the kinetics and to determine simultaneously all the possible side reactions and by-products, especially under industrially relevant conditions. One of the aims of section 1.3 of this chapter is to show how this deficiency can be overcome.

A somewhat older review by Butler[42] still contains some useful information on suitable methods for the diazotization of heterocyclic aromatic amines. Studies that have appeared up to now and that offer useful kinetic results and mechanistic interpretations have recently been reviewed by Zollinger.[33]

The most systematic study on the kinetics of the diazotization and related reactions of a number of aminopyridine derivatives has been published by Kalatzis et al.[43-46] A typical example of such a kinetic investigation is depicted in Figure 1.14. The mechanism of the diazotiza-

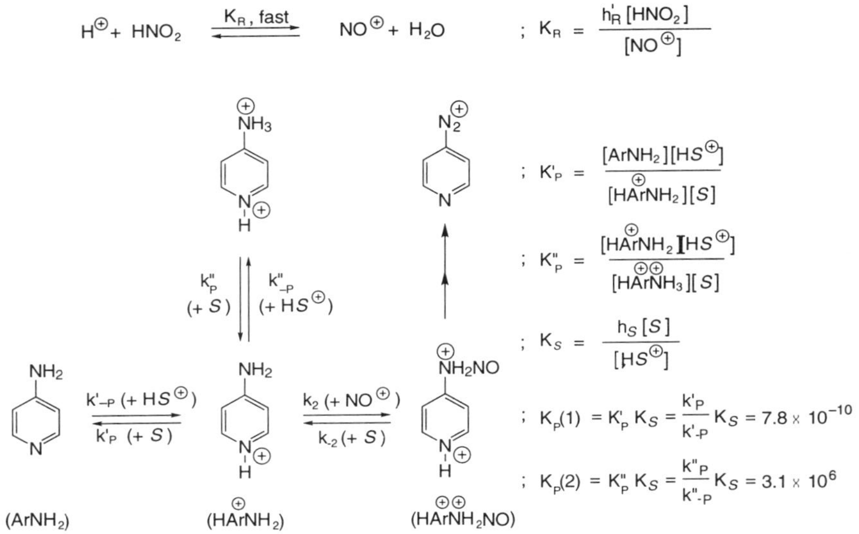

Figure 1.14 Diazotization of 4-aminopyridine.

tion of 4-aminopyridine at moderate acidities ($H_0 = 1$ to -6) is similar, but not equal, to that proposed for the diazotization of the aniline derivatives having electron-withdrawing groups within the same acidity range (see section 1.2.1.2).

From the experimentally determined kinetic equation (1.22) it must be concluded that the aminopyridinium ion reacts with the nitrosonium ion in a rate-limiting step. For such a reaction sequence (Figure 1.14), the rate equation (1.23) can be derived if steady state kinetics and the limiting cases $h_0 \gg K_P$ and $h'_R \ll K_R$ are assumed.

$$\text{Rate} = k_{\exp}[\overset{+}{\text{HArNH}_2}][\text{HNO}_2]_T 10^{-1.33 H_0} \quad (1.22)$$

$$\text{Rate} = k_{\exp}[\text{HNO}_2]_T [\text{ArNH}_2]_T \quad (1.23a)$$

$$\text{where } k_{\exp} = \frac{k_2}{K_R} h'_R = \frac{k_2}{K_R} 10^{-H'_R} \quad (1.23b)$$

Since under these conditions no significant amount of the double protonated 4-aminopyridine is detectable it can be assumed that $[\text{HArNH}_2] \cong [\text{ArNH}_2]_T$. Furthermore, for the considered moderate acidity range one can easily show[16] that, within experimental error, $H'_R = 1.3 H_0$. Thus, the experimentally measured kinetic behaviour displayed in equation (1.22) corresponds to the expected diazotization rate derived in equation (1.23).

The common mechanistic feature of both the diazotization of anilines and heteroaromatic amines such as 4-aminopyridine, is that in both cases the nitrosonium ion reacts with the protonated aromatic amine. However, in the first case the attack occurs at the aromatic π-electrons whereas in the second case the electrophile can directly react with the free electron pair at the amino group. As a consequence, it is not surprising that the rate-limiting steps for these two diazotizations are different.

Another noteworthy difference between the behaviour of the carbocyclic and the heterocyclic primary aromatic amines towards diazotization is apparent in the composition of the prototropic equilibrium mixture of N-nitrosamine and diazohydroxide formed. On one hand, the diazotization of ring-substituted anilines yields primary N-nitrosamines only in apolar, aprotic media. These fairly unstable compounds, which appear as transient species in the reaction path to the diazonium ions (Figure 1.4), have been detected[48] spectroscopically at low temperature ($-78\,°C$). It is reasonable to assume that in polar, protic reaction systems the lifetime for primary N-nitrosamines formed from carbocyclic aromatic amines is too short to enable detectable amounts to be accumulated. On the other hand, a large number of heterocyclic primary aromatic amines, when diazotized even in polar, protic media, give rise to particularly stable primary N-nitrosamines.[42] Some of these N-nitrosamines can readily be isolated from dilute acid solutions and characterized.[49,50] Their rather surprising stability has been attributed to internal hydrogen bonding.[42] As a consequence, these stable N-nitrosamines do not show any electrophilic property for azo coupling reactions in very dilute acid solution. Only after treating these intermediates with stronger acid do they form electrophilic diazonium ions. In this context, it must be noted that the N-nitrosamine is only one of the possible prototropic isomers formed as an intermediate in the diazotization of heteroaromatic amines. A typical example is depicted in Figure 1.15 for the nitrosation products of 5-amino-3-R-1,2,4-thiadiazole. For a more detailed discussion on the properties of N-nitrosamines, Williams,[13] Zollinger,[33] Butler,[42] Müller and Haiss,[48] and Goerdeler and co-workers[49,50] should be consulted.

For a better understanding of the diazotization mechanism it is also necessary to consider the problem of the reversibility of the reaction, especially in highly acidic reaction media. Whereas it seems to be widely accepted that the reaction sequence for the N-nitrosation of heterocyclic primary aromatic amines leading to the corresponding diazonium ions is reversible (Figure 1.4), there is still some controversy concerning the position of the equilibrium state. The reversibility has been shown experimentally, e.g. for the 5-phenyl-1,3,4-thiadiazole-2-diazonium ion in dilute acid solution[52] and for the 4-chloro-3-cyano-5-formyl-2-thiophene-diazonium ion by means of ^{15}N-isotope exchange in 92–96% sulphuric acid.[36] Further evidence for the reversibility is given in other experimental studies.[49,53,54] However, so far as the position of the equilibrium

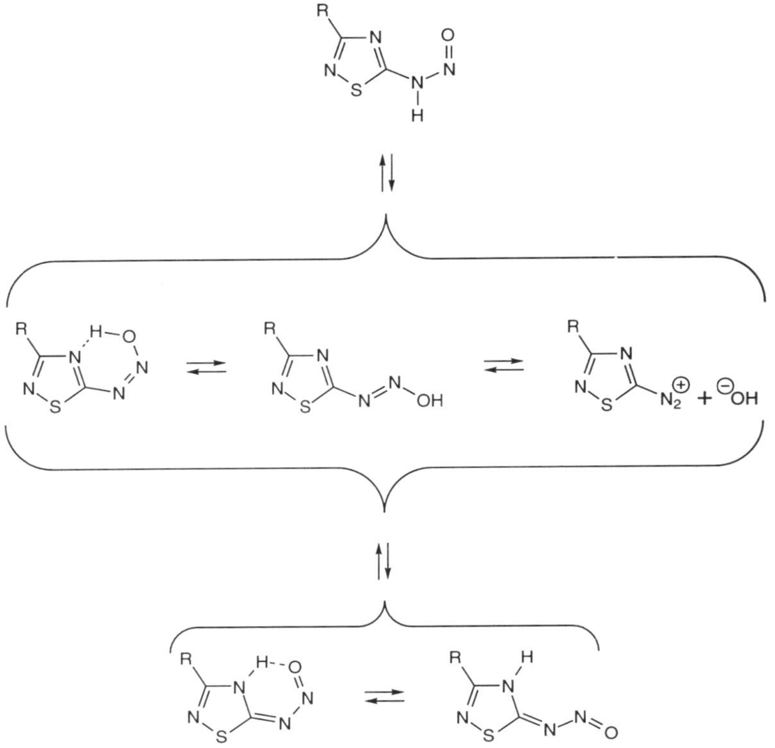

Figure 1.15 Prototropic equilibrium of 5-nitrosamino-3-R-1,2,4-thiadiazole.[49]

of the diazotization is concerned, there appears to be no definitive and clear-cut experimental evidence which would enable the equilibrium position to be measured quantitatively. Indeed, it has only recently been claimed[55,56] that the equilibrium concentrations of thiazole-2-diazonium ion in the diazotization of 2-aminothiazole with nitrosyl sulphuric acid in 70% sulphuric acid could be determined using UV/Vis spectroscopy. However, the conclusions obtained by this method are by no means totally convincing, since the exact composition of the product mixture could not be evaluated. Our own studies on the diazotization of 2-aminothiazole have revealed[36] that more than one competitive consecutive as well as parallel reaction occurs, thus consuming additional nitrosating agent. In addition, the present author acknowledges having serious difficulties following and understanding the arguments as well as the formula for calculating the equilibrium constant used in the above mentioned study.[56]

1.2.3 Diazotization of cyano-substituted aromatic amines

Many weakly basic aromatic amines contain the cyano group as a strong electron-withdrawing acidifying substituent. Its ability to undergo acid-catalyzed hydrolysis places severe limits on the reaction conditions under which these cyano-substituted amines can be diazotized efficiently. The ideal diazotization conditions must bring about a sufficiently high rate of diazotization together with the highest possible ratio of diazotization to hydrolysis. In addition to these requirements, the conditions chosen must also ensure that the stability of the diazonium ion formed is adequate for a two-stage diazotization/azo coupling process. The various requirements may run counter to each other in that a good design with respect to one requirement may be poor with respect to the other. In cases where these often conflicting demands do not allow an economic compromise, more sophisticated and unconventional operating conditions must be devised. Some interesting examples will be discussed in section 1.3.8.2.

To the best of the author's knowledge, there have been up to now no investigations of the competitive acid-catalyzed hydrolysis of the cyano group occurring during the diazotization of cyano-substituted aromatic amines. The present understanding of the mechanism of the acid-catalyzed hydrolysis of nitriles is based on a few kinetic studies[57-60] involving molecules such as benzonitrile and acetonitrile. In dilute acid solutions, the hydrolysis of nitriles yields exclusively the corresponding carboxylic acids. Under highly acidic conditions, however, the product distribution is temperature dependent. At low temperature, nitriles are selectively converted to the corresponding primary amides. A further conversion to the carboxylic acids can only be achieved at fairly high temperatures. Kinetic studies[60] have revealed that in concentrated sulphuric acid, several pathways for the hydrolysis of nitriles exist, often taking place concurrently, the relative contribution of each depending on the experimental conditions (Figure 1.16). Under the assumption of steady state kinetics related to the concentration of the protonated nitrile, the kinetic equation (1.24) can be derived.[36]

$$\text{Rate} = \frac{k_1[\text{R}-\text{CN}]h_0 a_W(k_2 + k_3/\gamma)}{k_{-1} + a_W(k_2 + k_3/\gamma)} \quad (1.24)$$

where a_W denotes the activity of water and $\gamma = a_{HSO_4^-}/a_W$ denotes the ratio of the activities of hydrogen sulfate and water.

For the limiting case of

$$k_{-1} \ll a_W(k_2 + k_3/\gamma)$$

the experimentally measured rate constant becomes

$$k_{\text{exp}} = k_1 h_0 \quad (1.25)$$

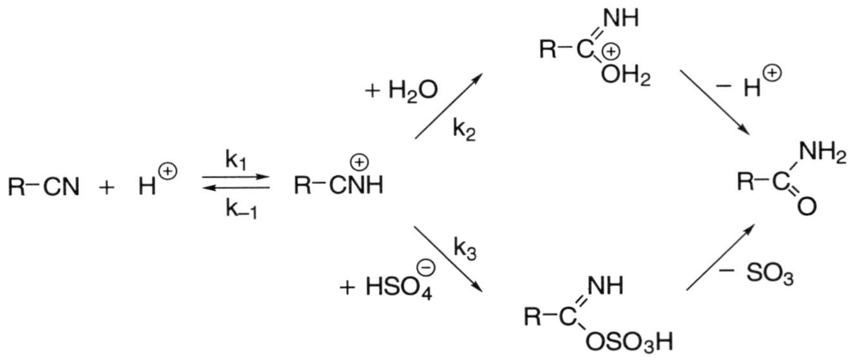

Figure 1.16 Hydrolysis of a nitrile group in concentrated sulphuric acid at room temperature.[60]

In the case where
$$k_{-1} \gg a_W(k_2 + k_3/\gamma) \quad \text{and} \quad k_2 \gg k_3/\gamma,$$
the expression for k_{exp} is

$$k_{exp} = \frac{k_1 k_2}{k_{-1}} h_0 a_W = k''' h_0 a_W \tag{1.26}$$

Thus

$$\log k_{exp} = -H_0 + \log a_W + \log k''' \tag{1.27}$$

Since $H'_R = H_0 + \log a_W$, equation (1.27) becomes

$$\log k_{exp} = -(2H_0 - H'_R) + \log k''' \tag{1.28}$$

It has been shown previously[16] that within the high acidity range (> 65% sulphuric acid) the acidity function H'_R corresponds approximately to $1.5 H_0$, and that equation (1.28) converts to equation (1.29).

$$\log k_{exp} \approx -0.5 H_0 + \log k''' \tag{1.29}$$

The acidity dependence predicted by the two relationships expressed by equations (1.25) and (1.29) has been observed for the hydrolysis of the cyano groups of 2-cyano-4-nitroaniline in 87–97% sulphuric acid at temperatures between 5 °C and 30 °C[36] and of 2-amino-4-chloro-5-formyl-3-thiophenecarbonitrile in 90–98% sulphuric acid at temperatures between 20 °C and 30 °C.[36] The hydrolysis of a cyano group in sulphuric acid concentrations of 100% or higher is more complex. Aromatic nitriles can, for example, form cyclic intermediates with sulphur trioxide[61] (Figure 1.17). An investigation by NMR spectroscopy has also revealed[35] (see section 1.3.4) that at concentrations of sulphuric acid at which the value

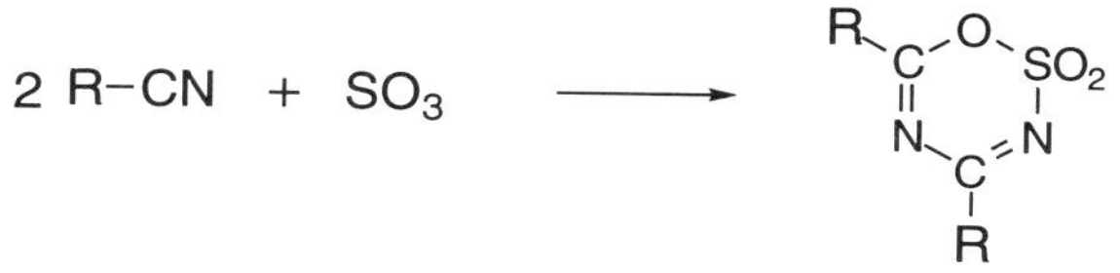

DIAZOTIZATION OF WEAKLY BASIC AROMATIC AMINES

Figure 1.17 Cyclic intermediate formed in 100% sulphuric acid and oleum.[61]

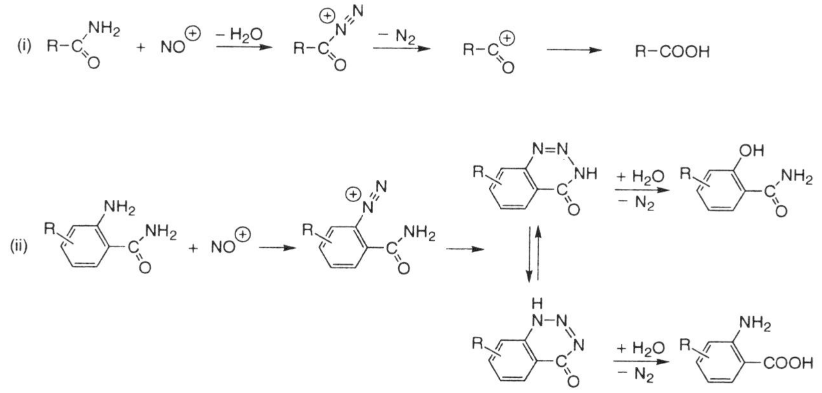

Figure 1.18 Reactions of amides in nitrosyl sulphuric acid.

for γ becomes greater than one, the hydrolysis passes through a transient hydrogen sulfate species as shown in Figure 1.16.

In the case where the hydrolysis of a cyano group takes place as a competitive reaction to diazotization, the formation of the corresponding carboxylic acid from the intermediate amide is often catalyzed by nitrosonium ions. This is depicted in Figure 1.18 for an isolated amido group (i), and for an amido group *ortho* to a diazonium group (ii). The formation of a triazine intermediate from the latter type amides has been verified.[36,62]

1.3 Diazotization under industrial reaction conditions

Until now, what has been known about the diazotization mechanism of weakly basic aromatic amines has been obtained from kinetic studies in dilute solutions having amine concentrations in the range of $10^{-3}\,\text{mol}\,\text{l}^{-1}$ to $10^{-4}\,\text{mol}\,\text{l}^{-1}$. It has proved, however, to be of some practical importance that the kinetic behaviour measured in such dilute solutions and the mechanistic conclusions drawn therefrom cannot simply be transferred to the reaction conditions prevailing at concentrations utilised in industrial processes. Such an extrapolation leads unquestionably to false predictions. This may be the reason why only few papers have appeared in the scientific literature pertaining to the subject of the diazotization of weakly basic aromatic amines. Two main causes for the difference in kinetic behaviour at low and high amine concentrations can be traced. On the one hand, the H_0 value of a given sulphuric acid solution is appreciably lowered, if amine is dissolved at high concentrations, thus decreasing or increasing the rates of the hydrolysis of a cyano group, if present, or of the diazotization step, respectively. Since the concentration of the amine decreases as diazotization proceeds, one should expect an increase of the H_0 value of the reaction mixture with time. However, this effect seems to be, at least partly, counter-balanced by the simultaneous formation of water.[35,36] On the other hand, according to equation (1.19) one must expect that amines, if present at high concentrations, can also act as Brønsted base catalysts. The reaction rate then becomes second order with respect to the free base form of the amine and, this has been proved experimentally.[35,36]

1.3.1 Analytical methods and practical units

It is common practice to determine the end point of the diazotization reaction with the help of the potassium/iodide/starch paper test. A fast colour change of the test paper indicates that there is still nitrosating reagent present and consequently that the diazotization has not yet been completed. Often this test is also used for determining the completion of the diazotization in concentrated sulphuric acid. However, as the test procedure involves a strong dilution of the highly acidic reaction sample, and as the diazotization is much faster in diluted acid solutions, the result obtained is not indicative of what has taken place in the concentrated solution.

Normally, a better way to determine whether the diazotization is complete and that the coupling reaction can be started is to analyze the reaction mixture by UV/Vis spectroscopy.[24,25] However, this direct method is not very suitable for reaction mixtures containing the high amine concentrations employed commercially. It has recently been shown[35,36] that NMR spectroscopy is by far the best analytical tool to use in investigating diazotization mixtures generated under industrial reaction conditions. The remainder of this chapter is devoted to a presentation of the potential utility of this powerful method in providing additional kinetic information as well as new insights into the mechanism of diazotization in highly acidic solutions. The examples chosen are not meant to be comprehensive, but rather illustrative. Furthermore, as is a common practice in industrial work, practical units are used, such as $mol\,kg^{-1}$ and min. Although it is not correct from the physico-chemical standpoint of view to use kg instead of l, the advantage, is evident. By using the unit mol kg^{-1}, the change in the density during the course of the reaction can be ignored. Certainly, the insuing kinetic expressions are not quite correct as the probability of an encounter between the reacting species is proportional to the volume and not to the mass of the reaction mixture. Nevertheless, the error introduced by using the mass unit instead of the volume unit is less than 10%.

1.3.2 2,6-Dichloro-4-nitroaniline

The kinetics of the diazotization of 2,6-dichloro-4-nitroaniline ($pK_A = -3.3$) under highly acidic conditions (80–97% sulphuric acid) within a temperature range of 15–30 °C and at industrial amine concentrations (0.39–0.87 mol kg^{-1}) have been measured by using ^{1}H-NMR.[35] The method is illustrated in Figure 1.19. The ^{1}H-NMR spectrum obtained is very simple as both ring protons are equivalent. The signal at higher field corresponds to the ring protons of the protonated amine. Its intensity decreases with time. Synchronously, the intensity of the ring protons of the diazonium ion at lower field increases as the diazotization proceeds. The diazonium ion has been confirmed as the only product formed[35] by

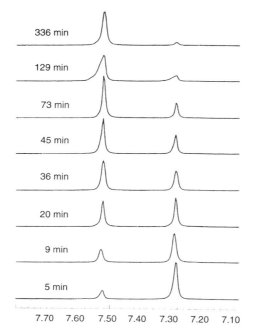

Figure 1.19 ^{1}H-NMR determination of the diazotization rate of 2,6-dichloro-4-nitro-aniline.[35]

performing a diazotization experiment using ^{15}N nitrosyl sulphuric acid and monitoring the reaction by ^{15}N-NMR spectroscopy, observing ^{15}N signals at 385 ppm and at 328 ppm for the nitrosonium and the diazonium ion, respectively. Due to the fact that at high concentrations the aniline lowers the acidity of the medium by being protonated and also acts as a Brønsted base in the rate-determining deprotonation step, the kinetic order with respect to the amine is greater than 1 and is time dependent.[35] As a consequence, at aniline concentrations relevant for diazotizations in industry (~ 0.5–$1.0\,\text{mol}\,\text{kg}^{-1}$), the half-lifetime of the diazotization is much lower than the half-lifetime determined by kinetic measurements at low amine concentrations (10^{-3}–$10^{-2}\,\text{mol}\,\text{kg}^{-1}$). Depending on the sulphuric acid concentration and on the reaction temperature, these half-lifetimes can differ by a factor of 50 to 200! Under the reaction conditions investigated, no by-products have been detected.[35] Examples of diazotization rate curves at low and at high aniline concentrations are depicted in Figures 1.20 and 1.21, respectively.

1.3.3 2-Bromo-4,6-dinitroaniline

The diazotization kinetics of 2-bromo-4,6-dinitroaniline ($pK_A = -6.7$) were elucidated within the same acidity, temperature and high amine concentration range as was employed in the evaluation of the diazotization kinetics of 2,6-dichloro-4-nitroaniline (section 1.3.2). A typical time

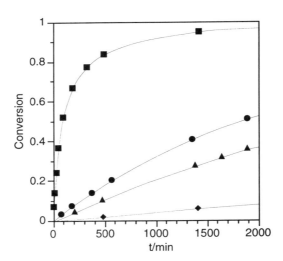

Figure 1.20 Diazotization rate curves for 2,6-dichloro-4-nitroaniline at 25 °C and low ($10^{-2}\,\text{mol}\,\text{kg}^{-1}$) amine and nitrosyl sulphuric acid concentrations.[35] Analytical method: UV/Vis spectroscopy. ■, 84.9% H_2SO_4; ●, 90.0% H_2SO_4; ▲, 92.2% H_2SO_4; ◆, 94.3% H_2SO_4.

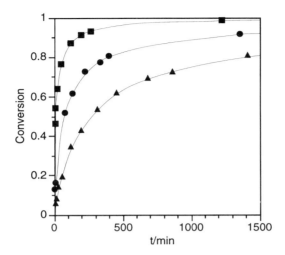

Figure 1.21 Diazotization rate curves for 2,6-dichloro-4-nitroaniline at 25 °C and high (0.8 mol kg^{-1}) amine and nitrosyl sulphuric acid concentrations.[35] Analytical method: ^{1}H-NMR spectroscopy. ■, 92.9% H_2SO_4; ●, 95.2% H_2SO_4; ▲, 96.9% H_2SO_4.

dependence of the ^{1}H-NMR spectrum representing the course of the diazotization reaction is shown in Figure 1.22. Both the 2-bromo-4,6-dinitroanilinium ion and the corresponding diazonium ion contain two non-equivalent ring protons in the 3- and 5-positions. The pair of NMR signals at higher field correspond to the aromatic protons of the protonated amine, whereas the two ^{1}H-NMR signals for the ring protons of the diazonium ion appear at lower field. The *meta*-coupling of the signals is not detectable, because there was purposely no deuterated solvent used to stabilize the magnetic field. The kinetic measurements have revealed that, as in the case of 2,6-dichloro-4-nitroaniline, the half-lifetime of the diazotization is up to 200 times shorter at high than at low aniline concentrations (Figures 1.23 and 1.24). The explanation for this remarkable finding was covered in section 1.3.2.

1.3.4 2-Bromo-6-cyano-4-nitroaniline

The diazotization of 2-bromo-6-cyano-4-nitroaniline ($pK_A = -8.7$) is accompanied by the competitive acid-catalyzed hydrolysis of the cyano group. Thus, the design of an industrial process affording the highest possible yield and selectivity for the diazotization of such cyano-substituted amines requires a very thorough evaluation of the optimal reaction and process conditions. The various competitive reactions occurring in highly acidic diazotization media (90% to 100% sulphuric

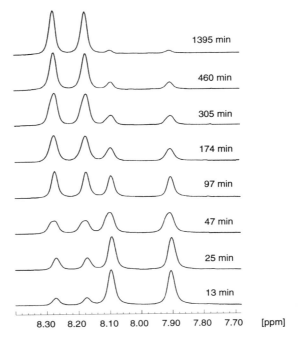

Figure 1.22 ¹H-NMR determination of the diazotization rate of 2-bromo-4,6-dinitroaniline.[35]

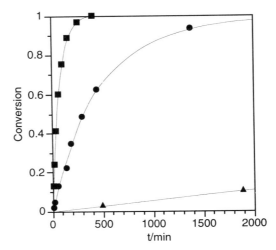

Figure 1.23 Diazotization rate curves for 2-bromo-4-6-dinitroaniline at 25 °C and low (10^{-2} mol kg^{-1}) amine and nitrosyl sulphuric acid concentrations.[35] Analytical method: UV/Vis spectroscopy. ■, 87% H_2SO_4; ●, 90% H_2SO_4; ▲, 95% H_2SO_4.

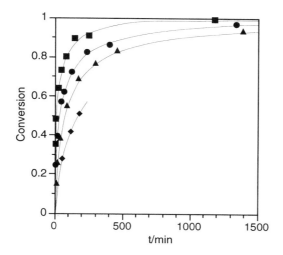

Figure 1.24 Diazotization rate curves for 2-bromo-4,6-dinitroaniline at 20 °C and high (0.8 mol kg^{-1}) amine and nitrosyl sulphuric acid concentrations.[35] Analytical method: ^{1}H-NMR spectroscopy. ■, 93.8% H$_2$SO$_4$; ●, 94.9% H$_2$SO$_4$; ▲, 95.8% H$_2$SO$_4$; ◆ 96.8% H$_2$SO$_4$.

acid) are depicted in Figure 1.25. These reactions have been characterised using NMR spectroscopy.[35]

The composition of the diazotization reaction mixture and its temperature and its strong acidity dependence can be traced from Figure 1.26 and Table 1.1. Using the appropriate experimentally determined kinetic expressions for the various competitive reactions, a computer-aided evaluation of the optimal reaction and process conditions can be performed.[36]

1.3.5 2-Aminothiazole

Today, there is no longer doubt as to whether 2-aminothiazole (pK_A = 5.39;[63-66] pK_A = 5.28[33,56]) can be diazotized in good yield. The diazonium ion formed can be identified by ^{15}N-NMR spectroscopy.[36] The rate of its formation under industrial reaction conditions has been determined using the ^{1}H-NMR technique.[36] As the chemical shifts of the ^{1}H-NMR signals for the CH-protons do not change over the entire acidity range from 2% to 100% sulphuric acid, it is safe to assume that 2-aminothiazole is only present as the monoprotonated species. There is evidence[36,56,67] that the protonation occurs at the heterocyclic nitrogen atom and not at the amino group. The latter is not even protonated in trifluormethane sulphonic acid, which has an H_0 value[68] of about −13. The rate and selectivity of the diazotization of 2-aminothiazole is very

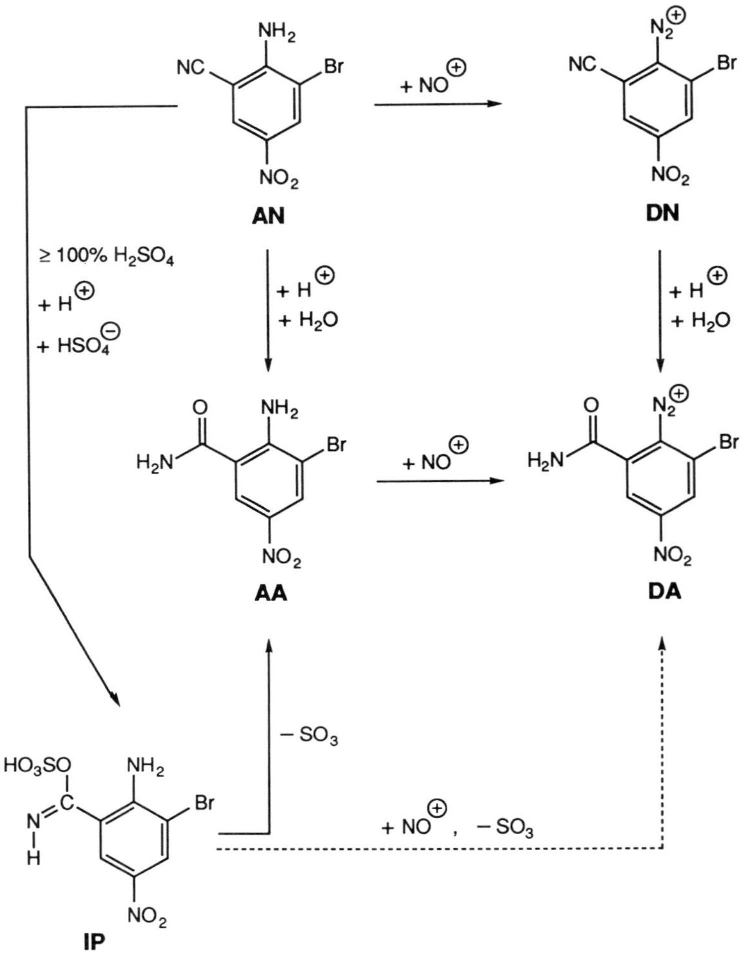

Figure 1.25 Competitive reactions occurring during the diazotization of 2-bromo-6-cyano-4-nitroaniline in 90–100% sulphuric acid.[35]

sensitive to the acidity of the reaction medium. For an optimum yield of the 2-thiazolediazonium ion it is necessary to determine this acidity dependence separately for every aminothiazole concentration and reaction procedure used commercially. For example, at an equimolar concentration of the reactants of $0.5\,\mathrm{mol\,kg^{-1}}$, the maximum yield of 2-thiazolediazonium hydrogensulfate is 15% in 95%, 55% in 92% (Figure 1.27) and 95% in 85% sulphuric acid. As has been discussed in section 1.2.2, there is no evidence that the yield of the diazotization depends on the equilibrium constant of the N-nitrosation step, as has been reported

Table 1.1 Maximum yield of the cyano-substituted diazonium ion (DN) generated in the diazotization of 2-bromo-6-cyano-4-nitroaniline.[35]

t_{max}* (min)	$\dfrac{[DN]_{max}}{[AN]_0}$	T (°C)	$\dfrac{[AN]_0}{(mol\,kg^{-1})}$	H_2SO_4 (%)
270	0.48	15	0.5	96.2
50	0.48	30	0.5	96.2
–	–	15	0.5	92.8
30	0.81	30	0.5	92.8
50	0.94	15	0.5	90.5
–	–	30	0.5	90.5

*t_{max}: reaction time for obtaining maximum yield of DN.

elsewhere.[33,55,56] The yield is rather modulated by the competitive parallel formation of by-products.[36] Although the structures of the by-products formed have not yet been determined, it is certain that their formation is not accompanied by an evolution of molecular nitrogen. In Figure 1.28, several possible reaction products from the diazotization of 2-aminothiazole are depicted.

1.3.6 2-Amino-4-chloro-5-formylthiazole

2-Amino-4-chloro-5-formylthiazole behaves differently from 2-aminothiazole in nitrosyl sulphuric acid. After three hours, a sample of this amine dissolved in a highly acidic diazotization mixture (96% sulphuric acid) shows no trace of reaction, not even at a high amine concentration of 1.1 mol kg^{-1}. However, one observes that the nitrosyl sulphuric acid signal in the ^{15}N-NMR spectrum is shifted towards lower fields and is considerably broadened, while the ^{1}H- and ^{13}C-NMR spectra remain unaffected.[36] An explanation for this observation was presented in section 1.2.1.2 of this chapter. In addition, UV/Vis spectra of 2-amino-4-chloro-5-formylthiazole in solutions covering the acidity range from 3% to 100% sulphuric acid reveal the existence of three isobestic points at acidities corresponding to 15%, 83% and 96% sulphuric acid.[36] These three protonation steps were explained with the aid of ^{1}H- and ^{13}C-NMR spectroscopy, and are shown in Figure 1.29.

Azo dyes obtained by a coupling reaction with diazotized 2-amino-4-chloro-5-formylthiazole cannot be synthesized by the classical two-stage procedure in which the diazotization step and subsequent coupling reaction are carried out in two separate reaction vessels. This constraint is due to the fact that there exists no acidity region in which the diazotization of 2-amino-4-chloro-5-formylthiazole occurs sufficiently fast and, in which the diazonium ion formed is stable enough for its solution to be transferable to the coupling reaction mixture.[36] The way in which this dilemma can be overcome is described in section 1.3.8.2.

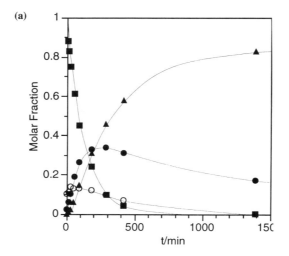

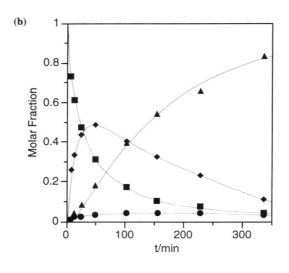

DIAZOTIZATION OF WEAKLY BASIC AROMATIC AMINES

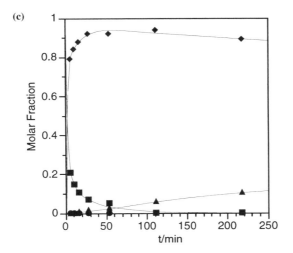

Figure 1.26 Competitive diazotization and hydrolysis of 2-bromo-6-cyano-4-nitroaniline.[35] (a) 99.8% H_2SO_4, 30 °C; (b) 96.2% H_2SO_4, 30 °C; (c) 90.5% H_2SO_4, 15 °C. ■, AN; ●, AA; ▲, DA; ◆, DN; ○, IP.

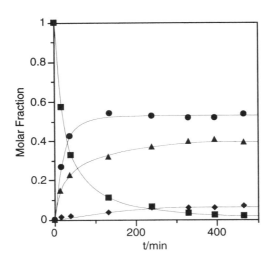

Figure 1.27 Diazotization rate curves for 2-aminothiazole at 20 °C and equimolar high concentrations (0.5 mol kg^{-1}) of amine and nitrosating agent (nitrosyl sulphuric acid) in 92% sulphuric acid.[36] Analytical method: ^{1}H-NMR spectroscopy. ■, 2-aminothiazole; ●, 2-thiazolediazonium ion; ▲, by-product 1; ◆, by-product 2.

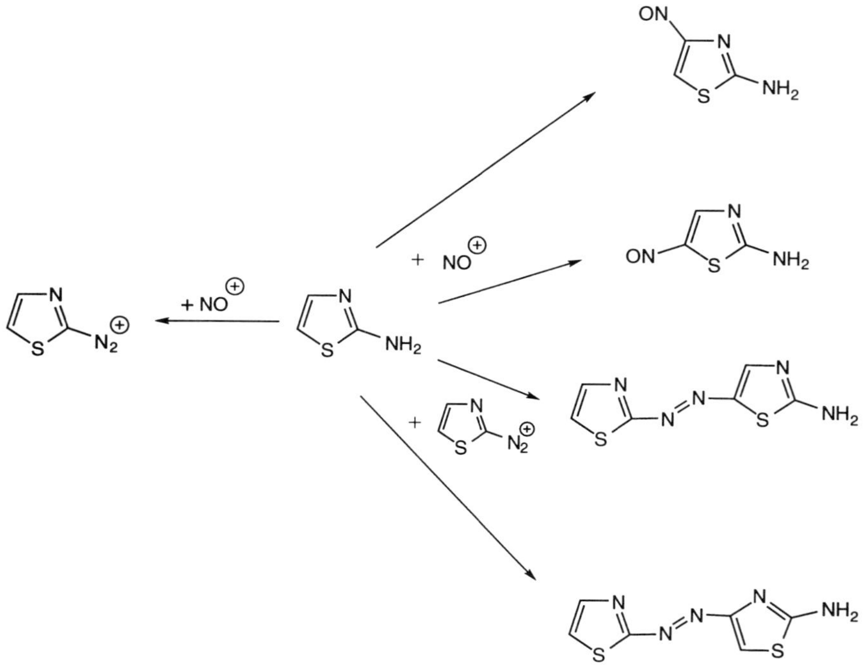

Figure 1.28 Possible by-products formed during the diazotization of 2-aminothiazole.

1.3.7 2-Amino-4-chloro-5-formyl-3-thiophenecarbonitrile

The various competitive reactions occurring during the diazotization of 2-amino-4-chloro-5-formyl-3-thiophenecarbonitrile are depicted in Figure 1.30. Most of these reactions have been characterised by NMR spectroscopy.[36]

In contrast to diazonium ions derived from weakly basic anilines, the diazonium ions shown in Figure 1.30 are not very stable and undergo dediazoniations. The maximum yield of the diazotized 2-amino-4-chloro-5-formyl-3-thiophenecarbonitrile and its temperature as well as its acidity dependence is shown in Table 1.2. How the optimum reaction and process conditions can be evaluated from experimental data at various concentrations, temperatures and acidities is briefly discussed in section 1.3.8.1 of this chapter.

1.3.8 Evaluation of the optimum reaction and process conditions

In order to realize the highest possible yield from the manufacturing of an azo dye it is necessary to optimize the selectivity of the diazotization

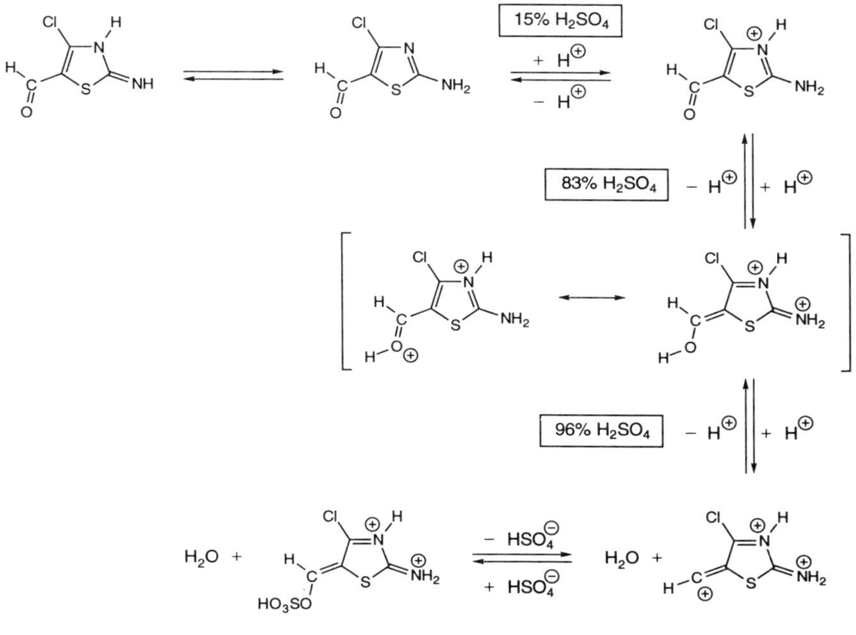

Figure 1.29 Possible protonation steps of 2-amino-4-chloro-5-formylthiazole in sulphuric acid.[36]

process, especially when weakly basic aromatic amines are involved. The remainder of this section is devoted to a brief presentation of such optimizing procedures. For the case where the diazotization step is clearly separated from the azo coupling process (section 1.3.8.1), a fairly accurate procedure has been developed[35,36] with the aim of satisfying the requirements for a wide application in industry. The procedure is kept as simple as possible and makes use of a computer-based methodology for regression analysis and experimental design, with non-linear algebraic and ordinary differential equation multiresponse models developed elsewhere.[69] For the case of a 'one-pot' dye production (section 1.3.8.2), the situation is much more complex, as the yield of the azo dye does not depend only on the chemical events but also on the hydrodynamic properties of the reaction mixture.

1.3.8.1 Diazotization and azo coupling as a two-stage process. The diazotization of 2-cyano-4-nitroaniline accompanied by the competitive acid-catalyzed hydrolysis of the cyano group is a typical example for which a very thorough evaluation of the optimum reaction and process conditions is required before designing an industrial process. One possible procedure for such an evaluation has recently been discussed in

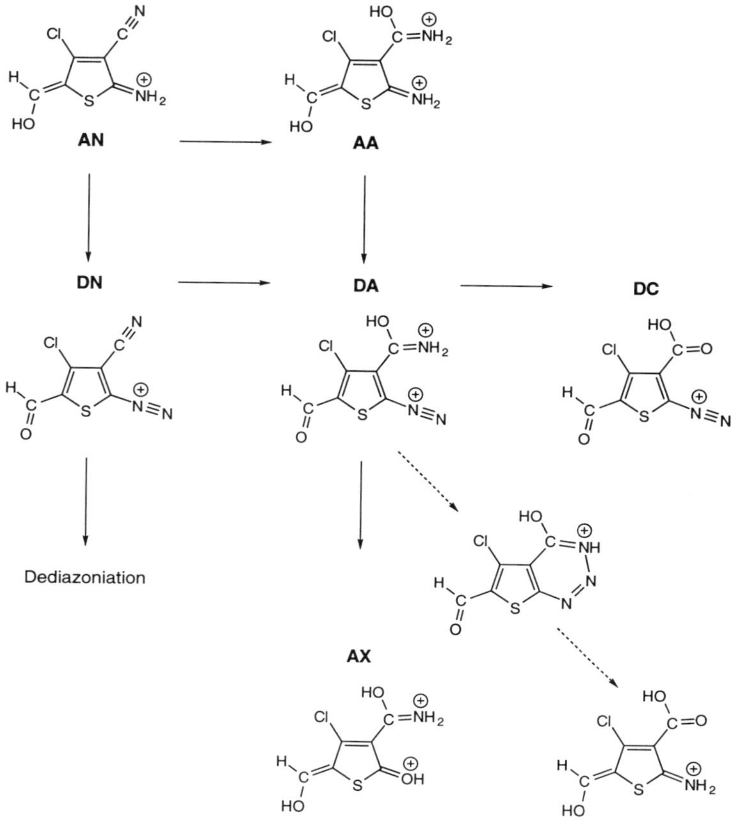

Figure 1.30 Competitive reactions occurring during the diazotization of 2-amino-4-chloro-5-formyl-3-thiophenecarbonitrile in 85% to 100% sulphuric acid.[36]

detail.[36] The following brief summary is an attempt to provide an indication of the utility of such a procedure. First, the kinetics and their dependence on the temperature, on the acidity of the reaction medium and on the initial concentration applied for all the competitive processes, such as the diazotization, the hydrolysis of the cyano group and other reactions possible, must be elucidated. This can be done with the help of NMR spectroscopy, as has been demonstrated earlier in this chapter. For the case of 2-cyano-4-nitroaniline, as an example, this procedure has led to a general reaction scheme and the corresponding rate equations summarized in Figure 1.31.

The dependence of the rate constants on the acidity and on temperature are defined in equations (1.30) and (1.31), with k'_i and k'_j being the rate constants at the standard conditions of $H_0 = -8$ and $T = 298$ K. For the acidity dependence of the hydrolysis rate constant formulated in

Table 1.2 Maximum yield of the cyano-substituted diazonium ion (DN) in the diazotization of 2-amino-4-chloro-5-formyl-3-thiophenecarbonitrile.[36]

t_{max} (min)	$\dfrac{[DN]_{max}}{[AN]_0}$	T (°C)	$\dfrac{[AN]_0}{(\text{mol kg}^{-1})}$	H_2SO_4 (%)
450	0.25	15	0.50	96
100	0.25	30	0.50	96
20	0.95	30	1.00	96
30	0.25	30	0.25	92
250	0.55	15	0.50	92
70	0.55	30	0.50	92
50	0.82	30	0.75	92
200	0.90	15	0.50	88
30	0.92	30	0.50	88
100	0.95	20	0.25	85
20	0.98	20	0.50	85
70	0.92	30	0.50	85
40	0.99	30	1.00	85

$\dfrac{[DN]_{max}}{[AN]_0}$ is the maximum yield of DN. t_{max} is the reaction time for obtaining maximum yield of DN.

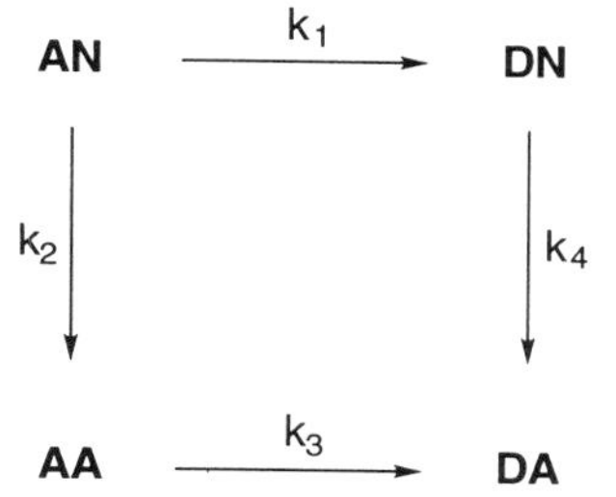

$$\text{AN} \xrightarrow{k_1} \text{DN}$$
$$\text{AN} \xrightarrow{k_2} \text{AA} \xrightarrow{k_3} \text{DA}$$
$$\text{DN} \xrightarrow{k_4} \text{DA}$$

$$-\frac{d[AN]}{dt} = k_1 [AN]\left([NO^{\oplus}]_0 - [DN] - [DA]\right) + k_2 [AN]$$

$$-\frac{d[AA]}{dt} = -k_2 [AN] + k_3 [AA]\left([NO^{\oplus}]_0 - [DN] - [DA]\right)$$

$$-\frac{d[DN]}{dt} = -k_1 [AN]\left([NO^{\oplus}]_0 - [DN] - [DA]\right) + k_4 [DN]$$

$$-\frac{d[DA]}{dt} = -k_4 [DN] - k_3 [AA]\left([NO^{\oplus}]_0 - [DN] - [DA]\right)$$

Figure 1.31 Competitive diazotization and hydrolysis of 2-cyano-4-nitroaniline.[36] AN is 2-cyano-4-nitroaniline, DN is 2-cyano-4-nitrobenzenediazonium ion, AA is 2-amino-5-nitrobenzamide and DA is 2-carbamoyl-4-nitrobenzenediazonium ion.

equation (1.31), the limiting conditions leading to equation (1.25) apply.

$$k_i = k'_i 10^{E_i} 10^{2(H_0+8)} \quad \text{(for } i = 1 \text{ and } 3\text{)} \tag{1.30}$$

$$k_j = k'_j 10^{E_j} 10^{-(H_0+8)} \quad \text{(for } j = 2 \text{ and } 4\text{)} \tag{1.31}$$

$$E_i = \frac{E_{ai}}{R}\left(\frac{1}{298} - \frac{1}{T}\right) \quad \text{and} \quad E_j = \frac{E_{aj}}{R}\left(\frac{1}{298} - \frac{1}{T}\right)$$

The dependence of the acidity function H_0 on the temperature and on the initial concentration $[AN]_0$ of the amine has been determined[35] using literature data.[70] The minimum number of experiments required for the evaluation of the entire competitive reaction system, as well as the position of these experiments within the parameter field, can be determined by existing computer-based optimization programs. The experimental kinetic data were evaluated by means of a non-linear regression program.[69] The results are summarized in Table 1.3.

Making the system of the kinetic equations in Figure 1.31 dimensionless reveals that the maximum yield of the cyano-substituted diazonium ion (DN) of 2-cyano-4-nitroaniline depends on the following five dimensionless quantities, equation (1.32).

$$\frac{[DN]_{max}}{[AN]_0} = f(\Phi_1, \Phi_2, \Phi_3, \alpha, \tau) \qquad (1.32)$$

where $\Phi_1 = k_1[AN]_0/k_2$ is the rate ratio of the diazotization of AN to its hydrolysis,

$\Phi_2 = k_3/k_1$ is the ratio of the rate constants of the diazotizations of AN and AA,

$\Phi_3 = k_4/k_2$ is the ratio of the rate constants for hydrolysis of the nitrile group of AN and DN,

$\alpha = [NO^+]_0/[AN]_0$ is the mole ratio of the nitrosating agent and the cyano-substituted amine and

$\tau = k_2 t$ is the dimensionless time.

The quantity Φ_1 can be calculated from the initial reaction conditions according to equation (1.33).

$$\Phi_1 = \Phi_1'[AN]_0 10^{3(H_0+8)} \exp\left[\frac{\Delta E}{R}\left(\frac{1}{298} - \frac{1}{T}\right)\right] \qquad (1.33)$$

The standard quantity Φ_1' is the quantity Φ_1 under the standard conditions of $H_0 = -8$, $T = 298$ K and $[AN]_0 = 1$ mol kg^{-1}. Its value has been determined to be that shown in equation (1.34)

$$\Phi_1' = 3 \times 10^5 \qquad (1.34)$$

The units applied in equation (1.33) are mol kg^{-1} for $[AN]_0$ and degree Kelvin for T. The term ΔE is the difference in activation energies, equation (1.35)

$$\Delta E = E_{a1} - E_{a2} \qquad (1.35)$$

Table 1.3 Characteristic kinetic parameters for the competitive diazotization and hydrolysis of 2-cyano-4-nitroaniline.[36]

k_i'; k_j'	k_1' (kg mol^{-1} min^{-1}) 2.7	k_2' (min) 9.2×10^{-5}	k_3' (kg mol^{-1} min^{-1}) 2.1	k_4' (min) 3.7×10^{-4}
E_{ai}; E_{aj}	E_{a1} (kJ mol^{-1}) 34	E_{a2} (kJ mol^{-1}) 158	E_{a3} (kJ mol^{-1}) *	E_{a4} (kJ mol^{-1}) 138

*Not determined as not relevant for the selectivity.

It is safe to assume that the quantities Φ_2 and Φ_3 are practically independent of acidity and temperature and their values proved to be essentially constant over a wide range of reaction conditions. It is also worth noting that the selectivity proved to be essentially independent of temperature despite the fact that the activation energies E_{a1} and E_{a2} are quite different (Table 1.3). The reason for this is the additional temperature dependence of the acidity function H_0. If a rate constant is proportional to 10^{H_0}, an increase in temperature increases the rate constant by a factor corresponding to an apparent activation energy of 39 kJ mol^{-1}.[35,36] Taking all these factors into account, one can show that the maximum yield of the diazonium ion DN for a given value of α is essentially determined by the value of Φ_1. This is convincingly shown in Figure 1.32.

It remains to be discussed what influence the reactor has on the maximum yield of the diazonium ion DN. Figure 1.33 shows this maximum yield as a function of the quantity Φ_1 for a batch and an ideal stirred tank (IST) reactor. The residence time in the IST reactor is chosen such that the maximum yield is reached. Furthermore, the curves are mean curves calculated for the four sets of reaction conditions used in Figure 1.32. For instance, if a selectivity of more than 99% is desired, the value for Φ_1 must be about 3.0×10^4 for a batch reactor, but 1.4×10^7 for an IST reactor.

The conclusion drawn from experimental results and from model calculations is that a high selectivity towards the diazotization cannot be

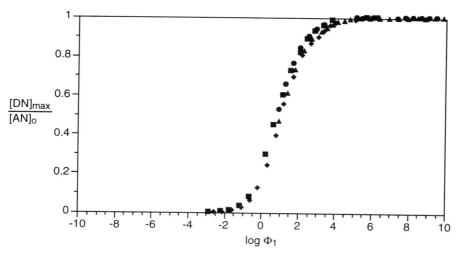

Figure 1.32 Maximum yield of the cyano-substituted diazonium ion (DN) from the diazotization of 2-cyano-4-nitroaniline, as a function of the quantity Φ_1, for a mole ratio of $\alpha = 1.02$.[36] ◆, [AN] = 0.25 mol kg^{-1}, $T = 5\,°C$; ●, [AN] = 1 mol kg^{-1}, $T = 30\,°C$; ■, [AN] = 0.25 mol kg^{-1}, $T = 30\,°C$; ▲, [AN] = 1 mol kg^{-1}, $T = 5\,°C$.

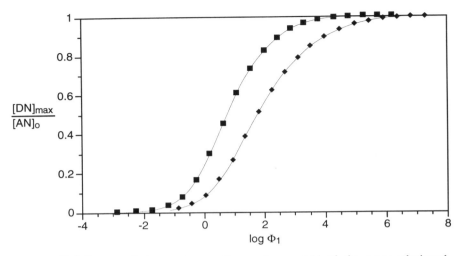

Figure 1.33 Influence of the reactor on the maximum yield of the cyano-substituted diazonium ion (DN) in the diazotization of 2-cyano-4-nitroaniline for a mole ratio of $\alpha = 1.02$.[36] ■, batch reactor; ◆, constant flow ideal stirred tank reactor (IST).

achieved unless the acidity is strongly decreased. For example, 2-cyano-4-nitroaniline and 2-bromo-6-cyano-4-nitroaniline or other weakly basic cyano-substituted aromatic amines can be diazotized in 60% to 80% sulphuric acid reaction mixtures with an excellent selectivity. However, at industrially relevant concentrations the amines are not fully dissolved. Thus, the reaction rates become limited by the rates of the dissolving process of the amines. A practical limit for accomplishing a diazotization under such conditions is often also the high viscosity of these suspensions. In certain cases, however, it is possible to obtain well agitatable fine suspensions for a successful diazotization process in 70% to 80% sulphuric acid.[36]

1.3.8.2 Diazotization and azo coupling as a one-stage process. There are several weakly basic aromatic amines such as 2-amino-4-chloro-5-formylthiazole, for which no acidity range can be found that brings about a diazotization rate high enough to be acceptable for an industrial process while simultaneously affording adequate stability to the diazonium ion formed. For cyano-substituted amines, the following additional requirement must be satisfied. The acidity of the diazotization solution must also ensure the highest possible ratio of the diazotization to the hydrolysis rate. In cases where the acidity for an acceptable diazotization rate has to be decreased into a region within which the diazonium ion formed rapidly decomposes, the diazotization and the coupling reaction must take place in the same reaction vessel. This can be achieved in the following way.

First, the amine is dissolved in concentrated sulphuric acid and the appropriate amount of nitrosyl sulphuric acid is added. The acidity is chosen to be high enough to prevent the onset of diazotization. Afterwards, this solution is added slowly under controlled stirring conditions to a well cooled aqueous solution of the coupling component. Within the contact zone between both solutions steep acidity, amine and nitrosyl sulphuric acid gradients are formed by molecular diffusion. The coupling component forms a gradient in the opposite direction. The more the acidity decreases towards the coupling solution, the faster the diazotization, as well as the decomposition of the diazonium ion, becomes. The contact zone turns now into a reaction zone where all the competitive chemical events occur faster than the mixing process of both solutions. If the coupling component present in the reaction zone is reactive enough, the decomposition process can successfully be surpassed by the coupling reaction. In certain cases, this 'one-pot' procedure can lead to surprisingly high yields of the azo dyes desired. However, fine tuning of the mixing conditions and of the local concentration of the amine in the reaction zone is often needed.

1.4 Conclusions

It is the aim of this chapter to demonstrate the great utility of NMR spectroscopy in determining the kinetics of diazotization in highly acidic media and at amine concentrations relevant to industrial processes. The evaluation of the kinetics of the various competitive reactions taking place in diazotization reaction mixtures is essential to achieve an optimum yield in the diazotization step and also in the subsequent azo-coupling process. In some cases, the application of the presented analytical method and of the optimizing procedure to processes already implemented in practice, brought about a considerable improvement in the existing yield and selectivity. In other cases, it was surprising that no reaction took place in the diazotization vessel in spite of a positive test with potassium/iodide/starch paper. Reaction occurred only after the diazotization mixture was added to the coupling solution.

Finally, the reader should be warned that improper handling of concentrated diazotization mixtures, especially suspensions, might cause these reaction systems to explode.

Acknowledgements

I thank my Ph.D. students T. Ernst and C. R. C. Fleury for the examples presented. I am extremely thankful to F. Bangerter, who did – and is still doing – an excellent job on the NMR spectrometers. And last but not least, I am much indebted to Dr. A. J. Klaus and Mrs. V. Sperb for helping me to write this chapter.

References

1. Zollinger, H., *Color Chemistry*, 2nd ed., VCH, Weinheim, 1991, Chapter 7.7.
2. Schwander, H. R., *Dyes Pigm.*, 3 (1982) 133.
3. Weaver, M. A. and Shuttleworth, L., *Dyes Pigm.*, 3 (1982) 81.
4. Dawson, J. F., *J. Soc. Dyers Colour*, 99 (1983) 183.
5. Dawson, J. F., *Rev. Progr. Coloration* 14 (1984) 90.
6. Hallas, G., in *Developments in the Chemistry and Technology of Organic Dyes*, Griffiths, J. ed., Blackwell Scientific, Oxford, 1984.
7. Annen, O., Egli, R., Hasler, R., Henzi, B., Jakob, H. and Matzinger, P., *Rev. Progr. Coloration*, 17 (1987) 72.
8. Singer, K. and Vamplew, P. A., *J. Chem. Soc.*, (1956) 3971.
9. Bayliss, N. S., and Watts, D. W., *Austral. J. Chem.*, 9 (1956) 319.
10. Bayliss, N. S., Dingle, R. and Watts, D. W., *Austral. J. Chem.*, 16 (1963) 933.
11. Bayliss, N. S. and Watts, D. W., *Austral. J. Chem.*, 16 (1963) 927, 943.
12. Seel, F. and Winkler, R., *Z. Phys. Chem.*, 25 (1960) 217.
13. Williams, D. L. H., *Nitrosation*, Cambridge University Press, Cambridge, 1988.
14. Allan, Z. J., *Collect. Czech. Chem. Commun.*, 15 (1950) 904.
15. Kobayashi, II., *Nippon Kogaku Kaishi*, (1976) 387.
16. Deno, N. C., Jaruzelski, J. J. and Schriesheim, A., *J. Am. Chem. Soc.*, 77 (1955) 3044.
17. Cox, B. A. and Yates, K., *Can. J. Chem.*, 61 (1983) 2225.
18. Hammett, L. P. and Deyrup, A. J., *J. Am. Chem. Soc.*, 54 (1932) 2721.
19. Macháčková, O. and Štěrba, V., *Collect. Czech. Chem. Commun.*, 37 (1972) 3313.
20. Jahelka, J., Macháčková, O. and Štěrba, V., *Collect. Czech. Chem. Commun.*, 38 (1973) 706.
21. Beránek, V., Štěrba, V. and Valter, K., *Collect. Czech. Chem. Commun.*, 38 (1973) 257.
22. Littler, J. S., *Trans. Faraday Soc.* 59 (1963) 2296.
23. Bersier, P., Valpiana, L. and Zubler, H., *Chem.-Ing.-Tech.*, 43 (1971) 1311.
24. Konecny, J. and Wenger, H., *Helv. Chim. Acta*, 55 (1972) 127.
25. Röbisch, G. and Süptitz, H., *J. Prakt. Chem.*, 320 (1978) 1047.
26. Hughes, E. D., Ingold, C. K. and Ridd, J. H., *J. Chem. Soc.*, (1958) 65.
27. Hughes, E. D. and Ridd, J. H., *J. Chem. Soc.*, (1958) 70.
28. Hughes, E. D., Ingold, C. K. and Ridd, J. H., *J. Chem. Soc.*, (1958) 77.
29. Hughes, E. D. and Ridd, J. H., *J. Chem. Soc.*, (1958) 82.
30. Hughes, E. D., Ingold, C. K. and Ridd, J. H., *J. Chem. Soc.*, (1958) 88.
31. Ridd, J. H., *Quart. Rev.*, 15 (1961) 418.
32. Ridd, J. H. *J. Soc. Dyers Colour.*, 81 (1965) 355.
33. Zollinger, H., *Diazo Chemistry I*, VCH, Weinheim, 1994.
34. de Fabricio, E. C. R., Kalatzis, E. and Ridd, J. R., *J. Chem. Soc. B*, (1966) 533.
35. Ernst, T., *Beiträge zur Kinetik der Diazotierung schwach basischer Aniline und Nitrosylschwefelsäure*, Ph.D. Thesis No. 8348, ETH Zürich, 1987.
36. Fleury, C. R. C., *Diazotierung schwach basischer aromatischer und hetero-aromatischer Amine in Nitrosylschwefelsäure: Eine ^{1}H- ^{13}C- und ^{15}N-NMR-spektroskopische Untersuchung*, Ph.D. Thesis No. 9470, ETH Zürich, 1991.
37. Challis, B. C. and Ridd, J. H., *Proc. Chem. Soc.*, (1960) 245.
38. Cox, R. A., *J. Am. Chem. Soc.*, 96 (1974) 1059.
39. Giauque, W. F., Hornung, E. W., Kunzler, J. E. and Rubin, T. R., *J. Am. Chem. Soc.*, 82 (1960) 62.
40. Brønsted, J. N. and Pederson, K., *Z. Phys. Chem.*, 108 (1924) 185 (see also ref. 41).
41. Eigen, M., *Angew. Chem.*, 75 (1963) 489.

42. Butler, R. N., *Chem. Rev.*, **75** (1975) 241, and the references therein.
43. Kalatzis, E., *J. Chem. Soc. B*, (1967) 277.
44. Kalatzis, E. and Mastrokalos, C., *J. Chem. Soc. Perkin 2*, (1974) 498.
45. Kalatzis, E. and Papadopoulos, P., *J. Chem. Soc. Perkin 2*, (1981) 248.
46. Kalatzis, E. and Mastrokalos, C., *J. Chem. Soc. Perkin 2*, (1983) 53.
47. Rochester, C. H., *Acidity Functions*, Academic Press, London, 1970.
48. Müller, E. and Haiss, H., *Chem. Ber.*, **96** (1963) 570.
49. Goerdeler, J. and Deselaers, K., *Chem. Ber.*, **91** (1958) 1025.
50. Ginsberg, A. and Goerdeler, J., *Chem. Ber.*, **94** (1961) 2043.
51. Hammett, L. P., *Physical Organic Chemistry*, McGraw-Hill, 1970.
52. Kaválek, J., Janák, K. and Štěrba, V., *Collect. Czech. Chem. Commun.*, **44** (1979) 3102.
53. Chmátal, V. and Štěrba, V., *Collect. Czech. Chem. Commun.*, **32** (1967) 2679.
54. Butler, R. N., Lambe, T. M., Tobin, J. C. and Scott, F. L., *J. Chem. Soc., Perkin Trans. 1* (1973) 1357.
55. Diener, H., *Fünfgliedrige heterocyclische Amine: Mechanismus der Diazotierung und Reaktivität der Diazoniumverbindungen*, Ph.D. Thesis No. 7693, ETH Zurich, 1984.
56. Diener, H., Güleç, B., Skrabal, P. and Zollinger, H., *Helv. Chim. Acta*, **72** (1989) 800.
57. Liler, M. and Kosanovic, D., *J. Chem. Soc.*, (1958) 1084.
58. Olah, G. A. and Kiovsky, T. E. *J. Am. Chem. Soc.*, **90** (1968) 4666.
59. Deno, N. C., Gaugler, R. W. and Wisotsky, M. J., *J. Org. Chem.*, **31** (1966) 1967.
60. Hyland, C. J. and O'Connor, C. J., *J. Chem. Soc., Perkin 2*, (1973) 223.
61. Weidinger, H. and Kranz, J., *Chem. Ber.*, **96** (1963) 2070.
62. Sauter, F. and Deinhammer, W., *Monatsh. Chem.*, **104** (1973) 91.
63. Albert, A., Goldacre, R. and Phillips, J., *J. Chem. Soc.*, (1948), 2240.
64. Coursot, P. and Wadsö, I., *Acta Chem. Scand.*, **20** (1966) 1314.
65. Phan-Tan-Luu, R., Surzur, J.-M., Metzger, J., Aune, J.-P. and Dupuy, C., *Bull. Soc. Chim. Fr.*, **9** (1967) 3274.
66. Haake, P. and Bauscher, L. P., *J. Phys. Chem.*, **72** (1968) 2213.
67. Toth, G. and Podanyi, B., *J. Chem. Soc., Perkin 2*, (1984) 91.
68. Olah, G. A., Prakash, G. K. S. and Sommer, J., *Superacids*, Wiley, New York, 1985.
69. Klaus, R., *A Computer-Based Methodology for Regression and Experimental Design with Nonlinear Algebraic and Ordinary Differential Equation Multi-Response Models*, Ph.D. Thesis No. 6866, ETH Zurich, 1981.
70. Johnson, C. D., Katritzky, A. R. and Shapiro, S. A., *J. Am. Chem. Soc.*, **91** (1969) 6654.

2 The heterogeneous kinetics of reactive and disperse dyeing

C. M. BRENNAN and J. F. BULLOCK

2.1 Introduction

The study of the kinetics of dyeing has long been of academic and practical interest. In this chapter we do not intend to revisit the many thorough previous treatments of the area (e.g. Vickerstaff,[1] Rattee and Breuer[2] and Johnson[3]); rather we intend primarily to concentrate on developments which have been published in recent years, highlighting those contributions which stress the heterogeneous nature of the process. Both the dyeing of polyester by disperse dyes, and of cotton by reactive dyes, are truly heterogeneous in nature, involving a liquid (dyebath solution and/or dispersion) and a solid (fibre). At a fundamental level, the reaction between a liquid and a solid involves a number of elementary steps, each of which may or may not be applicable to different specific reactions. These may be described as follows (*cf.* also Figure 2.1):

(i) Dissolution of dye molecules from the surface of any solid particle present and diffusional transport through a boundary layer to the bulk dyebath solution.
(ii) Transport through bulk solution to near the surface of the fibre.
(iii) Diffusional transport through any boundary layer surrounding the fibre and partition to the immediate surface of the fibre.
(iv) Diffusion within the fibre.
(v) Chemical reaction at an active site in the fibre.

2.2 The heterogeneous kinetics of dyeing polyester

2.2.1 Introduction

In this section, for simplicity, we confine ourselves to the batchwise (i.e. exhaust) dyeing of polyester at high temperature (typically 130 °C) in pressurised dyeing vessels. Having said that, it is true that the exhaust dyeing of other hydrophobic fibres, such as polyamide, acrylic and cellulose acetate, with disperse dyes is likely to follow the same series of heterogeneous processes.

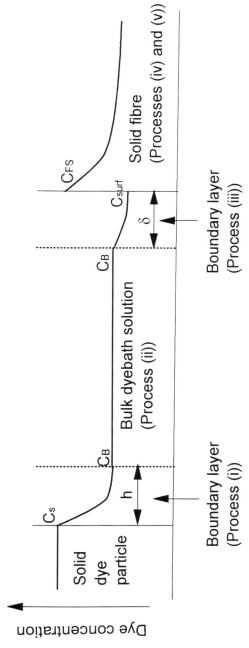

Figure 2.1 Schematic representation of heterogeneous dyeing processes.

Many mathematical treatments of the dyeing process are derived from those of Hill,[4] Crank[5] and Wilson.[6] These in turn derive from Fick's second law of diffusion and treat the system as effectively heterogeneous, with a single diffusion coefficient describing the process. Wilson's equation, for example, refers to an efficiently stirred finite dyebath in which the fibre is considered to be a cylinder of infinite length. The uptake of dye at time t can be written:

$$\frac{M_f(t)}{M_f(\infty)} = 1 - \sum_{n=1}^{\infty} \frac{4\alpha(\alpha + 1)}{4 + 4\alpha + \alpha^2 q_n^2} \exp\left(\frac{-D_f t q_n^2}{r^2}\right) \quad (2.1)$$

where $M_f(t)$ is the concentration of dye in the fibre at time t, D_f is the diffusion coefficient of the dye in the fibre, r is the fibre radius, α is equal to $1 - 1/E_\infty$, E_∞ is exhaustion of the dyebath at equilibrium and q_n is the nth root of the equation:

$$\alpha q_n J_0(q_n) + 2J_1(q_n) = 0 \quad (2.2)$$

where J_0 and J_1 are zero and first-order Bessel functions. Hill's equation describes the situation where the dyebath is infinite (i.e. the bath is saturated and presents a constant concentration of dye to the fibre throughout the dyeing process), and can be written as follows:

$$\frac{M_f(t)}{M_f(\infty)} = 1 - \sum_{n=1}^{\infty} \frac{4}{q_n^2} \exp\left(\frac{-D_f t q_n^2}{r^2}\right) \quad (2.3)$$

These equations can be solved graphically or by using a computer program, to derive D_f from a rate of dyeing curve; however, Shibusawa[7] has derived relatively simple polynomials which approximate both to Wilson's equation:

$$\frac{M_f(t)}{M_f(\infty)} = \frac{A_1 X + A_2 X^2}{1 + B_1 X + B_2 X^2} \quad (2.4)$$

($X = (D_f t/r^2)^{0.5}$ and A_1, A_2, B_1 and B_2 are simple functions of E_∞) and also to Hill's equation:

$$\frac{M_f(t)}{M_f(\infty)} = 1 - \left(\frac{1 - c_1 X}{1 + c_2 X + c_3 X^2 + c_4 X^3 + c_5 X^4}\right) \quad (2.5)$$

(c_1 to c_5 are constants). Values of the diffusion coefficient obtained from such analyses are given by the author. For the application of C.I. Disperse Violet 1 to polyester at 100 °C, D_f was found to be approximately 3×10^{-11} cm^2/min. An earlier approximation to Wilson's equation[8] has also been fitted to experimental data obtained from dyeing nylon at 40 °C with a disperse dye in the presence of an anionic dispersing agent. The rate of dyeing was retarded in the presence of the dispersing agent, and this was ascribed to a decrease in the apparent partition coefficient, K. In this example a value of D_f of 3.05×10^{-10} cm^2/min was obtained.

In addition to these mathematical treatments, a wide variety of empirical equations have been developed to describe the process of disperse dyeing. Most of these treatments yield a rate constant, from which a diffusion coefficient may be derived (e.g. Cegarra and Puente,[9] Militký[10]). These have been summarised and compared by Shibusawa[11] who concluded that many are applicable when the diffusional boundary layer (see section 2.2.5 below) plays an appreciable part in introducing a delay to the dyeing process.

It is well established that transport of dye molecules from dispersed solid particles in the dyebath to a 'solid solution' in the polyester occurs via a dyebath solution phase, even though the dye may have sparing solubility in the dyebath. For a true heterogeneous treatment, it is necessary to break down the disperse dyeing process into its constituent steps, which may be described as shown in Figure 2.1. Dye–fibre reaction may, of course, be ignored.

In the following sections these processes are described in more detail.

2.2.2 Description of the dissolution process

The process of dye dissolution is largely ignored by mathematical treatments of the aqueous disperse dyeing of polyester. The assumption is usually made that dye dissolution is relatively rapid, and that a saturated solution of dye is presented in a well stirred dyebath until such time that dye uptake by the polyester is sufficient and that all solid dye particles have dissolved.[12] In these treatments, the kinetics of dyeing are, therefore, determined only by those processes occurring after dissolution. On the other hand, a few authors have addressed the issue of dissolution and its influence on dyeing kinetics. For instance, Merian and Lerch[13] state that 'If the solubility is insufficient, not enough of the dye molecule is available for dyeing in the time specified, and the diffusion rate is no longer the phenomenon determining the rate of adsorption'. The rate of adsorption of dye onto acetate and nylon is shown to correlate well with the product $D_f\sqrt{C_s}$. Rattee and Breuer[2] state, without further discussion, that the dissolution rate of disperse dyes is of importance, while Braun[14] discusses in more detail the importance of dispersed particle size on both absolute thermodynamic solubility and dissolution rate.

The rate of dissolution of a solid particle is generally considered to depend on its surface area and, hence, on its linear dimension, as well as on the concentration gradient of dissolved solute between the surface of the particle and the bulk solution (see below). As the magnitude of this gradient depends on saturation solubility,[15] solutes having a low equilibrium saturation solubility will also tend to dissolve at a low rate. It is worth remembering that, although we generally define all disperse dyes as being 'sparingly soluble', equilibrium solubility can vary by over two orders of magnitude from dye to dye. Braun[14] states that a low dissolution rate may lead to misleading conclusions during a determination of the solubility of sparingly soluble solids such as hydrophobic fluorescence brightening agents and disperse dyes. It is, therefore, reasonable to assume that, on occasion, over the short period of time (about 1 h) associated with the batchwise disperse dyeing of polyester, the dyebath may not have become saturated with respect to the dye. Indeed, Odvárka and Huňková[16] have presented experimental evidence that the rate of dyeing of polyester fibre by the anthraquinone dyes C.I. Disperse Blue 56 and C.I. Disperse Red 60 can be limited by the rate of dissolution. However, it should be noted that the mean particle sizes quoted by these authors (6–10 μm) are rather larger than those found in commercial dye dispersions, and that the experiments were conducted in the absence of dyeing auxiliaries. One might expect such auxiliaries to enhance the rate of dye dissolution. This means that published evidence for the importance of dissolution rate dictating the overall kinetics of disperse dyeing is quite weak, although the topic appears to be of current interest. To quote Aspland:[17] 'If during dyeing the rate at which dye can be taken from solution by the fibre (the intrinsic rate of dyeing) is higher than the rate at which dye can be dissolved from the solid and released from the micelles, then the overall rate of dyeing will effectively be controlled by the rate at which dye molecules dissolve. To ensure that low rates of dissolution do not act as a bottleneck, the disperse dyes are ground to as fine a particle size as is reasonably possible ... because the rate of dissolution is related to $1/d$, where d is the particle diameter'.

For a mathematical treatment of the kinetics of dissolution one must look outside the dyeing literature. A recent review of dissolution kinetics for organic compounds is given by Grant and Higuchi[15] and is illustrated schematically, for instance, by Navratil.[12] The dissolution rate for many organic compounds can be described by the Diffusion Layer theory. In the simplest version of this model, the concentration of solute at the solid surface is equal to the saturation solubility (C_S), and this concentration decreases to that of the bulk solution (C_B) over a distance h (the thickness of the diffusion layer). The rate of dissolution (dm/dt) is then proportional to the surface area of the solid (A), to the diffusion coefficient of the solute molecule in the solvent (D_{Aq}), and to the concentration gradient across the diffusion layer, $(C_S - C_B)/h$, i.e.:

$$\frac{dm}{dt} = AD_{Aq}\frac{(C_S - C_B)}{h} \quad (2.6)$$

To determine the initial rate of dissolution in real systems (i.e. initial bulk concentration $C_B = 0$), the quantities A, D_{Aq} and C_S must be determined. A number of methods exist to determine the 'aqueous solubility' of disperse dyes.[12,18,19] This is a quantity somewhat more difficult to determine than might at first seem to be the case, due to the requirement to make the measurement under the 'real' dyeing conditions of high temperature and pressure in the presence of formulating agents and other auxiliaries. It must also be noted that the apparent solubility depends on the type and concentration of such additives in the dyebath, and much has been made of the distinction between dissolved dye and dye 'solubilised' in surfactant micelles or aggregates.[20] An estimate of the mean particle size or the particle size distribution may yield a value for A, either by assuming spherical particles of radius x, or by introducing a shape factor, if such information is known (e.g. from electron microscopy). A further complicating factor may be introduced if one takes into account that C_S depends on x, at least at sub-micron particle sizes,[14] according to the Ostwald–Freundlich expression:

$$\ln \frac{C_S(r_1)}{C_S(r_2)} = \frac{2M\sigma}{RT\rho}\left(\frac{1}{x_1} - \frac{1}{x_2}\right) \quad (2.7)$$

where $C_S(x_1)$ and $C_S(x_2)$ are the solubility of particles of radius x_1 and x_2, respectively, M is the molecular weight of the solid, σ is the specific free energy of the interface between solid particle and solution and ρ is the density of the solid.

It may be possible to measure h or to estimate its value relative to x, by using the Nusselt number, Nu.[22] The aqueous diffusion coefficient, D_{Aq} may be measured[15] or, more easily, estimated roughly by using a correlation such as that given by Wilke and Chang,[20] i.e.:

$$D = 7.4 \times 10^{-8} \frac{(yM)^{1/2}T}{\eta V^{0.6}} \quad (2.8)$$

V is the molal volume (cm^3/mol), T is the absolute temperature (K), M is the solvent's molecular weight, y is an association parameter used to account for hydrogen bonding in the solvent (= 2.6 for water), and η is the solvent viscosity (cp).

2.2.3 Practical factors affecting the rate of dissolution

In addition to the effects of particle size and surface area discussed above, a number of other experimental variables may influence the rate of dye dissolution. Different dyes will have different equilibrium solubilities and, hence, dissolution rates. Also, even for the same dye, different crystal polymorphs will have different equilibrium solubilities (through their different solid state free energies) and rates of dissolution. The effect of crystal polymorphism is illustrated by, for instance, the work of Biedermann,[23] where it is demonstrated not only that five polymorphs of the same dye achieve very different saturation values (C_f) in cellulose acetate, but also that the rate of achieving fibre saturation is more rapid for the high solubility (low stability) forms. It is not possible, however, to determine whether this rate effect is truly a dissolution rate effect or merely a manifestation of higher equilibrium solubility.

Numerous attempts have been made to predict the thermodynamic parameters of a dye (e.g. C_f, C_S and K) which can also determine the kinetic regime from a consideration of the molecular structure alone (for instance by using the Hildebrand solubility parameter), but it appears that, to date, no method has been entirely successful.[24]

The presence of formulating, dispersing or levelling auxiliaries in the dyebath will also affect the rate of dissolution, as partition of dye to any micellar or aggregate phase will reduce the amount of dye in 'true' bulk solution (C_B). If C_B is lower, the concentration gradient between the solid surface and the bulk solution will be maintained, thus enhancing the dissolution rate. It should be noted, however, that if dye is held in micelles or aggregates in the dyebath then the effective partition coefficient to the fibre surface will be reduced. In turn, this slows the rate of uptake of dye from bath to fibre and possibly also reduces the total colour yield.

The effect of temperature on dissolution rate is fairly obvious; increasing the temperature will more than likely increase the equilibrium solubility, C_S, as well as accelerate diffusion through the boundary layer by increasing D_{Aq}. It is worth remembering, although it is also difficult to model, that in real dyeing fibre is introduced to the bath at temperatures well below the final dyeing temperature, and that dissolution and transport to the fibre will commence before that temperature is reached.

The effect of efficient agitation is also obvious, as the width of any 'stagnant' diffusional boundary layer (h) surrounding the dye particles will be reduced, and the magnitude of the concentration gradient will be increased.

2.2.4 Finite or infinite kinetics?

The physical chemist is permitted the luxury, not always afforded to the practical dyeing technologist, of assuming that the dyebath is well agitated and that all of the fibre to be dyed is fully accessible to the dye in solution. Thus, we assume that the concentration of dye in bulk solution is uniform throughout the dyebath, although not of course necessarily uniform with time, apart from in those microscopic 'diffusional boundary' layers adjacent to the solid dye particle surface and to the fibre surface where concentration gradients exist. There remain, therefore, a number of processes to consider before the dye is absorbed by the fibre, i.e.:
(i) transport of the dye molecule across a concentration gradient from

the bulk solution to the fibre/solution interface, (ii) partition of dye from solution to the immediate fibre surface, and (iii) diffusion of the dye within the fibre.

At this stage, the concentration of dye in bulk solution as a function of time must be considered. Much has been made of the distinction between: (i) infinite dyebath kinetics (where solid dye is always present in excess and the dyebath remains saturated throughout the dyeing process), (ii) finite dyebath kinetics (where all dye is initially in solution and this bulk solution concentration drops throughout the dyeing process), and (iii) transitional kinetics (where all dye dissolves at some point during the dyeing process). McGregor and Etters[25] describe these regimes in some detail. The point at which the kinetic regime changes from infinite to finite depends on: (i) the liquor ratio (Λ) used in dyeing and (ii) the partition coefficient (K) of the dye between fibre and dyebath. At the point of transition, a discontinuity in the dyeing rate will be seen (Figure 2.2). The rate of dyeing will drop after dissolution is complete. In simple terms, the transition from saturated to unsaturated solution will provide a reduced driving force (through a reduced concentration gradient) for the uptake of dye by the fibre. McGregor and Etters compared rates of dyeing curves, calculated by numerical computer simulation, with simplified analytical expressions. The transition between infinite and finite kinetics occurs at

$$\frac{C_d}{C_f} = \alpha\left[\frac{C_B}{C_S} - 1\right] \quad (2.9)$$

where C_d is the concentration of dye in the fibre, C_f is the saturation value (saturation solubility) of dye in the fibre, C_B is the concentration of dye dissolved in bulk solution, C_S is the saturation solubility of the dye in the dyebath and $\alpha = \Lambda/K$. Alternatively, the condition for infinite kinetics can be expressed as:

$$C_0 \geq C_S(1 + (1/\alpha)) \quad (2.10)$$

where C_0 is the initial total dye concentration, both dissolved and as solid. Infinite kinetics will, therefore, be favoured by systems where the dye has low aqueous solubility (assuming that dissolution kinetics do not limit, as this would complicate the matter), where the liquor ratio is low or where the depth of shade is heavy.

A numerical simulation which takes into account this transition is described by Navratil.[12] In this computer model, the dyeing process is divided into a series of iterative steps. At each step, the concentrations of solid dye, dissolved dye, and dye in the fibre are all calculated and a revised distribution of the dye in the system is derived. Such a numerical treatment relies on a large number of assumptions and requires the

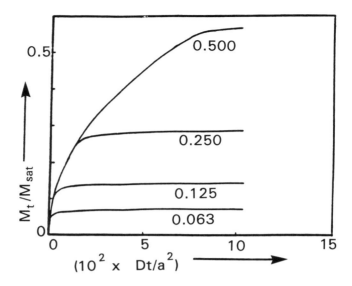

Figure 2.2 Dye uptake as a function of time under a 'transitional kinetic' regime. The respective lines refer to different depths of dyeing (as % dye on weight of fibre). Reproduced from McGregor and Etters[25] with permission.

measurement and/or estimation of a number of thermodynamic parameters for the system of interest. However, such a model appears to be very flexible in terms of the number of modifications which may be built in.

2.2.5 Transport to the fibre surface

Most treatments of the transport of dye from bulk solution postulate a 'boundary layer' which defines a region of concentration gradient. Such a layer is analogous to that described earlier for the dissolution process. Its existence is discussed by, for example, Etters[26] who states that it may become significant where dyebath agitation is inefficient, although its relevance to real systems is questioned by Burley.[27] Whether or not its influence is significant, it can be represented diagrammatically (Figure 2.3, from Jones[28]).

The rate of transfer across the boundary is then given by Fick's equation, so:

$$J = -D_{Aq}\frac{C_{Surf} - C_B}{\delta} \quad (2.11)$$

where J is the rate of change of mass per unit surface area of the fibre, C_{Surf} is the concentration of dye in solution at the fibre surface and δ is

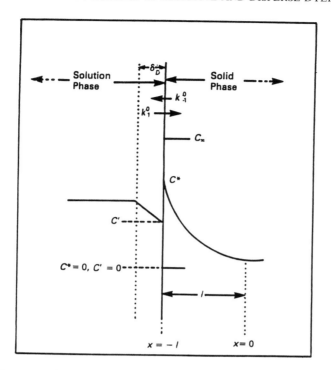

Figure 2.3 Representation of diffusion into a substrate through a diffusional boundary layer. Reproduced from Jones[28] with permission.

the width of the diffusional boundary layer. The concentration of dye in the fibre at the fibre surface can be related simply to that in the aqueous phase by the equilibrium partition coefficient, so:

$$K = \frac{C_{FS}}{C_{Surf}} \qquad (2.12)$$

where C_{FS} is the concentration of dye in the fibre at the fibre surface. This assumes that partition at the very surface of the fibre is, like dissolution at the surface of a solid dye particle, instantaneous. The partition coefficient is a thermodynamic parameter which describes the hydrophobicity of the dye molecule, and can be related to the thermodynamic affinity $-\Delta\mu^{\ominus}$ by:

$$-\Delta\mu^{\ominus} = RT \ln K \qquad (2.13)$$

Etters[26] concentrates on examining the effect on dye uptake when a significant diffusional boundary layer exists and the system deviates significantly from that described by Wilson's equation.[6] The author calculates the effect on the half-time of dyeing of the dyebath exhaustion E_∞ and of a parameter L which depends on δ as follows:

$$L = \left(\frac{D_{Aq}}{D_f K}\right)\left(\frac{r}{\delta}\right) \quad (2.14)$$

The data of Etters are presented graphically in Figure 2.4 and show that the half-time of dyeing is much more sensitive to L (and hence to δ) when the exhaustion is high.

Deviations from expressions such as Wilson's or Hill's equations have often been observed (e.g. Cegarra and Puente,[29] Narębska et al.[30] and Shibusawa[11]). These deviations manifest themselves as a delay in dye uptake early in the dyeing process and are variously explained by a desorption process at the fibre surface[29] or by a 'stagnant solution layer surrounding the fibres'.[11] The latter author states how Newman's equation[31] can be used to describe the uptake of dye at time t (i.e. $M_f(t)$) in an infinite dyebath in the presence of a diffusional boundary layer as follows:

$$\frac{M_f(t)}{M_f(\infty)} = 1 - \sum_{n=1}^{\infty} \frac{4L^2}{B_n^2(B_n^2 + L^2)} \exp\left(-B_n^2 \frac{D_f t}{r^2}\right) \quad (2.15)$$

(B_n is the nth non-zero root of $BJ_1(B) - LJ_0(B) = 0$, and J_0 and J_1 are zero and first-order Bessel functions respectively). As L decreases (i.e. δ increases) a delay in dye uptake is predicted. Shibusawa then went on to compare a number of published empirical rate of dyeing equations with Newman's equation. The empirical equation of Hirusawa and Matsumura[32] has the form:

$$\frac{M_f(t)}{M_f(\infty)} = 1 - \exp\left(-\frac{kt^{0.8}}{M_f^2(\infty)}\right) \quad (2.16)$$

and this was found to correspond well with Newman's equation when a value of $L = 8$ was used. The simple exponential form of a number of authors:[1,33-35]

$$\frac{M_f(t)}{M_f(\infty)} = 1 - \exp(-k't) \quad (2.17)$$

agrees with Newman's equation when the diffusional boundary layer is large (i.e. L tends to zero). The value of k' can be shown to depend on δ as follows (where A is the fibre surface area):

$$k' = A \frac{D_{Aq}}{\delta} \left(\frac{1}{K} + \frac{1}{\Lambda}\right) \quad (2.18)$$

HETEROGENEOUS KINETICS OF REACTIVE AND DISPERSE DYEING 55

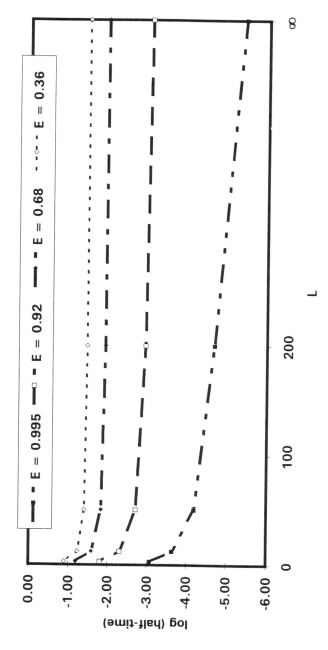

Figure 2.4 Transitional kinetics—dependence of half-time of dyeing on L (data from Etters[26]).

2.2.6 Practical factors which affect transport to the fibre surface

The effects of liquor ratio (Λ), partition coefficient (K) and aqueous solubility (C_s) on the transition between infinite and finite kinetics have been discussed above. Clearly these are all variables which can be manipulated to some extent to shift this transition point. Efficient agitation, as with dissolution, will increase the rate of uptake through reducing δ and increasing the magnitude of the concentration gradient. The partition coefficient K determines what proportion of dye in solution is taken up to the extreme surface of the fibre, and it may be manipulated practically (i) by changing to another dye having a different affinity or degree of hydrophilicity, (ii) by changing the concentration of solubilising auxiliaries, or (iii) by altering the temperature (a higher temperature will reduce K). The equilibrium thermodynamic quantities C_f (the maximum concentration of dye which can be achieved in the fibre), C_s and K for a particular dye are not independent, however, as $K = C_f/C_s$.

2.2.7 Description of the process of diffusion within the fibre

Diffusion of dye molecules within polyester fibre is most commonly represented by Fick's first law in which the flux (rate of mass transport per unit area), J is proportional to the concentration gradient. Thus:

$$J = D_f \frac{dC_f}{dx} \qquad (2.19)$$

The constant of proportionality, D_f, is the diffusion coefficient of the dye within the fibre, and generally has units of $[L]^2[T]^{-1}$. The determination of D_f from experimental measurements requires the solution of differential equations, and experimental methods for doing this are well summarised by earlier reviewers such as Rattee and Breuer[2] and Jones.[28] Such methods include (i) the measurement of the time lag required to achieve steady-state diffusion through a thin polyester film, (ii) the measurement of non-steady-state diffusion through parallel sheets or slabs of polymer, and (iii) direct measurement of the dye distribution in real fibres by the use of a microtome and a microdensitometer. If the diffusion coefficient is measured at more than one temperature, an activation energy for diffusion can be obtained[29,36] using an Arrhenius equation of the form:

$$D_f = D_0 \exp(-\Delta E/RT) \qquad (2.20)$$

Cegarra and Puente[29] quote activation energies for diffusion in polyester fibre of about 70 kcal/mol for C.I. Disperse Red 15 between 70 and 95 °C. Diffusion coefficients for typical disperse dyes in polyester at 130 °C are of the order of 10^{-10} cm^2/s.[37] The numerical computer model of Navratil[12] couples the description of diffusion within the fibre with the earlier process of partition from solution, and calculates iteratively the distribution of dye within the fibre as a function of time. This enables an analysis to be made of such phenomena as 'ring' dyeing and other inhomogeneous distributions of dye in the fibre.

Most treatments of dye diffusion within the fibre assume that the value of D_f is independent of concentration. Some evidence exists, however, showing that the diffusion coefficient of disperse dyes in polyester can increase considerably as the concentration of dye increases. Rattee and Breuer[2] summarise the free volume model of dye diffusion. At temperatures above the glass transition temperature (T_g) of the fibre (about 90 °C for polyester) dye can move through voids in the polymer (which is behaving like a very viscous liquid). Addition of a diluent such as dye molecules will alter the free volume (V_f) and T_g. The diffusion coefficient varies with V_f as follows:

$$D_f = RTa_1 \exp(-a_2/V_f) \qquad (2.21)$$

(a_1 and a_2 are constants) and, in principle at least, the variation of V_f with (i) dye volume fraction, (ii) the thermal expansion coefficients of dye and polymer, and (iii) T_g can be calculated to enable a prediction to be made of the variation of D_f with dye concentration.

Kojima and Ijima[38] quote results for a number of disperse dyes. C.I. Disperse Orange 3 at 100 °C, for instance, shows over a ten-fold increase in D_f as the dye concentration in polyester film increases from 0 to 1600 mol/kg. These authors then go on to calculate an intrinsic diffusion coefficient χ, given by:

$$\chi = \frac{D_f}{(1 - \phi_1)^3} \qquad (2.22)$$

Here, ϕ_1 is the volume fraction of the amorphous region of the polyester occupied by the dye. From the variation of ϕ_1 with the dye activity (or concentration) in the polyester, the thermodynamic diffusion coefficient, χ_T can be evaluated. The variation of $1/\ln(\chi_T/\chi_0)$ with $1/\phi_1$ was found to be linear, indicating that the free volume theory of Fujita et al.[39] can explain the concentration dependence of D_f.

In principle, a concentration dependent diffusion coefficient may be incorporated in a computer simulation such as that exemplified by Navratil.[12] In its simplest form it can be represented as:

$$D_f = D_{f,0} + BC_f \qquad (2.23)$$

where B is a constant determined experimentally for the system of interest. A recent mathematical treatment of concentration dependent diffusion has been given by Lin,[40] although it is primarily applicable to the adsorptive (pore) model of dye diffusion which normally applies to fibres such as cellulose and wool.

A combination of the free volume theory and the pore model to describe diffusion in polyester has been made by Hori et al.[41] primarily in order to study the process of solvent dyeing where pores and fibre swelling are important. For aqueous systems, Hori et al. conclude that a high T_g and strong dye–polymer interactions (i.e. the partition coefficient, K, is high) account for low values of D_f. Pore swelling is not very significant for aqueous dyeing. The authors derive an equation which describes D_f as follows:

$$\ln D_f = \ln \frac{\phi_0 \delta^2}{6} + \ln \frac{\theta}{\tau} - \ln K - \frac{\Delta H_H}{RT} - \frac{AB}{B + (T - T_g)} \quad (2.24)$$

The first term is a probability term, the second takes into account the pore model (θ is the pore volume and τ is the 'tortuosity' of the pore network), the third term describes dye–polymer affinity, the fourth describes the enthalpy for 'hole' formation in the polymer, and the final term is a free volume term.

2.2.8 Practical factors affecting the rate of diffusion in the fibre

The rate of dye diffusion in the fibre may of course be increased by increasing the diffusion coefficient of the dye. In the simplest approximation (see the correlation of Wilke and Chang[21] above) the diffusion coefficient is inversely proportional to the molecular volume to the power 0.6 and, therefore, smaller molecules will have a higher diffusion coefficient. However, since no dye molecule is spherical, it seems likely that molecular shape will also play a role. The above correlation is essentially hydrodynamic in nature and so ignores any specific interactions (i.e. dipole–dipole, hydrogen bonding, Van der Waals interactions) which may occur between dye molecule and polymer molecule. It is likely, then, that specific functionalisation of a dye molecule will alter its diffusion coefficient, even if it does not change its molecular volume.

Carriers have long been used to enhance the rate of dye uptake, particularly so that temperatures above the boil are not required. They are generally believed to act by swelling the fibre and so increasing the free volume, enabling more rapid diffusion as

$$D_f = RTa_1 \exp(-a_2/V_f) \quad (2.25)$$

The free volume also depends on temperature and the glass transition temperature:[2]

$$V_f(T) = V_f(T_g) + \alpha(T - T_g) \quad (2.26)$$

Therefore, D_f increases as we get further away from the glass transition temperature either by increasing the dyeing temperature or by using a less crystalline or modified polyester having a lower T_g.

2.2.9 The heterogeneous processes of levelling (migration)

Once a disperse dye has diffused into the fibre during the dyeing process, it does not become immobilised. In practical dyeing, because of the limitations of the dyeing machinery, because of instability of the dye dispersion or because of the properties of the dye molecule itself, it is often the case that the textile is dyed unevenly on a macroscopic scale. To some extent level dyeing can be promoted by control in application, by the use of levelling auxiliaries, or by a post-treatment of the dyed fibre. During a dyeing process (before or after dyebath exhaustion), level dyeing may be promoted by movement of dye from heavily dyed regions to regions of paler depth. This levelling or migration process is generally held to occur via the solution phase and so involves the following processes (Figure 2.5):

(i) Diffusion of dye molecules to the surface of the heavily dyed piece.
(ii) Partition at the surface (and diffusion through any aqueous boundary layer).
(iii) Transport in the bulk dyebath.
(iv) Partition to the fibre surface (preceded by diffusion through any boundary layer to the surface of the pale dyed fibre).
(v) Diffusion of dye molecules through the pale dyed fibre.

The driving force for levelling by these processes is an overall concentration gradient between the dye in the heavily dyed areas and the dye in less heavily dyed areas. The processes (with the exception of diffusion through the boundary layer) have been modelled by Navratil,[42] using a method similar to that used by the same author for the disperse dyeing process itself, as they are entirely analogous. Most of the quantities which must be measured or estimated are also required to model the dyeing process. On this occasion, however, it was possible to derive simpler approximations which describe the migration process. The simplest of these defines so-called 'standard' conditions of the dyed piece and the dyebath (liquor ratio, dyebath volume, temperature, etc.) and the concentration of dye in the piece to be levelled ($C_L(t)$, initially set to be zero) varies with time as follows:

$$C_L(t) = \frac{50K}{K+10}\left(1 - \exp\left(-\left(\frac{a_1 D_f}{r^2} + a_2\right)t + \frac{K}{(a_3 + a_4 K)}\right)\right) \quad (2.27)$$

where a_1 to a_4 are constants, (59.04, 9.17 × 10⁻⁴, 571.91 and 5.2295,

60 PHYSICO-CHEMICAL PRINCIPLES OF COLOR CHEMISTRY

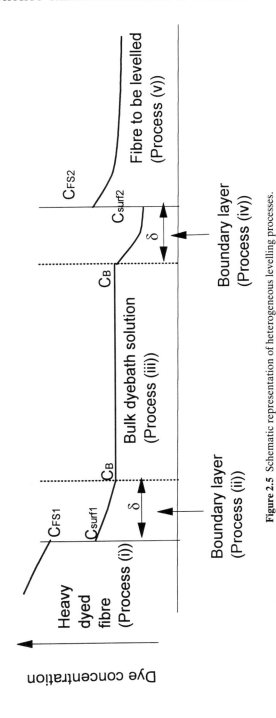

Figure 2.5 Schematic representation of heterogeneous levelling processes.

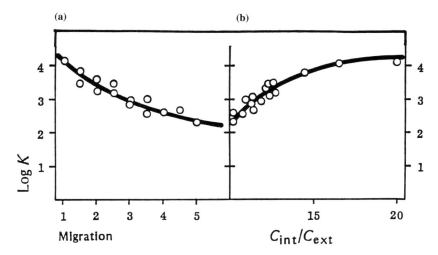

Figure 2.6 Relation between partition coefficient and (a) migration and (b) unevenness of dye packages. C_{int}, dye concentration on the internal surface of the package (g/kg); C_{ext}, dye concentration on the external surface of the package (g/kg); K, partition coefficient (C_f/C_s). Reproduced from Gerber[43] with permission.

respectively). Results from using these approximations are shown to be in good agreement with experiment.

The above equation shows that, all other conditions being equal, the rate of migration depends on a balance of two properties, K and D_f which are essentially properties of the dye molecule alone. From a consideration of the above five processes migration (i) will be favoured by a high value of D_f, (ii) will be favoured by a low value of K, (iii) may be ignored in a well stirred dyebath, (iv) will be favoured by a high value of K, and (v) will be favoured by a high value of D_f. Which of these processes dominates will depend on the dye molecule. Merian and Lerch[13] and Gerber[43] show data to demonstrate that good migration performance correlates well with a low value of K (Figure 2.6), indicating that in these cases process (ii) (partition to the dyebath) dominates.

2.3 The heterogeneous kinetics of dyeing cellulose with fibre reactive dyes

2.3.1 Introduction

On a molecular level cellulose consists essentially of poly-1,4-D-glucopyranose with a backbone of hydroxy groups which have an apparent pK_a of 13.74.[44] Under basic conditions these groups can ionise, to an extent determined by the local pH, to give the corresponding nucleophilic cellulosate anion. Following the invention of dyes possessing electrophilic

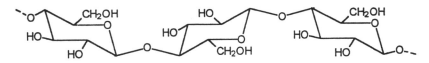

Structure 2.1 Structure of glucopyranose.

groups which could react with these nucleophiles, extensive effort was invested in understanding the dyeing process involved. These efforts can be split broadly into two areas: qualitative descriptions of actual dyeings as exemplified by times of half dyeing (relating to bulk dyebath concentrations) and more fundamental descriptions of the physical and chemical processes that lead to the observed variations in bulk concentrations. Whilst the former without doubt helps the dyer in the application process, it is only the latter of these two types of investigations that can lead to a rigorous understanding needed for the development of improved dyes and application processes.

Initial mathematical treatments treated the system as a competitive reaction, for the electrophilic dye molecule, between the cellulosate anion (fixation) and hydroxide (hydrolysis) whilst attempting to take account of the heterogenous nature of the reaction (Figure 2.7). The derived equations involve similar treatments, as outlined for disperse dyeing, but now have to take into account the chemical reactions involved. Adapta-

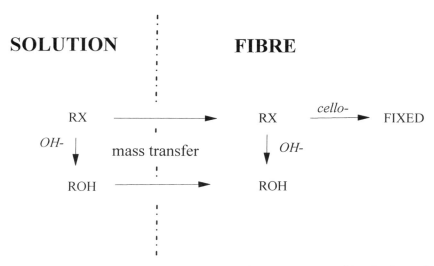

Figure 2.7 Schematic representation of reactive dyeing process: competitive fixation and hydrolysis in fibre phase.

tion of Ficks's second law to take account of coupled mass-transfer and chemical reaction gives the usual expression:

$$\frac{dC}{dt} = D\left(\frac{d^2C}{dx^2}\right) - kC \tag{2.28}$$

where k is the first-order rate constant for the reaction that produces immobilisation of the dye, D is the diffusion coefficient, and C the concentration of reactive species.

Application of suitable boundary conditions, assuming specific substrate geometries, allows the solution of the above equation in several different forms. For example, Danckwerts' treatment[45] of coupled diffusion–reaction into an infinite slab from a source of constant surface concentration gives

$$F_t = C_{FS}\left(\frac{D_{obs}}{k_f}\right)^{0.5}(k_f t + 0.5)\,\text{erf}\,(k_f t)^{0.5} + \left(\frac{k_f t}{\pi}\right)^{0.5}\exp\,(-k_f t) \tag{2.29}$$

where F_t is the quantity reacted in time t, C_{FS} is constant surface concentration of reactive dye, D_{obs} is the observed diffusion coefficient and k_f is the first-order rate constant for reaction with the cellulose. The most well-known simplification of this expression is that derived by Sumner and Weston[46] who showed that when $\text{erf}(k_f t) \sim 1$, i.e. $k_f t$ is large, the equation becomes

$$\frac{dF}{dt} = C_{FS}(D_{obs}k_f)^{0.5} \tag{2.30}$$

where dF/dt is the rate of fixation of reactive dye per unit area of surface. The competitive hydrolysis reaction can occur in both the bulk dyebath solution and within the internal pore solution, with an overall rate of hydrolysis given by

$$\frac{dC_{ROH}}{dt} = k_{OH_b}C_b C_{OH_b} + k_{OH_i}C_i C_{OH_i} \tag{2.31}$$

in which C_{ROH} represents the hydrolysed dye concentration, C_{OH} the hydroxide ion concentration, k_{OH} the second-order rate constant for reaction with hydroxide, and the subscripts b and i refer to the bulk and internal solution phases, respectively. The derived equations for coupled mass-transfer and chemical reaction are only applicable to first-order reactions and, so, it is traditional to also express the competitive hydrolysis as pseudo-first-order with $k_w = k_{OH}[OH^-]$. It must be noted that the pseudo-first-order rate constant for fixation, k_f, can also be expressed in terms of bimolecular reaction with the cellulosate anion, $k_f = k_{cello-}C_{cello-}$. The concentration of cellulosate anion under varying conditions can be calculated,[2] and this has been used to calculate the apparent second-order rate constant.

The total rate of disappearance of the active dye is the sum of the rates of the two competing reactions

$$-\frac{dC}{dt} = \frac{dF}{dt} + \frac{dC_{ROH}}{dt} = C_{FS}(D_{obs}k_f)^{0.5} + k_{w_b}C_b + k_{w_i}C_i \quad (2.32)$$

We, hence, have mathematical descriptions of the rate of production of the two products derived from fixation and hydrolysis and the total loss of the reactive dye. An assessment of the efficiency of the dyeing process is, hence, given by the ratio of the rate of the desired process (fixation) to the unwanted side reaction:

$$E = \frac{dF}{dC_{ROH}} = \frac{C_{FS}(D_{obs}k_f)^{0.5}}{k_{wb}C_b + k_{wi}C_i} \quad (2.33)$$

These equations have been utilised by several workers to examine the behaviour of various dyes and have been further expanded, most notably by Rattee and Breuer,[2] to take account of the differing amount of each phase and the influence of the surface potential of the anionic fibre on the distribution of the numerous charged species.

A more realistic solution to the coupled mass-transfer–chemical reaction equation is to set boundary conditions to describe diffusion–reaction into a cylindrical fibre. The solution becomes

$$F_t = 4\pi D_{obs} C_{FS} \sum_{n=1}^{\infty} \frac{k_f + D_{obs}\alpha_n^2 \exp[-t(k_f + D_{obs}\alpha_n^2)]}{k_f + D_{obs}\alpha_n^2} \quad (2.34)$$

where α_n are the successive roots of the Bessel function $J_0(r\alpha) = 0$.[2,47] Rys and Zollinger[48] used this type of analysis to derive an efficiency expression for a constant dyebath concentration that is divided into two components, E_r and E_d. The factor E_r arises when reactivities are the only factors, and E_d when fixation is diffusion controlled. Which of the two dominates depends on the magnitude of the ratio of the relaxation times of the diffusion and reaction process. When this is much less than unity, the efficiency is dominated by chemical reactions, and when the ratio of relaxation times greatly exceeds unity, the role of diffusion becomes dominant.

This type of treatment splits the overall mechanism into components dominated by either chemical reaction or the transport steps, but fails to address realistic dyeing conditions not adequately described by the limiting boundary conditions used. As with all heterogeneous processes, the rate of reaction can be dependent upon any of the steps involved in transferring the reactant species from the bulk solution to the reaction site, including the chemical reaction itself. The relative contribution of

HETEROGENEOUS KINETICS OF REACTIVE AND DISPERSE DYEING   65

each of the steps to the overall process is governed by dyestuff molecular structure, fibre structure, and application conditions. Furthermore, the dynamic nature ensures that these relative contributions will be time dependent.

The overall heterogeneous reaction can be subdivided into a number of elementary steps corresponding to the processes (ii)–(v) outlined in Figure 2.1, and each of these will be examined in the sections which follow.

2.3.2 *Uptake of dye from bulk solution to internal fibre*

In the above analyses, the concept of three distinct phases has been used: the bulk solution phase, the internal solution phase, and the fibre substrate. The usual definition of the process of adsorption from the bulk dyebath solution through an interfacial boundary layer is not as clear. For the discussion here, this microscopic diffusional boundary layer is described as that between the bulk dyebath and the internal phase which consists of the free pore solution and the fibre surface (Figure 2.8). Before this adsorption process can occur, the dye must migrate through the bulk solution to the interface. As dye is adsorbed, the adjacent liquor will become deficient in dye; the deficiency is made up by diffusion of

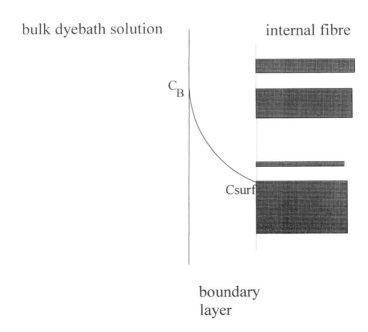

Figure 2.8 Representation of bulk solution–internal fibre interface.

more dye to the interface. It is generally assumed that this aspect can be ignored, provided that dyeing is carried out under conditions of adequate agitation.[49,50] This assumption would appear reasonable, given that a major part of dyeing machine technology has focused on it.

In order to probe the actual interfacial process, a knowledge of the true surface concentrations is required. In contrast, most dyeing kinetics research has monitored changes in the bulk dye concentration to make inferences about the surface processes.[15,51–54] Measurements of bulk concentrations provide little information on the surface reaction mechanism due to the remote and insensitive nature of the detection. Until recently, there have been few, if any, satisfactory methods of interrogating heterogeneous processes occurring between solution species and non-conducting solids, such as cotton. Compton and co-workers[55,56] have described the use of the channel flow cell (Figure 2.9) to investigate the reaction of dichlorotriazine-based dyes **1** and **2** with cotton.

The method involves pumping alkaline dye solution through a thermostatted cell, under modellable laminar flow conditions over the cotton substrate which forms part of one wall of the channel. The downstream 'detector electrode' monitors either an electroactive component of the dye molecule itself (e.g. anthraquinone or azo chromophore) or chloride

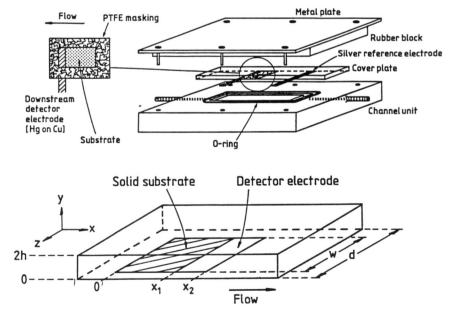

Figure 2.9 Channel flow cell used for the study of the heterogeneous kinetics of dyeing and schematic diagram of cell dimensions. Reproduced from Compton and Wilson[56] with permission.

Structure 2.2 Structures of dyes **1** and **2**.

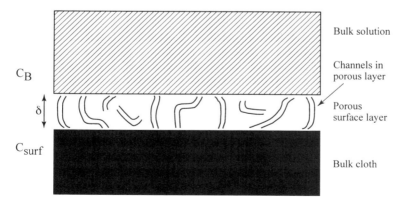

Figure 2.10 Diagram representing adsorption of dye into surface pore structure. Reproduced from Gooding[58] with permission.

ion concentration (liberated as a result of either hydrolysis or fixation). In this way, the actual concentrations of the active species at the interface are known and utilised in subsequent kinetic analyses. If the rate of solution hydrolysis is well defined under the experimental conditions, analysis of the rate of fibre adsorption and/or chemical reaction (depending upon detection technique used) is possible. Varying the flow rate enables the extent of the contribution from diffusion to the interface to be assessed. For both dyes **1** (detected for rate of adsorption) and **2** (detected for rate of reaction) the reaction is seen to be first-order in the surface concentration

$$J = k_1 C_{\text{surf}} \tag{2.35}$$

where J is the interfacial flux, C_{surf} the dye surface concentration and k_1 the observed first-order rate constant found to be 2.6×10^{-4} cm/s and 4.3×10^{-4} cm/s, respectively. The rate constants are constant with varying flow rates, suggesting that diffusion from the bulk solution can be ignored.

This method seems to be of general applicability to the study of the kinetics of reactive dyeing, especially since the detection methods have been generalised for non-electrochemical techniques such as spectroscopic methods.[57] Utilisation of the differing detection methods possibly provides differentiation between the physical and chemical steps involved after transfer to the fibre surface. The work on dyes **1** and **2** was carried out under typical fixation conditions. More recently Gooding[58] used this method to investigate the adsorption of non-reactive water-soluble dyes onto cotton under neutral conditions analogous to the primary exhaustion stage in exhaust dyeing of reactive dyes. The acid dyes Orange G and Edicol Sunset Yellow were used as simple models, at concentrations and conditions where aggregation in solution was known not to occur. In this instance, a variation in the rate of adsorption with flow rate is observed.

The mechanism of dye uptake is described as being via rapid adsorption into a porous surface layer, followed by slow diffusion into the fabric. Paradoxically, the rate increases with increased flow rate. Gooding[58] proposes several possible mechanisms to explain the results. It was observed that the higher the flow rate the lower the surface concentration, even though dye is brought more rapidly to the adsorbing interface. Furthermore, the rate of adsorption is lower at higher surface concentrations, varying in a linear manner as described by the equation:

$$J = \text{constant} \frac{(C_{\text{B}} - C_{\text{surf}})}{C_{\text{B}}} \tag{2.36}$$

It is suggested that the dye is adsorbed into a porous surface layer and then diffuses through the surface layer into the bulk fibre. Adsorption of dye into this surface layer blocks the channels through which dye is

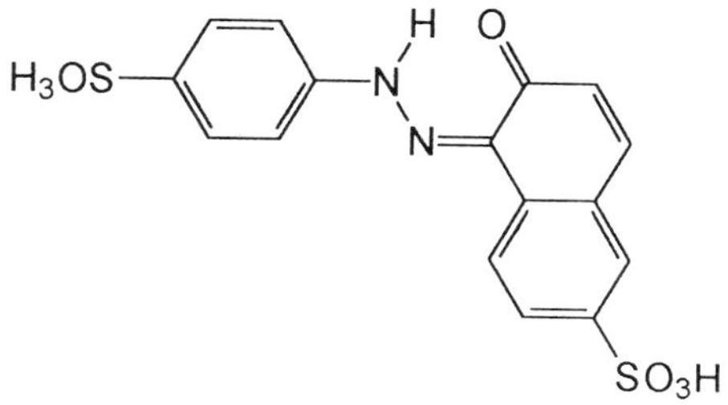

HETEROGENEOUS KINETICS OF REACTIVE AND DISPERSE DYEING 69

Edicol Sunset Yellow

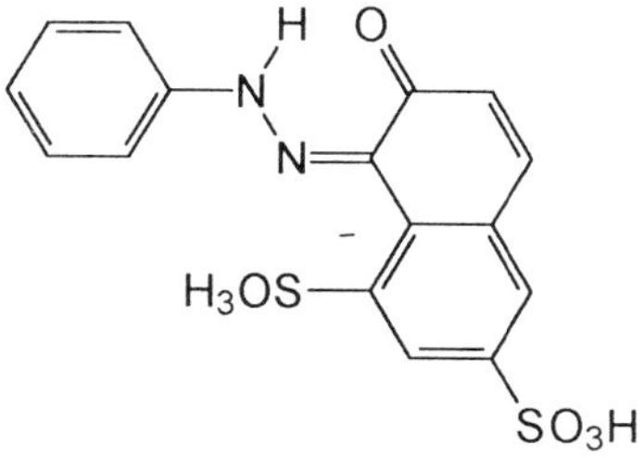

Orange 2

Structure 2.3 Structures of Edicol Sunset Yellow and Orange 2.

transported from solution to fibre bulk via some type of physical effect such as aggregation or precipitation (which is presumably concentration dependent). Thus, with progressive adsorption the rate of diffusion through the surface layer will decrease rather than increase, though presumably there is a maximum concentration where this effect takes over.

Assuming a Langmuir type adsorption isotherm, a rate equation is defined as

$$J = \frac{D_{\text{free}}}{\delta K}\left\{1 - \frac{C_{\text{surf}}}{C_B}\right\} = k_1 C_{\text{surf}} \qquad (2.37)$$

where D_{free} is the diffusion coefficient through the pores if no dye is adsorbed, δ is the thickness of the surface region, and K is the equilibrium constant for the adsorption/desorption process. The surface behaves as if it were reacting in a first-order manner with the dye, with a rate constant of uptake of dye into the bulk fibre from the surface region, k_1:

$$k_1 = \text{constant}\left\{\frac{1}{C_{\text{surf}}} - \frac{1}{C_B}\right\} \qquad (2.38)$$

For a given C_B, k_1 decreases as C_{surf} increases. As the flow rate increases, C_{surf} decreases and the rate of dyeing increases. The dyeing mechanism is a delicate balance between bulk dye concentration and the agitation in the dyebath. Increasing the rate is not as simple as increasing the bulk concentration, since this will result in an increase in the surface concentration and a decrease in the diffusion through the porous surface layer. In effect, this predicts that the dyeing process can be 'turned off', due to adsorption onto surface sites in the porous surface layer restricting passage of further material. Gooding[58] also looked at the adsorption of Orange G under non-steady-state conditions. A fast initial plateauing of dye adsorbed is observed, followed by a steady slow increase, consistent with the proposed mechanism.

A possible physical explanation for the observations of Gooding[58] is that although aggregation is not observed in solution, surface-induced association is occurring. Some evidence for this exists using Synchrotron X-ray Diffraction (XRD) to investigate the dye–fibre interface.[59] Similarly, Wilkinson[60] has used fluorescence to illustrate the possibility of cellulose surface-induced association.

Under the basic conditions employed for the reactive dyes, such surface saturation equilibria should be perturbed by the chemical reactions initiated. Furthermore, the physical processes and fibre structure will also be altered. The above type of analysis illustrates the importance of the fibre structure on the dyeing process and the variation in impact of fibre structure as a function of application conditions.

2.3.3 Fibre diffusion

The discussion in section 2.3.2 has introduced the importance of the porous structure of cotton to the dyeing process. The structure of cotton is well documented, although this documentation is often conflicting in minor details. At a basic level, it is believed that the individual polymer units form crystalline chains which bundle together to give microfibrils which in turn make up fibrils, and then the fibres themselves. This produces a porous network-type structure which is hydrophillic in nature and can absorb water up to about 45% by weight.[61] The presence of water is crucial to the diffusion process within the fibre,[62,63] swelling the fibre and providing a medium for mass and energy transport. Adsorption of the dye on cellulose will be controlled on a molecular scale by how it interacts with the polymer chain which is crystalline in nature. At a macroscopic level, however, the dye can only reach such polymer sites where the structure is overall amorphous or paracrystalline in nature. How the individual polymer chains interact to give single crystal units, and how these order to give microfibrils, fibrils and fibres with voids or pores between, will control the crucial diffusion process. This process has been described by two main modelling methodologies briefly mentioned in section 2.2.7: the porous matrix model[64-70] and the dynamic polymer chain model.[39]

The latter of these two models assumes that the water is dissolved in the polymer and is more suitable to fibres dyed above their glass transition temperature, T_g. For cotton, dyed below its T_g, the porous matrix model has been the most utilised. The basic principles involved are that (i) dye migration within the fibre occurs in the water-filled pores or interconnecting channels, (ii) an instantaneous dynamic reversible equilibrium exists between the migrating molecules and binding sites on the internal pore surface/walls, (iii) and when adsorbed the dyes are effectively immobilised (Figure 2.11). This temporary immobilisation via an internal partition effectively reduces the rate of migration. The observed diffusion coefficient (D_{obs}) will be altered relative to that in bulk solution (D_{H_2O}) by a factor dependent upon the isotherm describing the internal partition. For the simplest case, i.e. Nernst where $K = C_a/C_i$ (C_a refers to the dye adsorbed onto the pore walls, and the total reactive dye C, is the sum of C_a and C_i), D_{obs} is related to D_{H_2O} by the equation:

$$D_{obs} = \frac{D_{H_2O}}{1 + K} \qquad (2.39)$$

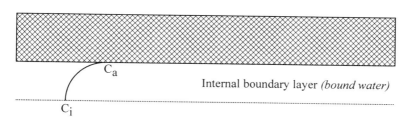
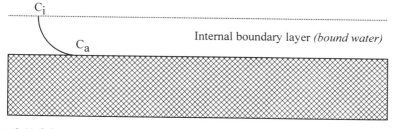

Figure 2.11 Schematic representation of water swollen pore with free and bound water, and equilibration of dye between internal solution and pore walls.

and

$$\frac{dC}{dt} = D_{obs}\frac{d^2C}{dx^2} \qquad (2.40)$$

If a more realistic adsorption isotherm is used, e.g. Freundlich $C_a = KC_i^n$, the diffusion equation becomes non-linear and not amenable to analytical solution. D_{obs} will then become concentration dependent, as observed in practice by numerous authors.[1-3]

The pore model further modifies the observed diffusion coefficient, and thus the rate of migration, by taking into account physical resistance to motion from the polymer matrix itself. This comprises the non-linear nature of the pores introducing a tortuosity factor (τ) to account for the meandering path the diffusing species must take, and a porosity factor (α) to describe the fraction of the substrate that is in fact accessible to the diffusant. The rate equation, thus, becomes

$$\frac{dC}{dt} = \frac{\alpha D_{H_2O}}{\tau(1+K)}\frac{d^2C}{dx^2} \qquad (2.41)$$

Typically D_{obs} is a few orders of magnitude less than D_{H_2O}.[3] This type of analysis has routinely been applied to direct and reactive dyes under neutral conditions. The starting point for any detailed analysis without α and τ becoming 'potentially compromising approximations' has to be some form of characterisation of the porous structure in terms of pore size and distribution (psd), etc. Traditionally, this has been achieved by methods such as nitrogen adsorption or mercury porosimetry, where the substrate is not in the form that it is during dyeing. Recently, other methods of measuring psd under more realistic conditions have been described. Brederbeck has used size exclusion methods[71,72] and then illustrated that the porosity or amount of free water is a factor when there are differences in the rate of diffusion of direct dyes. Grunwald et al.[73] have similarly used liquid chromatography methods with fibre packing material to assess the psd of cotton, and then analysed the effect on dyeing properties.[73] In a similar manner, Saafan[74,75] examined the interrelationship between pore structure and the thermodynamic dyeing properties for different types of cotton. With respect to direct dyes, this type of analysis has been carried out recently by several different workers.[76-82]

Zollinger and Hori[62,63,83,84] have summarised discussions of the pore model and other alternatives, concluding that for cellulosics the pore model best approximates to experimental observations. In effect, the model treats the internal fibre phase as consisting of the internal solution and solid substrate. The pores will contain an inner core of free liquid surrounded by a diffusion boundary layer situated adjacent to the walls, which means that the dye will not be free to diffuse through all of the water. It is thought that, under dyeing conditions, about half of the adsorbed water is intimately associated with the pore walls and the rest is present as free water within the pores.[85] This differentiation between the form the water is present in has been used by Gladden[86] to determine the pore size distribution profile of cotton in the water swollen state using NMR spectroscopy. Two distinct psds were observed centred around radii of 5 and 50 Å. Furthermore, dyeing was observed to alter the larger but not the smaller psd, indicating that dye only penetrates the larger pores and that these are altered.

The pore model has been criticized for simplicity and being misleading. Indeed, several problems and assumptions are inherent in the analysis. First of all, the water is assumed to diffuse ahead of both inorganics and dyestuff into the fibre. The dye, as it follows the water, is assumed to partition instantaneously on the dyeing timescale between the internal solution and fibre phase, following a simple linear isotherm, and this partition is deemed to be constant throughout. It is well known that the adsorption characteristics of anionic dyes onto an anionic cellulose surface do not follow linear isotherms and are dependent upon conditions and the extent of adsorption as sites are used up and negative charge builds up. Since the diffusion of the dyes through the cotton pores is dependent upon such an equilibrium, the diffusion will also be dependent upon the amount of dye present at a particular place and time. A single observed diffusion coefficient will not be capable of describing the whole process. Similarly, the pore structure is assumed to be constant during the dyeing process, which appears not to be the case. Modifications to the theory have appeared, and these have included variable sized pores[81] and the possibility of dual surface and pore diffusion.[79] The model does seem to start to take account of the actual nature of the substrate being dyed and probably needs to be adapted to further reflect this, more accurately taking into account changes in pore size, position, interconnectivity, and adsorption isotherm with time.

The chemical engineering literature is a valuable source of information with regards to diffusion within porous networks.[87] Several examples are present which show that such diffusion is not only dependent on the psd, tortuosity, porosity, but also on the interconnectivity of the pores. In coupled diffusion–reaction profiles this will also be the case.

2.3.4 Diffusion with simultaneous chemical reaction

Under basic fixation conditions, the migration of the dye throughout the cotton structure becomes a problem of coupled diffusion–reaction. All the considerations outlined in section 2.3.3 are still present, but now immobilisation via reversible adsorption on the pore walls can be accompanied by covalent bond formation. This chemical reaction is reversible (or subject to further hydrolysis), but on the dyeing timescale, the reversibility is generally slow and ignored. Effectively, we have an irreversible immobilisation process. To the best of the authors' knowledge, all attempts at quantifying the diffusion–reaction process in reactive dyeing have started from the equation

$$\frac{dC}{dt} = D_{obs}\left(\frac{d^2C}{dx^2}\right) - k_f C \qquad (2.42)$$

Analytical solutions to this equation are complex and require numerous assumptions and/or approximations. Most methods that have been applied assume (i) the dyebath and hence fibre surface concentration is constant, (ii) OH^- diffuses ahead of the larger dye molecule and equilibrates with the cellulose hydroxy groups so that pseudo-first-order kinetics can be applied at each point in the fibre, and (iii) the observed diffusion coefficient, D_{obs}, adsorption isotherm and general porous fibre structure are constant.

Initial work utilised Danckwerts[45,47] analytical solutions for the equation under specific boundary conditions (section 2.3.1) following the process via variation in bulk dyebath concentrations and the total amount of dye absorbed. As indicated in section 2.3.2, such remote detection can only provide limited information on the processes occurring within the fibre itself. An alternative method derives interpretations of the rate of dyeing by the extent and/or amount of dye penetration into a cylindrical film roll or sheet of substrate with time. In most instances, cellophane is used as a cellulose model for cotton. In this manner, Karasawa et al.[88] determined the diffusion coefficient of reactive dyes in cellulose by monitoring the concentration profile of the active species within successive layers of cellophane film. This type of analysis has been extended most notably by Motomura and Morita.[89] Starting from the equation describing coupled diffusion–reaction they have utilised Danckwerts[45] solution into an infinite slab (section 2.3.1) to derive expressions for the distribution of both the immobilised (fixed) and active (reactive) dye with time. The usual boundary conditions apply, i.e. no dye in substrate at $t = 0$, when $t > 0$ the surface concentration, C_{FS}, is constant.

Using cellophane rolled onto a glass rod, the concentration of dye in each layer is then followed with time. This is achieved by direct spectroscopic analysis of the dyed cellophane before scouring (to give total dye adsorbed) and after scouring to remove unfixed 'active' dye (to give the amount fixed). The concentrations are calculated from the optical densities of the UV/Vis transmission spectra relative to cellophane pieces calibrated with varying amounts of hydrolysed dye exhausted onto the substrate. It is assumed that the extinction coefficient for the reactive, hydrolysed and fixed dyes on cellophane will be constant within experimental error.

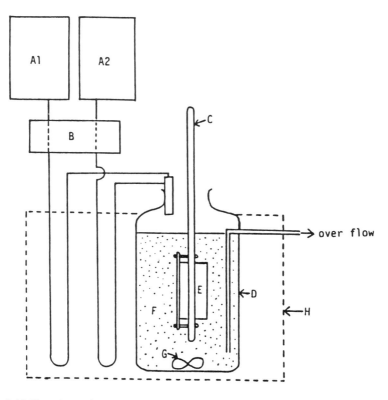

Figure 2.12 Experimental apparatus for cellophane roll method of investigative reactive dyeing: A1, A2, stock solutions of reactive dyes and buffers; B, quantitative pump; C, glass tubing; D, dyeing vessel; E, film roll; F, dyebath; G, stirrer; H, thermostat. Reproduced from Motomura and Morita[89] with permission.

From the concentration profiles obtained, the diffusion coefficient, D_{obs}, surface concentration, C_{FS}, and first-order rate constant for fixation, k_f, are calculated. This is achieved by first obtaining initial estimates of each. D_{obs} is estimated from that measured for the hydrolysed material, k_f from the rate of fixation in the surface layer, and C_{FS} by extrapolation. The latter of these is especially difficult to estimate with any certainty. The predicted profiles are then compared to the experimental ones and a non-linear regression analysis utilised to optimize D_{obs}, k_f and C_{FS}. Morita and Motomura's initial work concentrated on dichlorotriazine-based reactive dyes.[89–92] The fixed and unfixed profiles could be fitted at low pH values. At higher pH values, when fixation will become competitive and the equations describing the process realistic, agreement between theory and experiment for the unfixed dye could not be achieved. The dye apparently penetrated faster than could be explained by the theoretical curves. This was believed to be due to hydrolysis occurring within the fibre. Dye–fibre hydrolysis was counted out as being too slow to compete, and the experiment was designed so that the concentration of hydrolysed dye in the bulk dyebath was minimised and, hence, not available for significant adsorption. The original equation was modified to account for this,[91] and took the form:

$$\frac{dC}{dt} = D_{obs}\left(\frac{d^2C}{dx^2}\right) - (k_f + k_{w_i})C \qquad (2.43)$$

Even taking into account k_w, the distribution profile of the reactive dyes under basic conditions could not be adequately explained. Nevertheless, quantitative information on the relevant constants is derived. Variation in k_f with both temperature and pH is described. The pH variation for the dichlorotriazine dyes is similar to that observed for solution hydrolysis of the same dyes. This is explained as being due to both hydrolysis and reaction with the cellulosate anion occurring by the same aromatic nucleophilic substitution mechanism. Discussion of this mechanism with respect to dichlorotriazine electrophiles has been well documented previously.[2,3] From the temperature variation the activation energy for fixation, E_R, is calculated and a similar treatment for the derived diffusion coefficients gives an activation energy for fibre diffusion, E_D. Values for k_w within the substrate, i.e. k_{wi}, are also calculated and indicate that substantial hydrolysis occurs. Typical results obtained by Morita et al.[92] are shown in Figure 2.13, where $P = k_{wi}/k_f$.

The work was extended to monochlorotriazine,[93] vinylsulphone[94] and mixed bifunctional[95] reactive dyes. For the monochlorotriazine reactive dyes, a similar pH profile for k_f was observed as seen for the dichlorotriazine dyes. This is rather surprising considering the larger pK_a of the bridging imino group in the monochlorotriazines. Hydrolysis was estimated as being negligible in comparison to fixation, although it becomes slightly more noticeable at higher pH values. Similar conclusions were made for the vinylsulphone dyes studied, with $\log_{10} k_f$ varying linearly with pH, i.e. k_{cellO^-} is constant. In general, the authors observe that E_R is largest for dyes having monochlorotriazine electrophiles, followed by vinylsulphone and dichlorotriazine-based dyes.

Such an analysis provides detailed information on the chemical and physical processes operational within the substrate. Morita and Motomura point out that 'no precise value of k_w was determined, due to the insensitivity of the theoretical profiles for active and fixed species to the values of k_w or to the experimental profiles obtained'.[94] This highlights the inherent problem in trying to separate out the individual components associated with the simultaneous diffusion–reaction mechanism. The number of assumed constants (requiring a solution) invariably outnumbers the available variables. These constants are all related and their

HETEROGENEOUS KINETICS OF REACTIVE AND DISPERSE DYEING

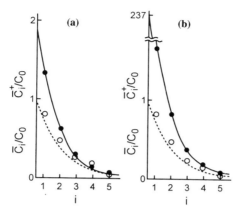

Values of $P(=k_w/k_{cell})$ at various ionic strengths, I, at 30 °C

C.I. Reactive	pH	Time (min)	Values of P		
			$I = 0.15$	$I = 0.5$	$I = 1.0$
Yellow 1	10.6	90	0.40	0.31	0.26
Yellow 4	10.6	90	0.65	0.62	0.60
Orange 1	10.6	90	1.68	1.51	1.30
	12.0	90	1.70	1.50	1.32
Red 1	12.0	60	0.51	0.39	0.38
Red 2	12.0	30	0.46	0.40	0.23
Red 8	10.6	90	0.70	0.69	0.61
Blue 1	10.6	90	0.54	0.48	0.46
Blue 4	10.6	90	0.90	0.88	0.83

Pseudofirst-order rate constants of reaction with cellulose, k_{cell} (min^{-1})

C.I. Reactive	Yellow 1			Red 2			Red 8	Blue 1	Blue 4
Temp. (°C)	25	30	35	25	30	35	30	30	30
pH (20 °C)									
8.4		0.0023	0.0029	0.0064	0.0080	0.0093		0.0020	
8.8	0.0048			0.016	0.020	0.023	0.0029		
9.0	0.0080	0.010	0.012				0.0047	0.0070	0.0079
9.2				0.038	0.048	0.057			
9.6	0.032	0.024	0.050		0.10		0.018	0.022	0.021
9.8		0.057							
10.0					0.16		0.031	0.046	0.053
10.6		0.088			0.25		0.075	0.070	0.084
11.0		0.093			0.47		0.10	0.081	0.089
11.7		0.093					0.11	0.083	0.091

Figure 2.13 Examples of concentration profiles and results obtained from the cellophane roll method on dichlorotriazine based reactive dyes. Concentration profiles for removed (○) and immobilized species (●) for C.I. Reactive Blue 4 at 30 °C, pH 9.0 and $I = 0.15$. Curves are the theoretical profiles for active (----) and fixed species (——) described for $D = 3.9 \times 10^{-7}$ cm^2/min, $k_{cell} = 0.0079$ min^{-1} and $P = 0.5$. Diffusion time: (a) $t = 240$ min, (b) $t = 300$ min. Reproduced from Morita et al.[92] with permission.

calculation is dependent upon reasonable initial estimations which is not always easy, especially taking into account the considerable changes associated with changes in application conditions. For example, Morita et al.[93] state that when the concentration of fixed dye in the outer layer becomes large, strong influences may be exerted on the adsorption of active species and hence C_{FS}. Furthermore, taking this point to an extreme, the number of available sites for reaction will decrease, pseudo-first-order kinetics will no longer be obeyed, and competition from hydrolysis should increase. The equations utilised will also no longer be valid and the 'constants' will vary with time. Rusznák and coworkers[96] have noted 'that the number of cellulose hydroxyl groups participating in the reactive dye–cellulose reaction is limited by accessibility. With increasing dye sorption during the fixation process, the number of accessible hydroxyl groups decreases at a higher rate than that corresponding to the degree of substitution. The phenomenon is due to the high space requirement of the dye molecule, and gives rise to an effective saturation value'.

These latter workers have utilised a similar procedure to look at the mechanism of reaction of monochlorotriazine dyes.[96] The surface concentration of reactive dye, C_{FS}, is again used and attempts are made to take account of changes in D_{obs} with time and concentration. Hydrolysis in the fibre is assumed to be negligible and the fixation is affected equally by rate of diffusion and chemical reaction, i.e. the process proceeds in the transitory region. This agrees with the model described by Rys and Zollinger[97] which takes the adsorption immobilisation effect into consideration not only on the diffusion but also the chemical reactions involved i.e. hydrolysis and fixation. This is not taken into account in the work of Motomura and Morita 'the diffusion coefficient of reactive dyes decides the concentration distribution in the substrate but has no relation to the fixation ratio',[91] presumably because of the pseudo-first-order nature assumed. They do make the point that strictly speaking, the values of D_{obs}, k_f, k_w and C_{FS} may change with concentration of the species involved (and hence with time).[93]

In their work on monochlorotriazine reaction dyes, Rusznák et al.[96] looked at differences when the dye is first allowed to migrate into the substrate under neutral conditions and then base is added to initiate chemical reaction. This mimics the exhaust dyeing situation but invalidates the assumption that base diffuses ahead of the dye thus allowing pseudo-first-order kinetics to be invoked. The value for k_f was found to be five to ten times greater under these conditions.

Morita et al.[98] have recently looked at the stability and reversibility of the dye–cellulose bond for vinylsulphone reactive dyes. The analysis provides an extension to their earlier work on the forward fixation reaction.

The work of Compton and co-workers[55–57] mentioned in section 2.3.2, investigated the interfacial kinetics associated with the fixation process for dichlorotriazine based reactive dyes **1** and **2**. Results show that under fixation conditions the mechanism is first order in dye surface concentration and controlled by a process at the interface or in the fibre, but do not provide explicit information that can be used to separate out the coupled diffusion–reaction processes. This work utilised the solution interfacial concentration, C_{surf}, which will be related to C_{FS} (as used by the workers using concentration profiles in the substrate) by the appropriate adsorption isotherm. Determining this relationship under the relevant perturbed equilibrium conditions is, however, not easy and will not be the same as that found under neutral conditions. This is one of the key problems that needs to be addressed further. If realistic dyeing conditions are to be quantified, variable surface concentrations, modification of the adsorbate by the adsorbant (including the porous structure), and dye–dye interactions will need to be included.

A discussion on the effect of electrolyte has not been included above. This has been well reviewed previously, indicating how charge balance in the separate phases affects the concentration of the individual charged species in each phase.[1–3] The basic conclusions will not be affected as long as the particular dyeing conditions used take this into account.

2.4 Summary

The dyeing of both cellulose with reactive dyes and of polyester with disperse dyes are true heterogeneous physical and chemical processes. As such, they may be described and understood in the same way as other heterogeneous chemical processes traditionally are. The heterogeneous nature introduces elements of complexity into what, at first sight, appear to be rather simple chemical or physical processes. Depending on the system being studied, any one of the constituent steps in the dyeing process may limit the uptake of the dye by the fibre. Thus, for the designer of new dye molecules or for the designer of new dyeing processes, it is very important to understand which are the key steps involved in transferring dye from the dyebath to the fibre and fixing it there. It is also important to understand how these steps depend on the physical properties of both the dye and the fibre, and on how these physical properties may be manipulated by modification of the molecular structure of the dye or by altering the variables of the dyeing process.

One area usually neglected is the influence of the second reactant, the fibre, on this process. A consequence of considering the dyeing mechanism as a detailed heterogeneous process is that our knowledge of the fibre structure and properties is also improved.

In the last 10–15 years there has been a slow but steady trend towards a heterogeneous rather than homogeneous description of fibre dyeing. If this continues then our understanding of this long established, but far from fully understood, physico–chemical problem will steadily grow.

References

1. Vickerstaff, T., *The Physical Chemistry of Dyeing*, Oliver and Boyd, 1954.
2. Rattee, I. D. and Breuer, M. M., *The Physical Chemistry of Dye Adsorption*, Academic Press, 1974.
3. Johnson, A. (ed.), *The Theory of Coloration of Textiles*, Society of Dyers and Colourists, 1989.
4. Hill, A. V., *Proc. R. Soc.*, **104B** (1928) 65.
5. Crank, J., *Phil. Mag.*, **39** (1948) 362.
6. Wilson, A. H., *Phil. Mag.*, **39** (1948) 48.
7. Shibusawa, T., *J. Soc. Dyers Colour.*, **101** (1985) 231.
8. Shibusawa, T., *J. Soc. Dyers Colour.*, **96** (1980) 293.
9. Cegarra, J. and Puente, P., *Text. Res. J.*, **37** (1967) 343.
10. Militký, J., *J. Soc. Dyers Colour.*, **95** (1979) 327.
11. Shibusawa, T., *J. Soc. Dyers Colour.*, **104** (1988) 28.
12. Navratil, J., *Melliand Textilber.*, **62** (1981) 333 and 403 (English ed. 419 and 515).
13. Merian, E. and Lerch, U., *Am. Dyest. Rep.*, **19** (1962) 695.
14. Braun, H. H., *Rev. Prog. Color.*, **13** (1983) 62.
15. Grant, D. J. W. and Higuchi, T., *Solubility Behavior of Organic Compounds*, Wiley Interscience, 1990.
16. Odvárka, J. and Huňková, J., *J. Soc. Dyers Colour.*, **99** (1983) 207.
17. Aspland, J. R., *Text. Chem. Color.*, **24** (1992) 38.
18. Braun, H. H., *J. Soc. Dyers Colour.*, **107** (1991) 77.
19. Datyner, A., *J. Soc. Dyers Colour.*, **94** (1978) 256.
20. Datyner, A., *Rev. Prog. Coloration*, **23** (1993) 40.
21. Wilke, C. R. and Chang, P., *AIChE J.*, **1** (1955) 264.
22. Spiro, M. and Freund, P. L., *JCS Faraday Trans. I*, **79** (1983) 1649.
23. Biedermann, W., *J. Soc. Dyers Colour.*, **87** (1971) 105.
24. Biedermann, W. and Datyner, A., *Text. Res. J.*, **61** (1991) 637.
25. McGregor, R. and Etters, J. N., *Text. Chem. Color.*, **11** (1979) 202.
26. Etters, J. N., *J. Soc. Dyers Colour.*, **107** (1991) 114.
27. Burley, R., *J. Soc. Dyers Colour.*, **107** (1991) 219.
28. Jones, F., in *The Theory of Coloration of Textiles*, (ed A. Johnson) Society of Dyers and Colourists, 1989.
29. Cegarra, J. and Puente, P., *Text. Res. J.*, **41** (1971) 170.
30. Narębska, A., Ostrowskā-Gumbkowska, B. and Krzstek, H., *J. Appl. Polym. Sci.*, **29** (1984) 1475.
31. Newman, A. B., *Trans. Am. Inst. Chem. Eng.*, **27** (1931) 203.
32. Hirusawa, Y. and Matsumura, S., *Sen-i-Gakkaishi*, **31** (1975) T175.
33. Boulton, J. and Crank, J., *J. Soc. Dyers Colour.*, **68** (1952) 109.
34. Armfield, J., Boulton, J. and Crank, J., *J. Soc. Dyers Colour.*, **72** (1956) 278.
35. Parish, G., *J. Soc. Dyers Colour.*, **78** (1962) 109.
36. Patterson, D. and Sheldon, R. P., *Trans. Faraday Soc.*, **55** (1959) 1254.
37. Navratil, J., *J. Soc. Dyers Colour.*, **106** (1990) 283 and 327.
38. Kojima, H. and Ijima, T., *J. Soc. Dyers Colour.*, **91** (1975) 103.
39. Fujita, H., Kishimoto, A. and Matsumoto, K., *Trans. Faraday Soc.*, **56** (1960) 424.
40. Lin, S. H., *J. Chem. Tech. Biotechnol.*, **54** (1992) 387.
41. Hori, T., Sato, Y. and Shimizu, T., *J. Soc. Dyers Colour.*, **97** (1981) 6.
42. Navratil, J., *Melliand Textilber.*, **63** (1982) 224 (English ed. 225).
43. Gerber, H., *J. Soc. Dyers Colour.*, **94** (1978) 298.

44. Neale, S. M., *J. Text. Inst.*, **20** (1929) T373.
45. Danckwerts, P. V., *Trans. Faraday Soc.*, **46** (1950) 300.
46. Sumner, H. H. and Weston, C. D., *Am. Dyest. Rep.*, **52** (1963) 442.
47. Danckwerts, P. V., *Trans. Faraday Soc.*, **47** (1951) 1014.
48. Rys, P. and Zollinger, H., in *The Theory of Colouration of Textiles*, (ed A. Johnson) Society of Dyers and Colourists, (1989), Chapter 6, 428.
49. Alexander, P. and Hudson, R. F., *Text. Res. J.*, **20** (1950) 481.
50. McGregor, R. and Peters, R. H., *J. Soc. Dyers Colour.*, **81** (1965) 393.
51. Popescu, C. and Segal, E., *J. Soc. Dyers Colour.*, **100** (1984) 399.
52. Popescu, C., *J. Soc. Dyers Colour.*, **108** (1992) 534.
53. Sada, E., Kumazawa, H. and Ando, T., *J. Soc. Dyers Colour.*, **100** (1984) 97.
54. Saafan, A. A. and Habib, A. M., *Colloids Surf.*, **34** (1988) 75.
55. Compton, R. G., Unwin, P. R. and Wilson, M., *Chemistry in Industry*, 2 April, 1990.
56. Compton, R. G. and Wilson, M., *J. Appl. Electrochem.*, **20** (1990) 793.
57. Compton, R. G., Coles, B. A., Stearn, G. M. and Waller, A. M., *J. Chem. Soc., Faraday Trans. I*, **84** (1988) 2357.
58. Gooding, J. J., *The Study of Interfacial Processes*, DPhil Thesis, Oxford Univ., 1993.
59. Edwards, D. J., in *Advances in Color Chemistry*, vol. 4 (eds A. T. Peters and H. S. Freeman) this volume.
60. Wilkinson, F., Leicester, P. A., Ferreira, L. F. V. and Freire, V. M. M. R., *Photochem. Photobiol.*, **54** (1991) 599.
61. Ingamells, W., in *The Theory of Coloration of Textiles*, (ed A. Johnson) 2nd Ed., Society of Dyers and Colourists, (1989), 169.
62. Hori, T. and Zollinger, H., *Melliand Textilber.*, **68** (1987) 267.
63. Hori, T. and Zollinger, H., *Melliand Textilber.*, **68** (1987) 657.
64. Neale, S. M., *Trans. Faraday Soc.*, **31** (1935) 282.
65. Valco, E., *Trans. Faraday Soc.*, **31** (1935) 278.
66. Morton, T. H., *Trans. Faraday Soc.*, **31** (1935) 262.
67. Weisz, P. B., *Trans. Faraday Soc.*, **63** (1967) 1801.
68. Weisz, P. B. and Hicks, J. S., *Trans. Faraday Soc.*, **63** (1967) 1807.
69. Weisz, P. B. and Zollinger, H., *Trans. Faraday Soc.*, **63** (1967) 1815.
70. Weisz, P. B. and Zollinger, H., *Trans. Faraday Soc.*, **64** (1968) 1693.
71. Bredereck, K., Bader, E. and Schmitt, U., *Textilveredlung*, **24** (1989) 142.
72. Bredereck, K. and Bluher, A., *Melliand Textilber.*, **73** (1992) 652.
73. Grunwald, M., Burtscher, E. and Bobleter, O., *Textilveredlung*, **27** (1992) 45.
74. Saafan, A. A., *Melliand Textilber.*, **68** (1987) 678; and **65** (1984) 534.
75. Saafan, A. A. and Habib, A. M., *Melliand Textilber.*, **68** (1987) 845.
76. Chrastil, J., *Text. Res. J.*, **60** (1990) 413.
77. Alberghina, G., Amato, M. E., Fisichella, S. and Pisano, D., *Colourage*, **34** (1987) 19.
78. Yoshida, H., Kataoka, T., Maekawa, M. and Nango, M., *Chem. Eng. J. (Lausanne)*, **41** (1989) B1–B9.
79. Yoshida, H. *et al.*, *J. Appl. Polym. Sci.*, **32** (1986) 4185.
80. Ibe, E. C., *Acta Polym.*, **37** (1986) 549.
81. Morita, Z., Tanaka, T. and Motomura, H., *J. Appl. Polym. Sci.*, **30** (1985) 3697.
82. Nango, M. *et al.*, *Text. Res. J.*, **54** (1984) 598.
83. Hori, T. and Zollinger, H., *Text. Chem. Color.*, **18** (1986) 19.
84. Zollinger, H., *Textilveredlung*, **24** (1989) 133.
85. Saafan, A. A. and Abdel-Halim, S. T., *Melliand Textilber.*, **69** (1988) 511.
86. Gladden, L. F., Hollewand, M. P. and Brennan, C. M., *I. Chem. E. Research Event*, (1993) pp. 558–560.
87. Gladden, L. F. and Hollewand, M. P., *Chem. Eng. Sci.*, **47** (1992) 1761.
88. Karasawa, M., Choii, N., Sakai, H. and Seikido, M., *Sen-i Gakkaishi*, **29** (1973) T–14.
89. Motomura, H. and Morita, Z., *J. Appl. Polym. Sci.*, **21** (1977) 487.
90. Motomura, H. and Morita, Z., *Bull. Chem. Soc. Jpn.*, **51** (1978) 1332.
91. Motomura, H. and Morita, Z., *J. Appl. Polym. Sci.*, **24** (1979) 1747.
92. Morita, Z., Nishikawa, I. and Motomura, H., *Sen-i Gakkaishi*, **39** (1983) T485.
93. Morita, Z., Kawamura, G. and Motomura, H., *Sen-i Gakkaishi*, **42** (1986) T93.
94. Morita, Z. and Motomura, H., *Sen-i Gakkaishi*, **42** (1986) T627.
95. Morita, Z., Kai, T. and Motomura, H., *Dyes Pigments*, **17** (1991) 241.

96. Rusnák, I., Zobor, J. and Lavai, G., *Dyes Pigments*, **2** (1981) 285.
97. Rys, P. and Zollinger, H., *The Theory of Coloration of Textiles*, (eds C. L. Bird and W. S. Boston) Chapter 7, The Dyers Company Publications Trust, Bradford, 1975.
98. Morita, Z., Kim, I. H. and Motomura, H., *Dyes Pigments*, **18** (1992) 11.

3 Aggregation and lyotropic liquid crystal formation of anionic azo dyes for textile fibres

D. J. EDWARDS, A. P. ORMEROD, G. J. T. TIDDY, A. A. JABER and A. MAHENDRASINGHAM

3.1 Anionic azo dyes

The three most important classes of anionic azo dyes are the acid, direct and reactive types. These dyes are water soluble and contain at least one solubilising group, most commonly a sulphonic acid group.

3.1.1 Acid dyes

In general, acid dyes have molecular weights in the range 300–500. They are used mainly for dyeing protein and synthetic polyamide fibres. As their name suggests, they are applied under acid conditions (pH 2–6). They generally contain one azo group and between one and three solubilising sulphonic acid moieties (cf. **1**).

3.1.2 Direct dyes

The first direct dye, Congo Red (**2**), was discovered in 1844 by Bottingen. The name is derived from the fact direct dyes are able to dye cellulosic materials without the aid of mordants. Their molecular weights are usually in excess of 500. More than one azo group is usually required to give the dyes sufficient affinity for cellulose.

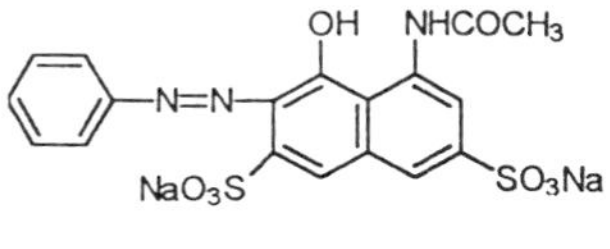

1 C.I. Acid Red 1.

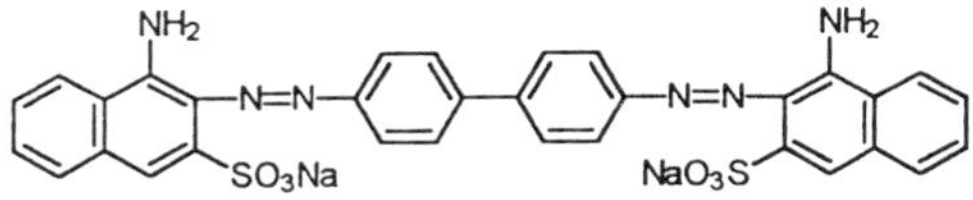

2 C.I. Direct Red 28 (Congo Red).

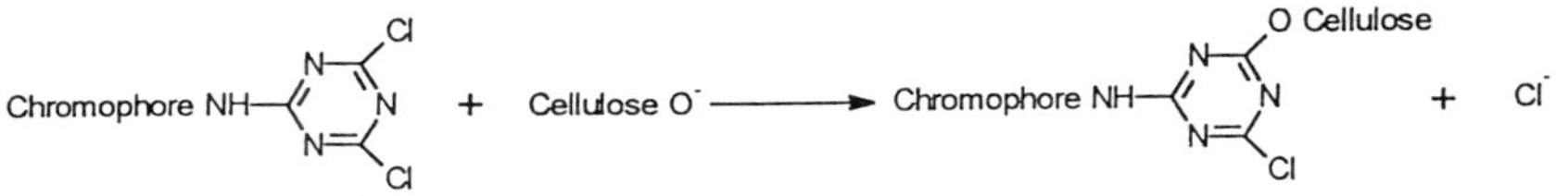

Scheme 3.1 Reactive dye fixation to cellulose.

3.1.3 Reactive dyes

Fibre-reactive dyes constitute one of the largest and fastest growing ranges of dyes. They contain one or more groups capable of forming a covalent bond with the fibre being dyed. Bonds are formed with the hydroxyl groups of cellulosic materials (Scheme 3.1), with the amino, hydroxyl and mercapto groups of protein fibres, and with the amino groups of polyamides.

3.2 Aggregation

3.2.1 Factors governing the formation and stability of aggregates

3.2.1.1 Hydrophobic effect. A significant proportion of the literature concerning the aggregation of dyes suggests that the hydrophobic effect is the major reason for the formation of dimers and larger aggregates[1-6]. This is no doubt primarily due to the belief that the dyes behave as amphiphiles and thus their aggregation is analogous to micellisation in surfactant systems.

The hydrophobic effect is the single major factor involved in the formation of micelles in surfactant/water systems. Below the Critical Micelle Concentration (CMC) the hydrophobic alkyl chains of the surfactant monomers are surrounded by a single layer of ordered water. At the CMC, stable micelles are formed with the hydrophobic chains at the centre, no longer in contact with the water. The driving force for this (micelle formation) is the reduction in water/hydrocarbon interactions on formation of the micelles. The water which was previously in a restricted range of orientations adjacent to the alkyl chains is now disordered (i.e. an entropy driven process occurs)[7].

It is unlikely that the hydrophobic effect plays as significant a role in anionic azo dyestuff aggregation. The aggregation process is a gradual one[8] with no sharp increases in aggregation number at a 'critical' concentration (i.e. a CMC). The hydrophobicity of planar aromatic ring systems is less than that of aliphatic hydrocarbons. This can be seen from the unitary free energy and entropy changes which occur on partitioning the hydrocarbons in water[9] (Table 3.1) and also from interfacial tension

Table 3.1 The changes in enthalpy, unitary free energy and unitary entropy on transferring various organic solvents from a pure liquid to an aqueous environment.[9]

Process	ΔS_u(e.u.)	ΔH(cal/mole)	ΔF_u(cal/mole)
Liq.propane to C_3H_8 in H_2O	−23	−1800	+5050
Liq.n-butane to C_4H_{10} in H_2O	−23	−1000	+5850
Liq.benzene to C_6H_6 in H_2O	−14	0	+4070
Liq.toluene to C_7H_8 in H_2O	−16	0	+4650
Liq.m- or p-xylene to C_8H_{10} in H_2O	−20	0	+5800

measurements; n-alkanes/water at 25 °C range from 49.0 dyn/cm ($n = 5$) to 53.3 dyn/cm ($n = 16$)[10], whilst benzene/water at 25 °C is much lower at 34.6 dyn/cm[11]. The addition of alkyl chains to anionic azo dyes will of course produce a more amphiphilic species and as a result, the hydrophobic effect will become more influential[1,6] (note the increased hydrophobic character of xylenes compared to benzene, Table 3.1).

3.2.1.2 Hydrogen bonding. The formation of hydrogen bonds between dye and/or dye–water molecules has been proposed as a possible factor in dimer and higher aggregate stabilisation. It has been suggested that dye molecules in an aggregate stack could be bound by means of a sandwiched water molecule[12,13]. Dimer binding energies for many dye/water systems have been calculated and their magnitude suggests association through hydrogen bonding[5,14]. However, firm spectroscopic evidence to support this is yet to be obtained.

3.2.1.3 π–π Interactions. The intermolecular spacing between molecules within aggregate stacks is consistently in the range 3.4–3.6 Å[15–18]. This fact seems to strongly support the case that π–π interactions play a major role in dyestuff self-association[19,20]. It is somewhat surprising then that many workers in the area have neglected to consider this interaction. It is also true to say that not only are adjacent dye molecules within aggregate stacks separated by a distance of c. 3.4–3.6 Å, but that they are also generally slipped or off-set with repsect to one another[21]. This arrangement appears to be the most favourable for minimising π–π repulsive whilst maximising π–σ attractive interactions[22].

3.2.1.4 van der Waals forces. Attractive van der Waals interactions have often been quoted as being responsible for dye aggregation, although usually in conjunction with other interactions such as the hydrophobic effect, hydrogen bonding or π–π interactions[3,14,22,23]. Of the three types of van der Waals interaction, dispersion or London forces[24–26] are the most significant. They play a lead role in the orientation/alignment of molecules within aggregate stacks. These forces are maximised when dye molecules are parallel and lying one above another. They act in conjunction with π–π interactions; and whilst the π–π appear dominant with respect to geometry, the dispersion forces contribute most to the net interaction energy[22].

3.2.1.5 Electrostatic repulsive forces. Anionic azo dyes possess ionic solubilising groups, commonly sulphonic acid groups present as their sodium salts. Consequently, the electrostatic repulsive forces between these charged species are a significant factor in dyestuff aggregation. These repulsive forces are reduced somewhat by hydration and counter-ion binding. The variation of electrostatic and steric interactions caused by changes in the number and position of ionic solubilising groups and their influence on aggregate structure and stability is discussed later. Electrostatic interactions are also particularly significant when considering the influence of dyeing auxiliaries (e.g. salt, urea) on dyestuff aggregation.

3.2.2 Measurement techniques

3.2.2.1 Introduction. Dyestuff aggregation can be expressed as either the particle weight, size or aggregation number (particle weight/molecular weight). A number of experimental techniques have been used to measure aggregation, including diffusion[27,28], conductivity[29,30], ultracentrifugation[31], light scattering[32,33], polarography[34,35], visible spectrophotometry[36–38], NMR spectroscopy[39–41] and X-ray scattering[42,43]. Most of these methods have been reviewed in some detail elsewhere[19,44,45]. This section will concentrate on the particular difficulties associated with measuring the aqueous solution behaviour of anionic azo dyes.

3.2.2.2 Difficulties associated with measuring anionic azo dyestuff aggregation. The literature on aggregation measurement is full of inconsistent data, different techniques giving rise to quite different values for the same dye system. One obvious, but important reason for this, is the size polydispersity of dye aggregate particles in many aqueous solutions. Diffusion-based methods, for example, give a number-average particle weight whilst light-scattering methods give a weight-average particle weight[46,47]. Certain methods emphasise low particle sizes, whilst in others high particle sizes and this must be taken into account when making cross comparisons.

Diffusion methods for determining aggregation behaviour include direct measurement techniques such as the diaphragm cell (steady state)[48], micro[49] and 'open-ended'[50] diffusion cells (non-steady state) and indirect measurement techniques such as polarography[34,35]. All involve using the determined diffusion coefficient to obtain the aggregate particle size from either the Stokes–Einstein equation (3.1):

$$D = RT/6\pi \eta r N \qquad (3.1)$$

where D = diffusion coefficient, R = gas constant, T = temperature, η = viscosity of medium, r = particle radius and N = Avagadro number, or from an empirical relationship between diffusion coefficients and molecular weights[6,19,35]. The Stokes–Einstein Equation describes the diffusion of uncharged spheres in a liquid. Dye aggregates are in most cases not spherical and are charged. The charge effects are often overcome by using large excesses of an electrolyte; this however, introduces an additional complication in that the electrolyte influences the very thing you are trying to measure, namely, aggregation.

Visible spectrophotometry has been used widely to determine the onset of self-association of dyestuff molecules in aqueous solution, as new electronic transitions and hence absorption bands often result. Unfortunately, for anionic azo dyes where the charges are localised, these metachromatic changes are less obvious and are often not observable over the concentration range of interest. Deviation from Beer's Law has, however, been used quite successfully to quantify dye–dye interactions[37]. This method is straight-forward and does not require the presence of a large excess of electrolyte.

^{1}H and ^{19}F NMR have been used to determine the most favoured orientation of azo dye molecules within aggregate stacks, by following chemical shifts as a function of dye concentration. For parallel stacking, a low field shift should occur on dilution, whereas for end-on aggregation, dilution should produce a shift to higher magnetic field. For ^{1}H NMR, however, the assignment of peaks is sometimes complicated and the change in chemical shifts is often small. On the other hand, the chemical shifts in ^{19}F NMR exist in the region from -450 to 550 ppm, and are highly sensitive to aromatic ring current effects and the proximity of carbonyl π-electrons[39–41]. Although qualitative, this method is relatively quick and easy.

Small Angle X-Ray Scattering (SAXS) is becoming more popular. It is capable of providing structural information describing heterogeneities of the order of 10–1000 Å in size. It can yield useful information about aggregate ordering, size, shape and separation (i.e. lyomesophase structure)[42,43].

Wide Angle X-Ray Scattering (WAXS) which probes the size range 1–50 Å can also be used to determine aggregation numbers. Aggregates, and in particular lyotropic liquid crystals, have certain similarities in structure with oriented fibres. Therefore, techniques and cameras developed for fibre diffraction studies are ideal for studying aggregate and lyomesophase structure[51]. One reflection common to all self-associating azo dye solutions is at 3.4–3.6 Å, and corresponds to the intermolecular spacing within the aggregate stack. The breadth of this reflection can be related directly to the length of the aggregate stack by the Scherrer-–Bragg approximation (3.2):

$$L = K\lambda/\beta_0 \cos\theta \qquad (3.2)$$

where $K = 1$, $\lambda =$ X-ray wavelength, β_0 is the half width of the intensity profile, and θ is the Bragg angle. An approximate aggregation number can be calculated from equation (3.3):

$$\text{Agg. N}° = L/\text{intermolecular spacing} \qquad (3.3)$$

3.2.3 Anionic azo dye aggregates

3.2.3.1 Acid dyes. Of all the anionic azo dye classes evaluated for aggregation and lyomesophase formation, acid dyes have been most widely studied. The work of Valko[52], Lenher and Smith[53] and Holmes and Standing[54] revealed that low molecular weight acid dyes existed as either small oligomers or were monomeric in aqueous solution.

Many workers have investigated the relationship between dyestuff aggregation and the number of sulphonic acid groups in the molecule[1,46,55–57]. In general as the number increases, electrostatic repulsive forces increase and so aggregation decreases. There are, however, some exceptions. The aggregation of dyes **3**, **5** and **6** (Figure 3.1) increase steadily on raising concentration, with dye **3** undergoing the highest level of aggregation and dye **6** the least. The exception was dye **4** which merely formed a stable dimer. This suggests, not surprisingly, that the position of the sulphonic acid group(s) in the molecule is also an important factor[1,57,58]. The variation in electrostatic and steric interactions caused by changes in number and position of sulphonic acid solubilising groups and the accompanying solvation and counter-ion effects influence significantly the molecular arrangement within the aggregate stack. A series of molecular models for simple acid dyes have been proposed from ^{1}H and ^{19}F NMR studies[39–41]. For example the favoured model for C.I. Acid Orange 1 (**7**)[40], is one in which the planes of the rings are stacked in an antiparallel array with an interplanar separation of 4.17 Å. The influence of alkyl chain length on aggregation has been investigated in some detail by Datyner, Flowers and Pailthorpe[46,47]. They have shown, for example,

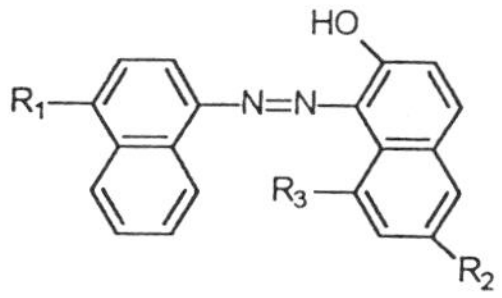

Dye	R_1	R_2	R_3
3	SO_3Na	H	H
4	H	SO_3Na	SO_3Na
5	SO_3Na	SO_3Na	H
6	SO_3Na	SO_3Na	SO_3Na

Figure 3.1 Some acid dyes of the anionic azo dye class.

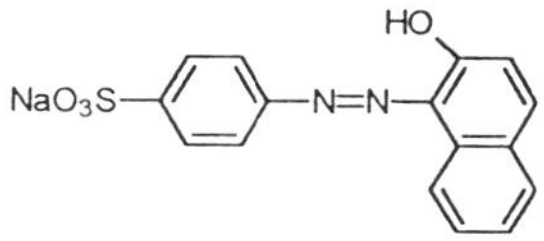

7 C.I. Acid Orange 2.

that the replacement of n-butyl by n-octyl, as the R group in **8**, dramatically increased both the degree and concentration dependence of the aggregation.

It seems reasonable then that introducing long alkyl chains into a dyestuff molecule could promote hydrophobic interactions which dominate over electrostatic repulsive forces and, hence, lead to an increase in aggregation, even on increasing the number of sulphonic acid groupings.

Herlinger and co-workers have suggested that the behaviour of aqueous solution acid dyes can be used as a guide to assess the physical state of these dyes in polyamide and hence can be related to fastness performance[59,60].

3.2.3.2 Direct dyes. The higher molecular weight direct dyes having proportionally fewer solubilising sulphonic acid residues, compared to acid dyes, aggregate to a much greater degree[52–54]. Congo Red, for

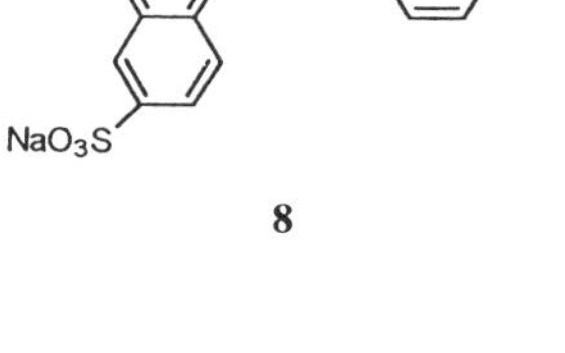

8

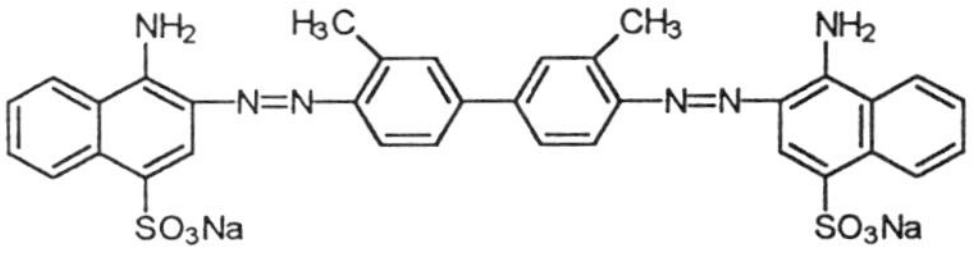

9 C.I. Direct Red 2 (Benzopurpurine 4B).

example, is known to aggregate strongly, particularly in the presence of electrolyte. Reported aggregation numbers for this dye include 20 000 (2.85×10^{-3} M[35]), 24 (7.4×10^{-5} M; 0.1 M NaCl[61]) and 45 (7.4×10^{-4} M; 0.2 M NaCl[48]). These differences reflect the varying conditions and measurement methods used.

Benzopurpurine 4B (C.I. Direct Red 2), 9, has been studied by a number of workers. In the absence of electrolyte, light scattering experiments suggested no appreciable aggregation. In the presence of 0.04 M KCl at 25 °C, Frank measured a weight-average molecular weight, M_w, of 2621 and a number-average molecular weight, M_n, of 552 for the aggregate particles[32,62].

3.2.3.3 Reactive dyes. There is relatively little published data on reactive dyestuff aggregation. However, it has been known for some time that the solution hydrolysis rates of reactive dyes can be reduced as a result of molecular self-association[63-65]. More recently it has been shown by Bredereck and Schumacher, for a series of azo reactive dyes based on H-acid, that aggregation significantly influences their fastness to hypochlorite[66]. Dramatic increases in the hypochlorite fading reaction rate are observed on disaggregation.

Harada *et al*. have investigated the relationship between the solution aggregation of C.I. Reactive Blue 19 and its dyeing performance on cellulose. The aggregation was studied using small angle x-ray scattering techniques. A stable ellipsoidal hexamer was present in the concentration range of $4.0-95.2 \times 10^{-3}$ mol/l at 20 °C which decreased to a trimer at 70 °C and increased to a 19 molecule unit in the presence of 0.3 M NaCl at 20 °C. Poor levelling was seen to accompany increased aggregation number[42,43]. Reactive dye aggregation will be covered further in sections 3.3 and 3.4.

3.3 Lyotropic liquid crystal formation

3.3.1 Liquid crystals

A liquid crystal represents a fourth state of matter, since it is neither an ordered solid nor an amorphous liquid, but portrays characteristics of both. A liquid crystal phase can be described as a liquid with a certain degree of order. Liquid crystal phases are also known as mesophases— from the greek, *mesos*, meaning middle. All mesophases possess orientational order, and the extent to which translational order is present gives rise to the many and varied types of mesophases that exist. The direction of orientation of the major axis of a liquid crystal phase is termed the director. A substance which is able to form liquid crystals is known as a mesogen.

There are two principal categories of liquid crystals; thermotropic and lyotropic. Thermotropic liquid crystal are produced when certain pure solids are heated. A range of mesophases may be formed as the temperature is raised, until the liquid phase becomes stable. The type of mesophase which may be formed is, thus, purely dependent upon temperature. Lyotropic liquid crystals are formed by the addition of a solvent to a solid. The mesophases (or lyomesophases) that are stable, therefore, depend upon composition as well as temperature.

The three common chromonic lyomesophases formed by anionic azo dyes in solution, as aggregates align with respect to one another, are the nematic, middle and P-phases.

3.3.2 Measurement techniques

3.3.2.1 Optical microscopy[67]. The practice of using optical microscopy to study liquid crystals commonly involves the sample being placed between crossed polarisers. If the sample is in the isotropic state, it will remain dark, because the orientation of the plane of polarisation is unaltered by the system. However, an anisotropic system will appear to be coloured except in four orientations, with respect to the polarisers, that are at 90° to each other. The sample appears to be dark in these extinction positions, due to the vibration directions being parallel to the polarisers. The colours are the brightest at an angle 45° to the extinction positions. These are called the interference or polarisation colours. A mesophase that is being observed through crossed polarisers will appear to be anisotropic over most of its area, though there may be some regions in which optical extinction occurs.

The extensive use of polarising optical microscopy in the study of liquid crystals is primarily due to the ease with which the mesophases can be identified by their optical textures. Consequently, observing the phase transitions is relatively straightforward during heating or cooling at a range of sample concentrations.

The nematic (N) mesophase actually derives its name from the optical texture that is observed under crossed polars. A number of mobile thread-like lines are commonly visible (Greek: *nematos*, meaning thread), which arise from structural discontinuities with the phase. Many nucleated domains are scattered throughout the phase; and, between crossed polars, dark branches can be seen joining one nucleus to another.

The middle (M) phase possesses its characteristic optical texture, due to the growth of small fan-like units which eventually form a composite fan-like texture. The initial units do not merge with each other when they meet, but instead from an irregular boundary. The fine ribs which constitute the fan-like design indicate that they are composed of narrow focal conic domains. The extinction arms that highlight the ribs, from the centre to the outer boundary, are typically straight. However, zig-zag alternations may develop once the rods which make up the phase become too long for the space that is available[68].

The P-type phase is characterised by the striated, tiger-skin texture. It has been suggested[18] that its formation is caused by the possible circular cross-section of the aggregates, which leads to the occurrence of bundles of columns. These are, then, able to lie above one another and any disclinations that may be present give rise to the tiger-skin optical texture.

3.3.2.2 NMR Spectroscopy[69,70]. When a nucleus having a magnetic moment (i.e. a spin) is placed in a magnetic field, there are a number of interactions which occur. The Zeeman interaction aligns the nuclear spins either parallel or anti-parallel to the magnetic field. A small magnetic field is induced by the electrons which circulate around the nuclei and the resultant magnetic field which is experienced by the nuclei causes all non-equivalent nuclei to have chemical shifts. Indirect spin–spin coupling can occur between nuclei. This is transmitted through chemical bonds and results in the resonance peaks being split into doublets, triplets, etc. with a splitting of a few hertz.

Dipole–dipole interactions can also occur between nuclei, and these produce much larger splittings of hundreds or thousands of hertz. The interaction, and hence the magnitude of the splitting, is proportional to $(1 - 3\cos^2 \theta)$, where θ is the angle between the liquid crystal director and the vector joining the two interacting nuclei. When the spin of the nucleus, I, is greater or equal to one, quadrupolar interactions can be observed between the nuclear electric quadrupole moment and the non-spherically symmetrical field gradient around the nucleus. The quadrupolar interactions are also orientation dependent, with the size of the splitting depending upon $(1 - 3\cos^2 \theta)$, where θ is the angle between the electric field gradient tensor and the director.

For a deuterium nucleus $I = 1$ and, therefore, two transitions of different energy are possible which produces a 1:1 doublet. By contrast, for a sodium nucleus $I = 3/2$ and hence three transitions of different energy can occur, thus causing the peak to be split into a triplet. No quadrupolar interaction is observed for a proton nucleus, because $I = 1/2$.

The dipole–dipole and quadrupolar interactions are responsible for the fundamental differences between solution and ordered state NMR spectroscopy. Much smaller splittings are observed in solution state spectra, because usually only indirect spin-spin couplings occur. The dipole–dipole and quadrupolar interactions are reduced to zero, due to the rapid tumbling motions of the molecules which cause all the values of θ to average to zero on the NMR timescale. However, liquid crystals are, by their nature, ordered and thus the tumbling motions of the molecules are not isotropic on the NMR timescale. Very often the director will align parallel or perpendicular to the magnetic field. Consequently, θ will have specific values and the dipole–dipole or quadrupolar interactions will be non-zero, and large resonance splittings will be observed. The magnitude of the splittings that are observed can be used to calculate the order parameters of the resonating nuclei, due to the orientation dependence of the spectra.

The order parameter, S, of a liquid crystal is the degree of alignment of the system with respect to the liquid crystal director. The order parameter of a particular nucleus is proportional to the dipole–dipole or quadrupolar interactions and hence, also the splitting.

In the study of lyotropic liquid crystals, the quadrupolar nuclei commonly belong to solvent molecules, for example, deuterium oxide. In such cases, a proportion of the solvent will be bound to the aggregates or micelles, whilst the rest will be free, unbound solvent. The motion of the bound water will be restricted, due to the lack of isotropic motion by the aggregates or micelles and, thus, quadrupolar interactions will occur. However, the free solvent will be able to rotate with typical isotropic, tumbling motion, thus reducing its quadrupolar interaction to zero. Since exchange between the two types of solvent is rapid with respect to the NMR timescale, an average spectrum is observed, with the splitting proportional to the bound fraction as well as the order parameter of the bound solvent. For the case of deuterium oxide, the exchange between different bound water molecules is also fast. Hence, for a single lyomesophase in deuterium oxide the single resonance peak of a deuterium spectrum will be split into a doublet, the magnitude of which depends upon the order parameter of the deuterons and also the fraction of bound water.

3.3.2.3 X-ray diffraction[51,70]. The application of small and wide angle X-ray scattering techniques to the study of aggregate and lyomesophase structure was covered in section 3.2. A combination of the two methods is usually required to fully characterise the lyomesophase structures formed by azo dyestuffs, as the repeat distances involved are typically in the range 1–100 Å (the most common unit cell for the ordered middle and P-phases is hexagonal). In general, the diffraction pattern for the ordered and well aligned middle and P-phases contains an arc along the meridian at 3.4–3.6 Å, up to four equatorial reflections and an off-axis reflection (cf. Figure 3.2). The amount of arcing of the meridional reflection (i.e. intermolecular spacing within aggregate stacks) is a measure of the degree of alignment of the stacks with respect to the director. The width of this arc is proportional to the length of the stack (see section 3.2.2). The off-axis reflection at $c.$ 7.0 Å suggests that the molecular stacks are helical in nature and comprised of dimer repeats. The four reflections along the equator indicate the degree of ordering of the stacks in the a/b plane of the two dimensional lattice. The spacings are in the ratio $1:\sqrt{3}:2:\sqrt{7}$ which is as expected for a hexagonal lattice.

3.3.3 Lyotropic liquid crystals

There are three major types of lyotropic liquid crystals: (i) amphiphilic formed by, for example, surfactants[71]; (ii) polymeric, e.g. surfactant or rigid polymers; and (iii) chromonic, e.g. dyes and drugs. The first lyotropic liquid crystal to be reported, though not identified, was in the nerve myelin/water system described in 1854 by Virchow[72].

3.3.4 Chromonic lyotropic liquid crystals

The first report of a lyotropic liquid crystal formed by a water soluble polyaromatic material was made in 1915 by Sandquist[73] who contacted crystalline 10-bromo-phenanthrene-3-(or 6-) sulphonic acid (**10**) with water.

This was followed, in 1927, by Balaban and King[74] who noted textures consisting of "waxy solids, limpid fluids possessing a sheen, clear gels, or

Figure 3.2 Schematic representation of the high angle X-ray diffraction pattern of the middle phase.

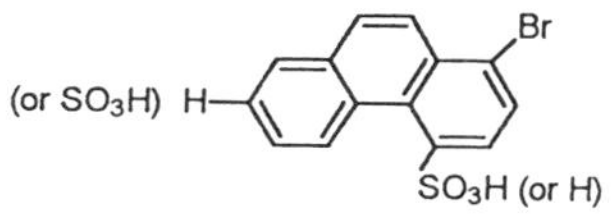

10 10-Bromo-phenanthrene-3-(or 6-)sulphonic acid.

worm-like growths which in most cases were strongly birefringent" when aqueous solutions of naphthylamine di- and tri-sulphonic acids were cooled.

In the early 1930s, Gaubert[75] discovered that aqueous solutions of the dyes tartrazine (**11**), croceine brilliant red, methylene blue and neutral red formed optically negative nematic and smectic-like phases when cooled. The mesophases of tartrazine were relatively unstable, with needle-like crystals forming, but, the phases produced by the other dyes were more stable.

A direct dye series having varying numbers and positions of sulphonic acid groups was studied by Jones and Kent in 1980[58]. Despite the tri-sulphonates crystallising out of solution, some of the mono- and di-sulphonates, **12**, were observed by polarising optical microscopy to form a 'lamellar' mesophase. A smectic-like mesophase was also observed on cooling from the isotropic liquid. Absorption spectrophotometry indicated that the process of aggregation was gradual. It was proposed that the aggregation of 'amphiphilic dyes' in aqueous solution involved stacks of dye molecules combining to produce a lamellar-like mesophase, as the dye concentration was raised. With the benefit of hindsight, the formation of a lamellar phase by such dyes appears to be unlikely and, in fact, the 'mesophase' that was reported was later defined as nematic[16].

In 1986, Attwood, Lydon and Jones[16] reported the aqueous solution behaviour for a further series of direct dyes. The mesophases observed

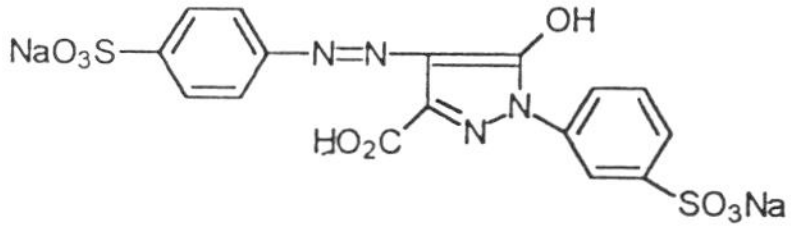

11 Tartrazine.

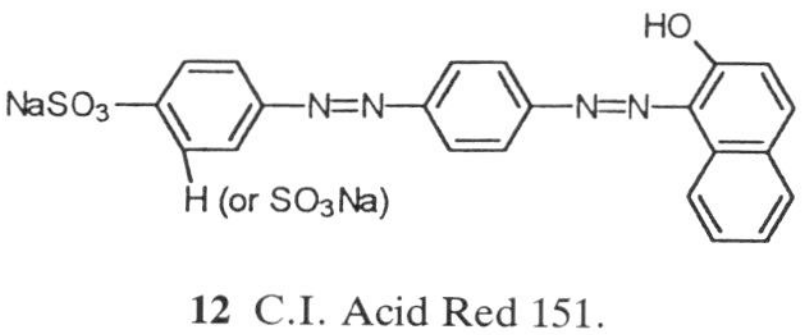

12 C.I. Acid Red 151.

were similar to the chromonic nematic and middle phases exhibited by the anti-asthmatic drug disodium chromoglycate (INTAL; **13**), in water[20,76,77]. (It is from this type of system that the name 'chromonic' originates, corresponding to the chromone structure). The middle phase was described as being made up of untilted stacks of mesogen molecules in a dynamic herringbone array in a water continuum, Figure 3.3. In this model, the apparent orthorhombic symmetry is reduced to hexagonal due to rotational disorder within the columns. Hence, it was suggested that the middle phase was analogous to the thermotropic S_B phase and the nematic to the S_A phase[78]. It was proposed that a phase similar to the S_E phase, with orthorhombic symmetry, may also occur. The results of x-ray diffraction studies on the nematic and middle phases exhibited by dyes **12** and **14** appeared to confirm the structural similarities. Optical microscopy, however, certainly showed the typical schlieren texture of the nematic phase but the middle phase appeared rather ill-defined and grainy. It had previously been observed that the addition of NaCl to the middle phase of the disodium chromoglycate/water system gave the same

13 Disodium Chromoglycate (INTAL).

14 C.I. Direct Brown 202.

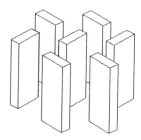

Figure 3.3 The proposed dynamic herringbone array for the chromonic middle phase.

texture. Since the preparation of anionic azo dyes often involves isolation of the solid from aqueous solution by salting out with sodium chloride, it was, therefore, assumed that residual salt had given rise to the poor structural definition in the optical micrographs. This was later confirmed in studies utilising a series of reactive dyes[79]. These studies[79] reported the formation of not only the two chromonic mesophases but also, at lower concentrations and with the addition of NaCl, a 'quasi-crystalline' form. This new solid form was birefringent with a fibrous morphology, though non-crystalline. It was shown that this quasi-crystalline form consisted of a columnar structure, with ordering along its length (i.e. similar to the chromonic phases). It was suggested that this new form had resulted from salt induced precipitation of the aggregates.

The occurrence of chromonic liquid crystals became more widespread, when research performed by Turner and Lydon[18] showed that a significant number of azo dyes formed these types of lyomesophases (i.e. the mesogens observed previously[16,20,76] were certainly not unique). Peripheral evaporation of many of the dye solutions produced a distinct boundary between the nematic and middle phases with no paramorphosis (i.e. a sharp first order transition had occurred). By contrast, a number of dyes produced a boundary which was very difficult to discern. This novel chromonic modification was termed a P phase. The explanation for this behaviour was based upon the cross-section of the stacks, Figure 3.4. It was proposed that in the conventional chromonic systems (N/M type) the stacks had a rectangular cross-section which resulted in a sharp transition from the nematic to the dynamic herringbone array of the middle phase. However, the aggregates in the P type systems possessed circular cross-sections which would lead to a gradual packing together of the stacks in the nematic-like region, until the hexagonal symmetry was achieved. Thus, it was suggested that the transition in the P type system

Figure 3.4 A schematic representation of the N/M and P type systems. The long axes of the columns are perpendicular to the plane of the paper with concentration increasing down the figure. The dotted line represents the sharp boundary between the nematic and middle phases.

was second order. The P phase possessed a characteristic optical texture which was similar to that of the nematic phase, although it also contained a striated/tiger-skin pattern of lines. The tiger-skin texture was explained by the formation of cylindrical bundles of the circular stacks, similar to a length of rope. Studies on the C.I. Reactive Red 3:1 (**15**)/water system revealed a number of unusual optical textures including one with orthorhombic symmetry, Figure 3.5.

The two extremes of columnar chromonic behaviour have been studied in some detail by various authors[51,70,80,81]. The dyes C.I. Food Yellow 3 (**16**) and C.I. Acid Red 266 (**17**), possess the most and least distinct boundary, respectively, between the nematic and middle (hexagonal) phases during peripheral evaporation of the solvent water[18] and, hence, represent the best examples of N/M and P type behaviour (cf. Figure 3.6).

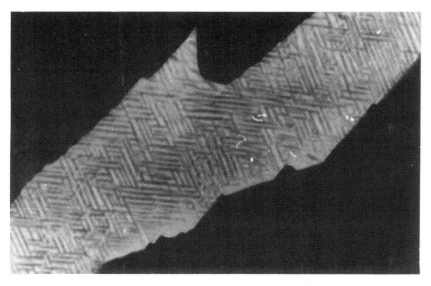

15 Reactive Red 3:1.

Figure 3.5 An unusual optical texture of the CI Reactive Red 3:1/water system which shows orthorhombic symmetry.

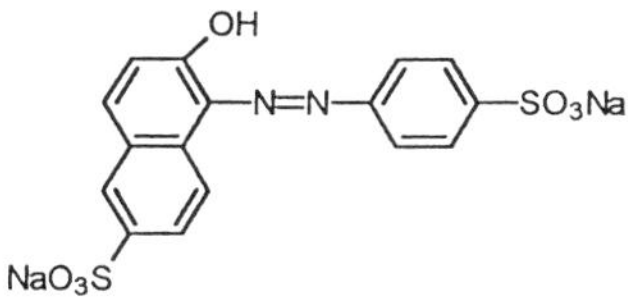

AGGREGATION AND LYOTROPIC LC FORMATION OF ANIONIC AZO DYES 99

16 C.I. Food Yellow 3.

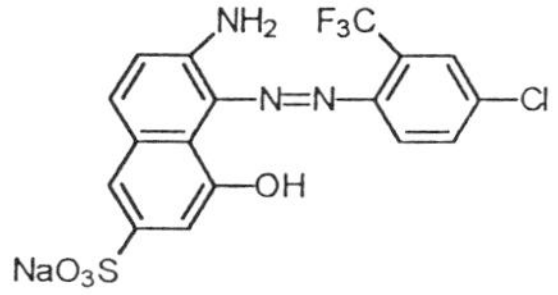

17 Acid Red 266.

The nematic and middle mesophases formed by C.I. Food Yellow 3 in aqueous solution, at 30% and 40% wt/wt dye, respectively, are stable over a wide temperature and concentration range (Figure 3.7). Deuterium NMR analysis of the C.I. Food Yellow 3/D_2O system shows an observable two-phase region between the nematic and middle phases (Figure 3.8). The order parameter changes for the aggregates during middle/nematic transition are shown in Table 3.2.

Typical high angle x-ray diffraction patterns of isotropic and aligned nematic and middle phases of C.I. Food Yellow 3 are shown in Figure 3.9. At 18 °C, a broad 3.34 Å ring present in the isotropic solution pattern (due to the intermolecular spacing of molecules within the aggregate stacks) gradually sharpens into a meridional arc, as the nematic phase forms ($c.$ 30 wt %) and an equatorial reflection at $c.$ 23–26 Å appears (corresponding to the ordering of the molecular stacks in the a/b plane). The hexagonal middle phase, which forms above 39 wt %, exhibits four reflections along the equator at 19.58, 11.33, 9.76 and 7.37 Å (i.e. in the ratio expected for a hexagonal lattice $1:\sqrt{3}:2:\sqrt{7}$). In addition off-axis reflections in the shape of an X appear at a spacing of 6.68 Å. These are believed to result from the helical nature of the molecular stacks where the molecules form dimer repeats with a twist angle of 180°. The refined lattice parameters for the hexagonal unit cell are $a = b = d_{100} \times 2\sqrt{3} = 23.04$ Å, $c = 6.68$ Å, $\alpha = \beta = 90°$, $\gamma = 120°$ (where d_{100} is the spacing of the (100) reflection). From the Scherrer–Bragg approximation, the aggregate stack size for the nematic phase (33 wt %, 18 °C) is 158 Å (or aggregation number = 46) and for the middle phase (46 wt %, 18 °C) is 185 Å (or aggregation number = 55).

The P phase that is formed by C.I. Acid Red 266 is stable even at

(a)

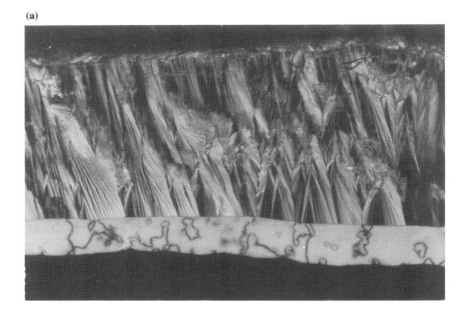

(b)

Figure 3.6 Optical micrographs of the peripheral evaporation of isotropic solutions of the dyes. (a) CI Food Yellow 3 (magnification ×80: low concn. at top). (b) CI Acid Red 266 (magnification ×80: low concn. at top).

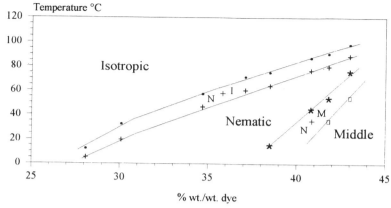

Figure 3.7 The phase diagram of CI Food Yellow 3 in aqueous solution.

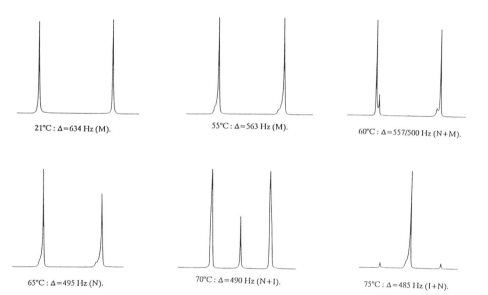

Figure 3.8 Deuterium spectra of CI Food Yellow 3/D$_2$O at variable temperatures. Fixed concn. (40% wt./wt. dye), M phase at room temperature.

concentrations of 0.25% wt/wt, which suggests that much longer rods are present than in the C.I. Food Yellow 3/water system. On peripheral evaporation of the isotropic solution a gradual increase in the viscosity of the mesophase can be observed. There does not appear to be a well-defined boundary between the hexagonal and nematic-like regions (*cf*. Figure 3.6). The splitting of the doublets in the ^{2}H NMR spectra of

Table 3.2 The order parameters of the naphthalene ring of CI Food Yellow 3 over a range of temperatures.

Order parameter (S)	Temperature (°C)	Mesophase
0.38	23	M
0.37	40	M
0.36	50	M
0.33	60	N + M
0.32	65	N

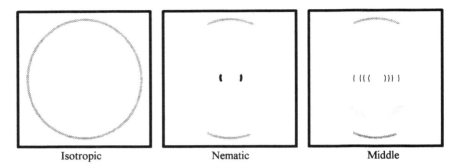

Isotropic   Nematic   Middle

Figure 3.9 Schematic representations of the high angle X-ray diffraction patterns of the CI Food Yellow 3/water system.

the C.I. Acid Red 266/D_2O system (Figure 3.10) are notably smaller than those observed with C.I. Food Yellow 3 due to the far greater water content in this case. By contrast, the splittings of the triplets in the ^{23}Na NMR spectra are much larger compared to C.I. Food Yellow 3 which indicates that the counter-ions possess a greater order parameter.

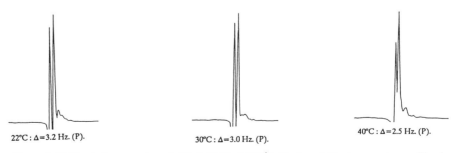

22°C : Δ=3.2 Hz. (P).   30°C : Δ=3.0 Hz. (P).   40°C : Δ=2.5 Hz. (P).

Figure 3.10 Deuterium spectra of CI Acid Red 266/D_2O at variable temperatures. Fixed concn. (0.5% wt./wt. dye), P phase at room temperature (side peaks due to field fluctuations; spectra gaussian enhanced).

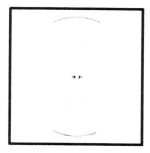

Figure 3.11 Schematic representations of the high angle X-ray diffraction pattern of the P phase of the CI Acid Red 266/water system.

The high angle x-ray diffraction pattern for the P-phase of C.I. Acid Red 266 is similar to that observed for C.I. Food Yellow 3 (Figure 3.11). The narrower arc observed for the P-phase along the meridian at 3.34 Å suggests a larger length aggregate.

The equatorial reflections at 62.5, 36.0 and 30.5 Å again indicate aggregate ordering in the a/b plane, although the observable streak of continuous scatter can be attributed to the irregularity of the hexagonal array. The lattice parameters were refined to the values of $a = b = 72.16$ Å, $c = 7.04$ Å, $\alpha = \beta = 90.0°$, $\gamma = 120.0°$. The length of the molecular stack in the P-phase (4 wt%, 18 °C) was found to be 321 Å (aggregation number = 94).

3.4 Textile dyeing

The role played by aggregation and lyomesophase formation in the dyeing of textiles fibres is a significant one, and the literature on the subject is extensive. A comprehensive review cannot be given here but a few general conclusions are drawn.

Anionic azo dyes are applied to textile substrates from, and are often formulated and sold as, aqueous solutions. If one considers probably the most complex of anionic azo dye application processes, the reactive dyeing of cotton, then dye–dye interaction can drastically influence each of the key physical chemical steps associated with this process, Figure 3.12. It can, for example, nucleate or retard dyestuff crystallisation, reduce the rate of solution hydrolysis, increase levels of adsorption and reduce fibre diffusion rates. Each of these effects influences the overall fixation efficiency (see chapter 2). The effect on crystallisation and hydrolysis is particularly important in the formulation of physically and chemically stable reactive dye solutions.

In turn, the application conditions used (e.g. dyestuff and salt

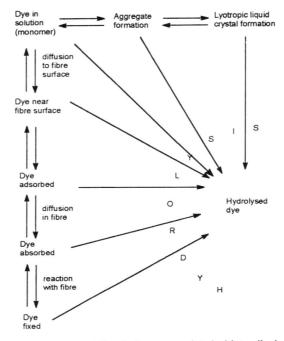

Figure 3.12 Key physical chemical steps associated with textile dyeing.

concentration, dyeing temperature and auxiliaries) also have a profound affect on the dyes solution behaviour and hence dyeing performance. The phase diagram shown in Figure 3.7 for C.I. Food Yellow 3 is characteristic of many acid, direct and reactive dyes and, hence, illustrates a common influence of concentration and temperature (i.e. as concentration increases and/or temperature decreases aggregation increases). Likewise Table 3.3 illustrates the typical effect of urea on anionic azo dye aggregation (i.e. aggregation decreases with increasing urea concentration).

Aggregation and lyomesophase formation will continue to receive a great deal of attention from dyestuff manufacturers and textile dyers for

Table 3.3 The effect of urea concentration on the aggregation of CI Food Yellow 3 in water (33 wt%, 18 °C). Calculated from high angle X-ray diffraction patterns using the Sherrer-Bragg approximation.

Urea concentration (wt%)	Stack length (Å)	Aggregation number
0	158	46
5	124	36
20	49	15
50	27	8

the reasons outlined above. It is still true to say that our understanding of the relationship between dyestuff molecular structure and aqueous solution behaviour is far from complete.

References

1. D. G. Duff, D. J. Kirkwood and D. M. Stevenson, *J. S. D. C.*, **93**, 303 (1977).
2. N. Boden, R. J. Bushby and C. Hardy, *J. Physique Lett.*, **46**, 325 (1985).
3. K. Bergmann and C. T. O'Konski, *J. Phys. Chem.*, **67**, 2169 (1963).
4. P. Mukerjee and A. K. Ghosh, *J. Phys. Chem.*, **67**, 193 (1963).
5. K. K. Rohatgi and G. S. Singhal, *J. Phys. Chem.*, **70**, 1695 (1966).
6. S. R. Iyer and G. S. Singh, *J. S. D. C.*, **89**, 128 (1973).
7. H. S. Frank and M. W. Evans, *J. Chem. Phys.*, **13**, 507 (1945).
8. D. R. Levin and T. Vickerstaff, *Trans. Faraday Soc.*, **43**, 491 (1947).
9. W. J. Kauzmann, *Adv. Protein Chem.*, **14**, 1 (1959).
10. C. O. Timmons and W. A. Zisman, *J. Coll. Inter. Sci.*, **28**, 106 (1968); R. Aveyard, B. J. Briscoe, J. Chapman, *J. Chem. Soc. Far Trans.*, **68**, 10 (1972).
11. K. A. Williams, *Oils, Fats and Fatty Foods*, 3rd edn., J&A Churchill Ltd. (1950).
12. S. E. Sheppard, R. H. Lambert and R. D. Walker, *J. Chem. Phys.*, **9**, 96 (1941); S. E. Sheppard, *Rev. Mod. Phys.*, **14** 303 (1942).
13. S. E. Sheppard and A. L. Geddes, *J. A. C. S.*, **66** 2003 (1944).
14. L. V. Levshin and V. K. Gorshkov, *Opt. and Spec.*, **10**, 401 (1961).
15. N. H. Hartshorne and G. D. Woodard, *Mol. Cryst. Liq. Cryst.*, **23**, 343 (1973).
16. T. K. Attwood, J. E. Lydon and F. Jones, *Liq. Cryst.*, **1**, 499 (1986).
17. N. Boden, R. J. Bushby, C. Hardy and F. Sixl, *Chem. Phys. Lett.*, **123**, 359 (1986); N. Boden, R. J. Bushby, L. Ferris, C. Hardy and F. Sixl, *Liq. Cryst.*, **1**, 109 (1986).
18. J. Turner, *Lyotropic Discotic Dye/Water Systems*, PhD Thesis, University of Leeds (1988).
19. E. Coates, *J. S. D. C.*, **85**, 355 (1969).
20. T. K. Attwood and J. E. Lydon, *Mol. Cryst. Liq. Cryst.*, **108**, 349 (1984).
21. H. L. Anderson, C. A. Hunter, M. N. Meah and J. K. M. Sanders, *J. A. C. S.*, **112**, 5780 (1990).
22. C. A. Hunter and J. K. Sanders, *J. A. C. S.*, **112**, 5525 (1990).
23. E. Rabinovitch and L. F. Epstein, *J. A. C. S.*, **63**, 69 (1941).
24. J. H. De Boer, *Trans. Far. Soc.*, **32**, 10 (1936).
25. J. Israelachvili, *Intermolecular and Surface Forces*, 2nd edn., Academic Press (1991).
26. F. London, *Z. Physik.*, **63**, 245 (1930); *Z. Physikal Chem.*, **11B**, 222 (1931).
27. B. Craven and A. Daytner, *J. S. D. C.*, **79**, 515 (1963).
28. A. Datyner, M. J. Delaney and H. Holliger, *Aust. J. Chem.*, **24**, 1849 (1971).
29. C. Robinson and J. L. Moilliet, *Proc. Roy. Soc. A*, **143**, 630 (1934).
30. Milicevic and G. Eigenmann, *Helv. Chim. Acta*, **47**, 1039 (1964).
31. O. Quensel, *Trans. Faraday Soc.*, **31**, 259 (1935).
32. P. Alexander and K. A. Stacey, *Proc. Roy. Soc. A*, **212**, 274 (1952).
33. R. B. Chavan and J. Venkata Rao, *J. S. D. C.*, **99**, 126 (1983).
34. W. U. Malik and P. Chand, *J. Electroanal. Chem. Int. Chem.*, **19**, 431 (1968).
35. P. J. Hillson and R. B. McKay, *Trans. Faraday Soc.*, **61**, 374 (1965).
36. R. Weingarten, *Melliand Textilber.*, **48**, 301 (1967).
37. D. Pugh, C. H. Giles and D. G. Duff, *Trans. Faraday Soc.*, **67**, 563 (1971).
38. U. Baumgarte, *Textilveredlung*, **15**, 413 (1980).
39. P. Skrabal, F. Bangerter, K. Hamada and T. Iijima, *Dyes and Pigments*, **8**, 371 (1987).
40. T. Asakura and M. Ishida, *J. Coll. Int. Sci.*, **130**, 184 (1989).
41. K. Hamada, M. Mitshuishi, M. Ohira and K. Miyazaki, *J. Phys. Chem.*, **97**, 4926 (1993).
42. N. Harada *et al.*, *Sen-'I Gakkaishi*, **47**, 528 (1991).
43. N. Harada, *Senshoku Kogyo*, **40**, 262 (1992).

44. A. H. Herz, *Adv. Coll. Int. Sci.*, **8**, 237 (1977).
45. B. C. Burdett, *Aggregation Processes in Solution* (eds E. Wyn-Jones and J. Gormally), Elsevier Science (1983).
46. A. Datyner, A. G. Flowers and M. T. Pailthorpe, *J. Coll. Int. Sci.*, **74**, 71 (1980).
47. A. Datyner and M. T. Pailthorpe, *J. Coll. Int. Sci.*, **76**, 557 (1980).
48. I. D. Rattee and M. M. Breuer, *The Physical Chemistry of Dye Adsorption*, Academic Press (1974).
49. R. H. Peters, *Textile Chemistry, Vol. III, The Physical Chemistry of Dyeing*, Elsevier (1975).
50. L. J. Gosting, *Adv. Protein Chem.*, **11**, 429 (1956).
51. A. A. Jaber, PhD Thesis, Physics Dept., Keele University, (1994).
52. E. I. Valko, *J. A. C. S.*, **63**, 1433 (1941).
53. S. Lenher and J. E. Smith, *J. A. C. S.*, **57**, 497, 504 (1935).
54. F. H. Holmes and H. A. Standing, *Trans. Faraday Soc.*, **41**, 542, 568 (1945).
55. A. M. Islam *et al.*, *Text. Dyer and Printer*, August, 41 (1973).
56. A. A. R. El-Mariah *et al.*, *J. Chin. Chem. Soc.*, **29**, 279 (1982).
57. K. Hamada *et al.*, *Dyes and Pigments*, **16**, 111 (1991).
58. F. Jones and D. R. Kent, *Dyes and Pigments*, **1**, 39 (1980).
59. H. Herlinger, B. Küster and D. H. Lämmermann, *Textil Praxis Int.*, September, 994 (1988).
60. H. Herlinger, B. Küster, E. Hindias and D. H. Lämmermann, *Textil Praxis Int*, October, 1105 (1988).
61. E. I. Valko, *Trans. Faraday Soc.*, **31**, 230 (1935).
62. H. P. Frank, *J. Coll. Sci.*, **12**, 480, (1957).
63. A. Datyner, P. Rys and H. Zollinger, *Helv. Chim. Acta.*, **49**, 755 (1966).
64. P. Rys, *Textilveredlung*, **2**, 95 (1967).
65. N. Bhattacharyya, *Text. Res. Journ.*, January, 54, (1987).
66. K. Bredereck and C. Schumacher, *Dyes and Pigments*, **21**, 23 (1993); *ibid*, 45.
67. N. H. Hartshorne and A. Stuart, *Practical Optical Crystallography*, 2nd edn., Arnold (1969).
68. N. H. Hartshorne, *The Microscopy of Liquid Crystals*, Microscope Publ. Ltd. (1974).
69. R. K. Harris, *Nuclear Magnetic Resonance Spectroscopy*, Pitman, 132 (1983).
70. A. P. Ormerod, MSc Thesis, Salford University, (1991).
71. G. J. T. Tiddy, *Physics Reports*, **57**, 1 (1980).
72. R. Virchow, *Virchow's Archiv.*, **6**, 571 (1854).
73. H. Sandquist, *Berichte*, **48**, 2054 (1915).
74. I. E. Balaban and H. King, *J. Chem. Soc.*, 3068 (1927).
75. P. Gaubert, *Compt. Rend.*, **197**, 1436 (1933); **198**, 951 (1934); **200**, 679 (1935).
76. J. S. G. Cox, G. D. Woodard and M. C. McCrone, *J. Pharm. Sci.*, **60**, 1458 (1971).
77. T. K. Attwood and J. E. Lydon, *Mol. Cryst. Liq. Cryst. Letts.*, **4**, 9 (1986).
78. G. H. Brown, *J. Opt. Soc. Amer.*, **63**, 1505 (1973).
79. D. E. Sadler *et al.*, *Liq. Cryst.*, **1**, 509 (1986).
80. A. P. Ormerod, PhD Thesis, Salford University, in preparation.
81. G. J. T. Tiddy, D. L. Mateer, A. P. Ormerod, W. J. Harrison and D. J. Edwards, *Langmuir*, **11**, 390 (1995).

4 Contribution of crystal form/habit to colorant properties
A. IQBAL, B. MEDINGER and R. B. McKAY

4.1 Introduction

This chapter deals with the contribution of crystal form and crystal habit to colorant properties. Crystal form will be considered to include crystal modification and crystallinity, whereas crystal habit will be treated in terms of the shape of such particles. Inseparably connected with form and habit is the size of particles and this will, therefore, be treated simultaneously. Although by definition colorants should also include dyes and inorganic pigments, discussion herein is consciously focused on *organic pigments*. Consideration will be given only to those colorant properties which relate to conventional pigmentation of paints, plastics, man-made fibres, and printing inks.

Organic pigments are finely divided crystalline solids, the primary function of which is to impart colour to paints, printing inks, plastics, man-made fibres, and various other materials. Unlike soluble dyes, they are essentially insoluble in application media and mechanical means have to be used to disperse these therein. In final application, the crystalline pigment particles thus are dispersed *within* a solid material, namely a dried paint or ink film, bulk polymer or fibre. This is their essential difference from dyestuffs, which are usually adsorbed from solution or dispersion by physical or chemical means onto the surface of a solid substrate, or sometimes are distributed via dissolution within application media.

The most important performance criteria of organic pigments are:

  (i) state of aggregation of primary particles
 (ii) application properties like dispersibility, dispersion stability, flocculation, flow properties, and levelling of dispersions,
(iii) colouristic properties of dispersions, such as hue/shade, chroma/saturation, lightness, hiding power (opacity, transparency), colour strength, dichroism, gloss, and
(iv) fastness properties, e.g. light, weather, heat, migration, chemical, and recrystallization stabilities.

Whereas the properties of dyes are determined almost exclusively by their chemical constitution, the overall pigmentary performance of a coloured particulate solid depends not only on intrinsic molecular properties, but also on molecular interactions within the solid state (crystal) lattice and at the interface with application media. Primary pigment particles are ordered assemblies of the molecules in a specific crystal lattice environment. Intrinsic molecular characteristics of relevance to pigments are light absorption, solubility, chemical stability and polarity/functionality, while crystal modification, molecular packing and crystallinity, particle size, shape, and surface energetics constitute the most important solid state characteristics.

This chapter is intended to provide some insight into how the above physical characteristics of organic pigments affect their application, colouristic, and fastness properties. Another aim is also to illustrate how this physical character can be controlled to give products of desired performance. A previous contribution from these laboratories has focused on dispersion properties of organic pigments.[1]

4.2 Molecular and crystal lattice characteristics of organic pigments and their effect on pigmentary performance

4.2.1 Molecular constitution

Most organic pigment molecules are characterized by certain common structural features:

(i) conjugated chromophoric system, embedded in relatively rigid planar or near-planar structural frame, enabling interplanar $\pi-\pi$ interaction, molecular stacking, resonance stabilization
(ii) van der Waals contacts, in some cases
(iii) functional groups, desirably capable of forming intermolecular hydrogen bonds, and
(iv) functional groups yielding insoluble salts or metal complexes.

The most important commercially available pigments possess one or more of the above molecular structural features. The main criteria which have led to their selection have been colour intensity, cleanliness of hue, fastness to light and weathering, heat, solvent, and chemical resistance, ecological and toxicological factors, raw material availability, and cost. The wide range of chemical constitutions, together with differing levels of performance, have led to classification into two broad groups, namely so-called classical pigments and high performance pigments. Chemical constitutions of organic pigments are detailed in the *Colour Index* published by the Society of Dyers and Colourists.[2] A number is assigned to each chemical type, for example, C.I. Pigment Yellow 13. This universally accepted nomenclature is used here. Comprehensive descrip-

CONTRIBUTION OF CRYSTAL FORM/HABIT TO COLORANT PROPERTIES 109

tions of the chemistry of organic pigments are given in Herbst and Hunger[3] and Zollinger.[4]

The term classical organic pigments is applied to a wide variety of monoazo and disazo pigments (mainly yellows, oranges, and reds), copper phthalocyanine pigments (blue and green), and basic dye complexes, all of which have been established for many years. Examples of their chemical constitution are given in Figure 4.1. The term azo pigments

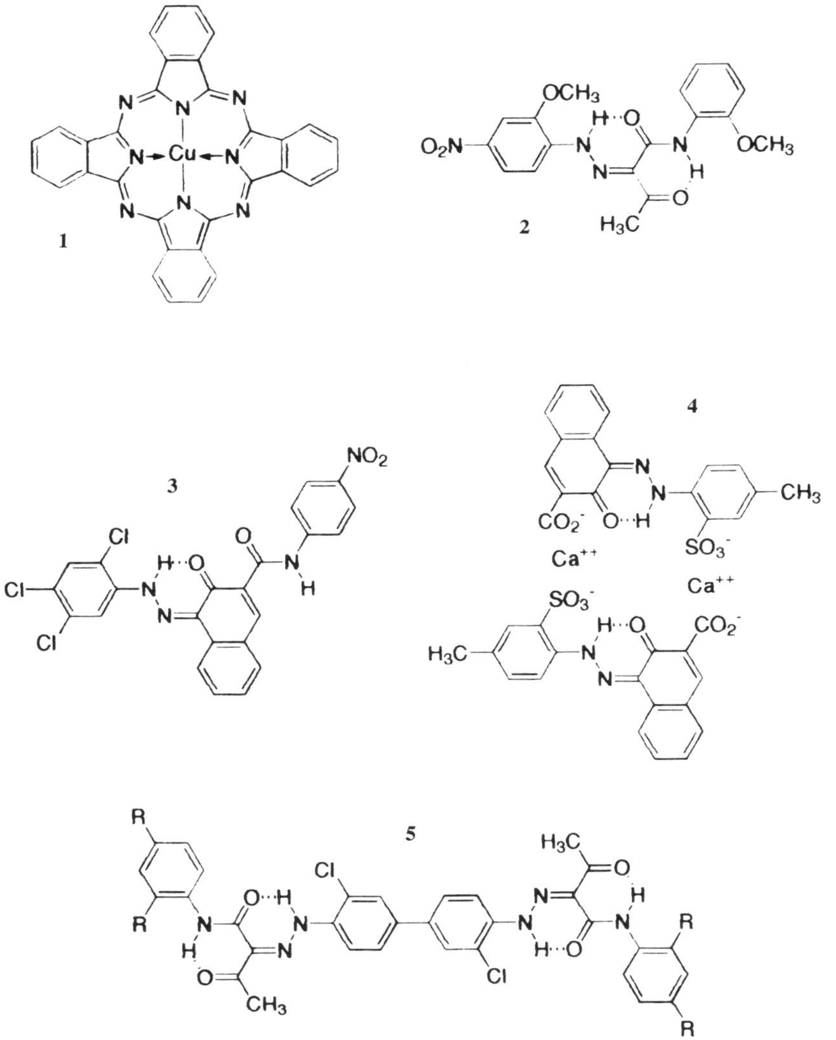

Figure 4.1 Chemical structures of some 'classical' organic pigments. (**1**) C.I. Pigment Blue 15; (**2**) C.I. Pigment Yellow 74; (**3**) C.I. Pigment Red 112; (**4**) C.I. Pigment Red 57:1; (**5**) R = CH_3 is C.I. Pigment Yellow 13, R = H is C.I. Pigment Yellow 12.

is a misnomer as their molecules tend to exist in the hydrazone form (as shown in Figure 4.1) stabilized by intramolecular hydrogen bonding.[5-7]

Copper phthalocyanine pigments have excellent fastness properties and are the main type of organic blue pigment for all applications. Monoazo and disazo pigments are of more limited applicability since they have generally less all-round fastness. Although a wide variety of yellow and red monoazo pigments and some yellow and orange disazo pigments find use in paints and plastics, the bulk of classical pigments are used in printing inks.

C.I. Pigment Yellows 12 and 13 (disazo pigments), C.I. Pigment Reds 57:1 and 53:1 (metal salts of sulphonated monoazos; commonly called calcium 4B toners and Lake Red C, respectively), and β-form copper phthalocyanine (C.I. Pigment Blue 15:3) are the most important for this application.

The term high performance pigments is applied to a group of more recently developed organic pigments with excellent all-round fastness properties. Being more costly to manufacture than classical pigments, they tend to be used for specialized and the more technically demanding applications, particularly automotive paints and constructional plastics, where exceptional light, weather, and heat fastness are prerequisites. Examples of their molecular structure are given in Figure 4.2.

There are two major chemical types of high performance pigments: azo and polycyclic pigments. Of the azo pigments the most important are yellow to red disazo condensation pigments (e.g. C.I. Pigment Red 144), yellow azo metal salts (e.g. C.I. Pigment Yellow 183), and yellow to red pigments based on benzimidazolone (e.g. C.I. Pigment Orange 36). Some azomethine-type pigments are important commercially; yellow isoindolinones (e.g. C.I. Pigment Yellow 110) and isoindolines (e.g. C.I. Pigment Yellow 139) are typical examples; azo pigments account for about 50% of world production of organic pigments, copper phthalocyanines and polycyclic pigments about 25% each.

The most important types of polycyclic pigments by far are quinacridones, including their various polymorphic forms (e.g. C.I. Pigment Violet 19), perylenes (e.g. C.I. Pigment Reds 179 and 224), and the recently introduced diketopyrrolopyrroles (C.I. Pigment Red 254). Two other types should be noted, the anthraquinoids (e.g. C.I. Pigment Red 177 and C.I. Pigment Blue 60) and dioxazines (e.g. C.I. Pigment Violet 23).

Properties of a pigment crystal, in particular its light reflectance, absorption characteristics, and chemical stability are to a large extent dependent on the structure of the pigment molecule. Fundamental principles and mechanisms of colour in a molecule have been extensively discussed elsewhere.[3,4,8-11] The position and form of the absorption bands in the visible and the ultraviolet part of the absorption spectrum are

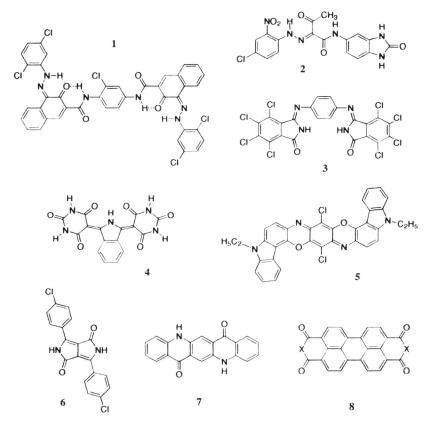

Figure 4.2 Chemical structures of some high performance organic pigments (**1**) C.I. Pigment Red 144; (**2**) C.I. Pigment Orange 36; (**3**) C.I. Pigment Yellow 110; (**4**) C.I. Pigment Yellow 139; (**5**) C.I. Pigment Violet 23; (**6**) C.I. Pigment Red 254; (**7**) C.I. Pigment Violet 19; (**8**) X = NCH_3 is C.I. Pigment Red 179, X = O is C.I. Pigment Red 224.

closely related to this structure. However, probably the most important property commercially is colour value. Colour value translates into strength and purity, both of which are dependent on the absorption spectrum. High maximum extinction coefficient with very sharp cut-off would represent the ideal profile. Any absorption outside the principal absorption band causes a reduction in purity of the colour. The position, the intensity and the shape of the bands very often may undergo a change upon transition from the molecular to the crystalline state. A certain correlation between the coefficient of maximum extinction in solution and in the solid state has also been established for a few pigment classes.[3]

Quantum mechanical calculations today allow us in many cases to predict the electronic spectra of a chromophore molecule with a fair degree of accuracy.[12] Such methods, however, have so far failed to provide us with a reliable prediction of the absorption spectrum of the crystalline aggregate of the same molecule, such as is present in a pigment crystal, and have been of limited value to pigment technology. Though extensively studied, the influence of molecular constitution on various pigment fastness properties still remains little understood. This is mainly due to the complexity of relationship between the earlier mentioned intrinsic molecular and solid state parameters. The chemical constitution of the individual molecule, clearly, will determine the crystal lattice, and hence also the fastness properties of the pigment in the solid state. The intrinsic (photo)chemical reactivity at the molecular level will also play as important a role here, as will electronic and steric factors. From case to case, one may observe certain correlations between molecular weight and substitution pattern of the pigment molecule and its light, weather, solvent and migration fastness.[3] This, however, cannot be generalized.

As a rule, the stability of pigment crystals is distinctly superior to that of the corresponding individual molecules. This can be substantiated particularly through appropriate comparative measurements on pigments in both crystalline (dispersed) and molecular (dissolved) state.

Apart from colouristic and fastness properties, the other application properties of a pigment are only indirectly influenced by the actual constitution of the pigment molecule. They derive primarily from its crystalline solid state and in particular from its particle surface characteristics. Although the structure of the pigment molecule itself does play an important role in determining the molecular packing and crystal surface, this role is often overwhelmed in practice by the presence of adsorbed impurities on the surface of the pigment crystals.

4.2.2 Crystal lattice properties

4.2.2.1 Molecule versus crystal. The importance of the packing arrangement adopted by pigment molecules is second only to their organic chemistry. Overall, the crystal lattice structure is a function of the chemical constitution of the pigment. By virtue of their structural features all pigment molecules exhibit a strong tendency to form highly ordered crystalline aggregates. This tendency is normally encountered also during commercial synthesis of pigments using wet methods. In fact, the extent of crystal growth can sometimes be so great that for optimum pigment performance it may become necessary to either reduce the size of resulting pigment particles by mechanical milling or inhibit growth by the incorporation of appropriate additives during the synthesis step.

An impressive demonstration of such crystal forming tendency is provided by the vapour deposition of copper phthalocyanine on aluminium, followed by solvent exposure of the deposited layer. Thus, Figure

4.3a shows the scanning electron micrograph (SEM) of a 0.15 μm thick layer of partially amorphous α-copper phthalocyanine before solvent exposure. Treatment of this layer with *liquid* xylene for only 1 min triggers the crystal growth process (Figure 4.3b). A more crystalline form of the same α-modification results, with simultaneous formation of cracks and partial peeling of the pigment layer.

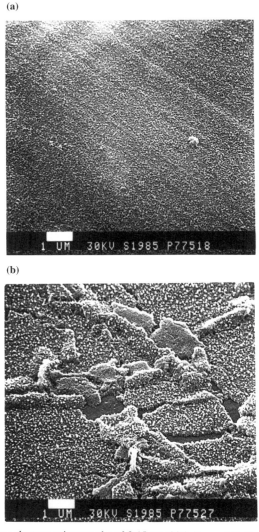

Figure 4.3 Scanning electron micrographs of 0.15 μm thick layer of CuPc (a) before and (b) after treatment with liquid xylene for 1 min. Reproduced from Mizuguchi and Rihs[13] and Mizuguchi and Wooden[20] with permission.

Surprisingly, the crystal forming forces are so strong that mere exposure of such a layer to xylene *vapour* at room temperature for 1 week is sufficient to cause the formation of 25–30 µm long needles (Figure 4.4a). After 4 weeks of xylene vapour treatment the entire layer is completely converted into extremely well-formed crystals which X-ray diffraction (XRD) measurements show to possess the β-modification. This phenomenon becomes even more conspicuous upon analogous treatment of a ten times thicker layer (Figure 4.4b).

The above mentioned transformation of the partially amorphous, less stable α-form of higher energy (Figure 4.3a) to the most stable, highly crystalline β-form of lowest energy of all the polymorphs (Figure 4.4a or b) takes place via the crystalline α-form of intermediate stability (Figure 4.3b), in conformity with Ostwald's law of successive reactions. A similar behaviour was also observed with an evaporated film of 1,4-diketo-3,6-diphenyl-pyrrolo-[3,4-c]-pyrrole (DPP).[13] When exposed to acetone vapour, the originally bright red film turns orange. The vapour treatment induces rearrangement of the DPP molecules within the crystal lattice. Figure 4.5 shows the XRD diagrams of evaporated DPP at room temperature before and after vapour treatment. The diffraction diagram before vapour treatment exhibits halo diffraction patterns, except for the peak at 2Θ = 6.4°, thus indicating a poor crystalline state. On the other hand, vapour treatment produces prominent diffraction peaks, corresponding to (001), (012), (111) and (110) plane spacings, thus indicating improved crystallinity.

Figures 4.6a and 4.6b show scanning electron microscope (SEM) photographs of evaporated DPP before and after vapour treatment. It is apparent that the crystal grains grow considerably upon vapour treatment to give rise to an improved crystalline state. In this connection it is worth mentioning that single crystals of linear unsubstituted quinacridone and 4,11-dichloroquinacridone have also been grown by the vacuum sublimation method.[14]

As mentioned earlier, the main driving forces for such crystal formation are intermolecular hydrogen bonding, π–π interaction between planar layers of molecules and van der Waals contacts. Hydrogen bonds, in particular, can be effective in two ways. Whereas *inter*molecular hydrogen bonding holds adjacent molecules together, *intra*molecular hydrogen bonding may also promote molecular planarity and hence π–π interaction and face-to-face stacking of molecules. The overall effect of such interactions is believed to lead to stabilization of the crystal lattice as well as of the molecule itself, through rearrangements of the molecular orbital energy levels. General consequences of this are improved fastness properties on the one hand, and a shift in wavelength of maximum absorption in the pigment solid state *vis-à-vis* in solution on the other. Any chemical and physico–chemical modification of the pigment crystal

CONTRIBUTION OF CRYSTAL FORM/HABIT TO COLORANT PROPERTIES 115

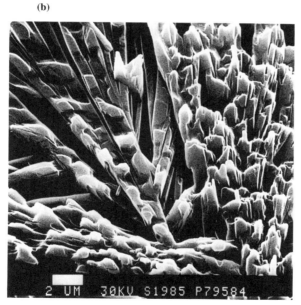

Figure 4.4 (a) Micrograph of a 0.15 μm thick layer of CuPc after 1 week of xylene vapour treatment (Differential Interference Contrast Microscopy by Nomarski); (b) Scanning electron micrograph of 0.15 μm thick layer of CuPc after exposure to xylene vapour for 4 weeks. (Please note the different magnification scales in Figures 4.4a and 4.4b). Reproduced from Mizuguchi and Rihs[13] with permission.

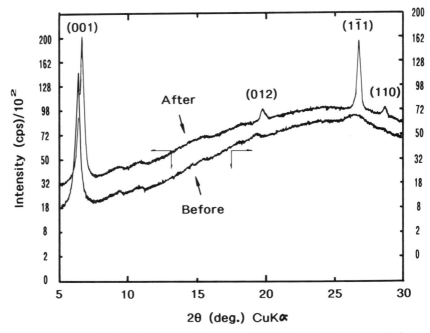

Figure 4.5 X-ray diffraction diagrams of evaporated DPP at room temperature before and after vapour treatment.

resulting in alteration of above intermolecular forces will consequently entail a corresponding change of pigment properties. For example, single crystal X-ray structural studies on DPP[13,15] clearly reveal the existence of chains of hydrogen bonds between the N–H of one molecule and the oxygen of another along the (110) direction (Figure 4.7a). As a consequence, DPP molecules align nearly in the same molecular plane, so that the lattice structure consists of a series of parallel planes with an interplanar spacing of 3.36 Å. There are also van der Waals contacts along the c-axis (Figure 4.7b), and significant interplanar π–π overlap of two molecules along the a-axis (Figure 4.8).

In solution, DPP fluoresces and appears pale greenish–yellow whereas in the solid state it is bright red. There is a significant resemblance in the absorption spectra of both states as shown in Figure 4.9. This would indicate that the basic absorption characteristics are well retained in the solid state and that the solid spectrum can roughly be pictured as the solution spectrum with a bathochromic displacement of about 40 nm (*ca.* 1400 cm^{-1}). An analogous bathochromic shift is also observed upon comparison of the visible spectrum of quinacridone in the solid state with that of a solution in N,N-dimethylformamide. A significant shift of 38 nm

CONTRIBUTION OF CRYSTAL FORM/HABIT TO COLORANT PROPERTIES   117

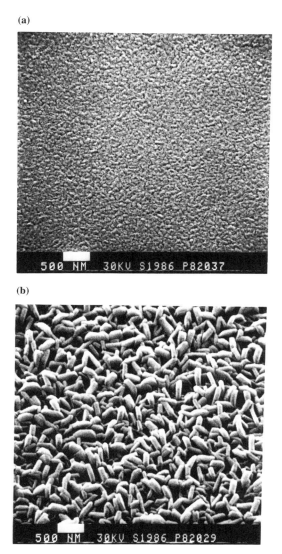

Figure 4.6 SEM pictures of evaporated DPP: (a) before and (b) after vapour treatment.

here can also be attributed to intermolecular hydrogen bonding in the solid state.[16,17]

On the other hand, differently substituted perylene-3,4:9,10-bis(dicarboxamide) pigments show closely related absorption spectra when molecularly dispersed (in solution). In the crystalline solid state, however, the absorption maxima of the same series of compounds spread over

(a)

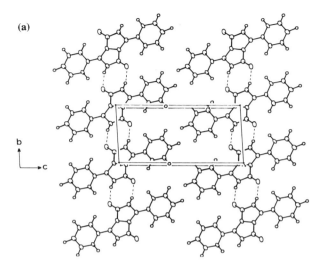

(b)

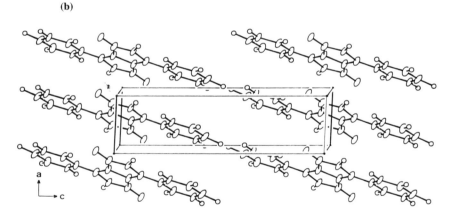

Figure 4.7 Projection of the DPP crystal structure onto the (a) (b,c) plane, and (b) (a,c) plane. The dotted lines represent the intermolecular hydrogen bonds between the N–H and the oxygen. The angle of the N–H···O is 173°, and the distances between the O/H, N/H and O/N are 1.82, 1.00 and 2.817 Å, respectively.

a range of more than 170 nm.[18] This crystallochromic solid-state effect is due to different packings of the differently substituted molecules in their crystal lattices—which feature face-to-face stacking of essentially planar molecules held together primarily by $\pi-\pi$ interactions. Similar steric and electronic effects of substitution on crystal geometry, molecular planarity and absorption properties could be rationalized by comparative three-

CONTRIBUTION OF CRYSTAL FORM/HABIT TO COLORANT PROPERTIES 119

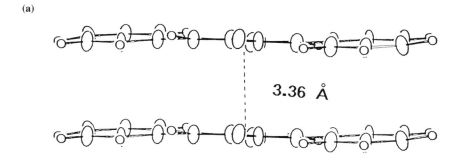

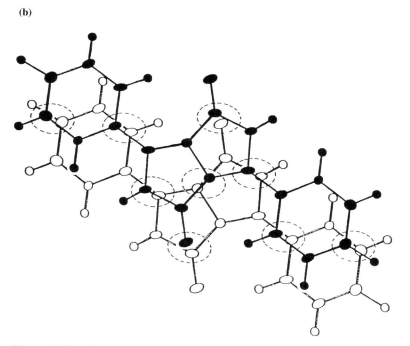

Figure 4.8 Overlap of the two DPP molecules along the *a*-axis: (a) as seen from the side of the molecular plane and (b) from the top. The overlap of interacting atoms is depicted by dotted ellipses.

dimensional XRD studies on red and brown representatives of the Naphthol AS pigment series; these pigments differed only by the presence or absence of one methoxy group in the anilide function of the coupling components.[3]

Polarized reflection spectra of single crystals of DPP (Figure 4.10), measured at various polarization angles on the (001) plane[13] show that

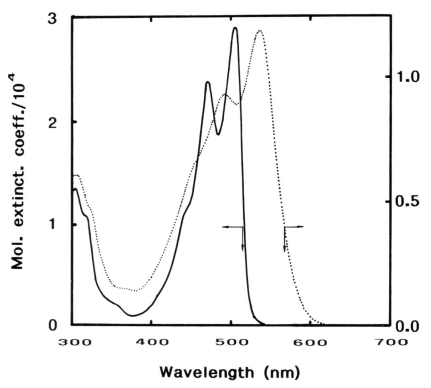

Figure 4.9 Solution spectrum of DPP in DMSO as well as the solid state spectrum of the evaporated DPP (film thickness: about 1200 Å). ——, Solution, · · ·, solid. Reproduced from Mizuguchi and Rihs[13] with permission.

the dominant reflection bands are most pronounced in the (110) direction along the intermolecular hydrogen bonds. Similar optical anisotropy of the pigment crystal may be held responsible for the pleochroic behaviour of β-copper phthalocyanine, showing changing shade as a function of particle shape.[19] Measurements on pigmented films in which the crystals had been oriented proved that the transmission colour is reddish blue in the direction of the long axis of the β-copper phthalocyanine and is greenish blue at right angles to the axis. As the axial ratio increases, the transmission colour along the axis of the rods exerts a greater effect on the colour of the films even if the particles are statistically oriented.

The intermolecular forces in the crystalline solid state may also contribute to heat and chemical stabilities of the pigment. This, for example, is borne out by the fact that cancellation of the intermolecular hydrogen bonding via N-methylation of DPP, yields not only more soluble products, in the order DPP ≪ N-monomethyl-DPP (MM-DPP) < N,N'-dimethyl-DPP (DM-DPP), but also less heat stable products. This

CONTRIBUTION OF CRYSTAL FORM/HABIT TO COLORANT PROPERTIES 121

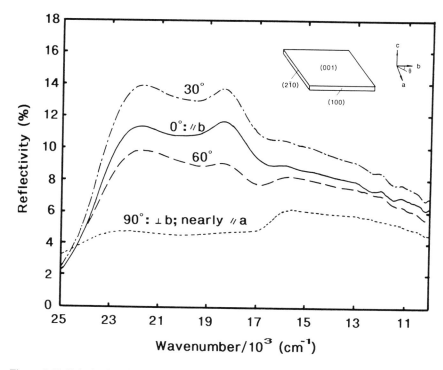

Figure 4.10 Polarized reflection spectra of the DPP single crystal measured on the (001) plane. The polarizer was rotated from $\Theta = 0°(\|b)$ to $90°$ ($\perp b$; nearly $\|a$) in 30° steps.

latter point is exemplified by Figure 4.11 where susceptibility of DPP, MM-DPP, and DM-DPP to weight loss by sublimation increases (indicating that intermolecular forces become weaker) in the same order.[20] Likewise N,N-dimethylquinacridone is alcohol-soluble,[21] because unlike quinacridone it presumably no longer possesses the required structural features for intermolecular hydrogen bonding.

Both insolubility and photochemical stability of organic pigments have been attributed to the presence of intermolecular hydrogen bonding.[22] Thus, 4,11-disubstituted quinacridones are generally more soluble and photochemically less stable than the parent compound.[16]

4.2.2.2 Polymorphism. The interplay of various forces of cohesion and repulsion between molecules of an organic pigment frequently leads to differently structured crystals. Depending on conditions of pigment synthesis and aftertreatment, molecules of a given pigment may sometimes be capable of forming more than one packing arrangement, commonly denoted as crystal modifications or polymorphs. Each of the polymorphs has its own characteristic lattice enrgy. Low energy barriers

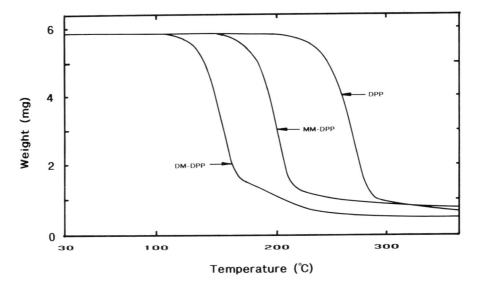

Figure 4.11 TGA measurements for DPP, MM-DPP and DM-DPP. Weight loss is plotted as a function of temperature. Reproduced from Mizuguchi and Rihs[13] with permission.

between such modifications of a pigment can allow easy phase transition during processing by pigment manufacturers, but can cause real problems for customers when transitions occur in uncontrolled circumstances. To be of value, polymorphs should offer useful and distinct property profiles, and sufficient stability to avoid uncontrolled transitions. Sometimes an unstable modification, if commercially attractive, can be stabilized by appropriate chemical and physico–chemical treatment.

A specific crystal modification of a pigment is generally characterized by its *powder* XRD pattern. Because of the extremely low solubility of organic pigments, only in a limited number of cases have crystals been grown large enough for lattice structure determination by three-dimensional XRD analysis of *single crystals*. Polymorphs can range from being almost indistinguishable in all properties except powder XRD, to differing so widely in their colouristic, application, and fastness properties, that they would not at first glance be considered even to be of the same chemical type. Variation in molecular packing modes of a pigment (determined by the intermolecular forces) may obviously be expected to lead to changes in intrinsic stability, light absorption, and surface characteristics of the crystal lattice. Crystal surface characteristics (functionality/polarity/energy) have a major influence on particle morphology (i.e. shape) with all relevant consequences. However, the possible correlation of polymorphic structure with crystal shape has not been widely investigated. Only in recent times have computational techniques been developed for predicting the morphology of crystals based on their surface energetics.[23]

In practice, polymorphic modifications of a given pigment are known to differ from one another with regard to their application, colouristic, and fastness properties. Copper phthalocyanine blue pigments can exist in no less than five modifications:[3,24–26] namely the α-form (red-shade blue), the thermodynamically stable β-form (green-shade blue) and the γ-, δ- and ε-forms (distorted α-forms). Only the first two have commercial significance and have quite distinct colours, showing equivalent light fastness but different solvent and heat resistance, the α-form tending to revert to the β-form.

The α-form *per se* (C.I. Pigment Blue 15) is of limited use in practice. The stabilized α-form (C.I. Pigment Blue 15:1) is commonly preferred; this is partially chlorinated, about 0.5 Cl per molecule being sufficient to prevent reversion to β-form. It is isomorphous with the α-form and little different in shade. Technological grades of the α-form usually consist of approximately brick-shaped crystals; β-form grades (C.I. Pigment Blue 15:3) feature either brick- or rod-shaped crystals.

Three distinct polymorphic forms have been reported for quinacridone[16,27] of which the α- and γ-modifications are red, while the β is violet. Thermodynamically, α is the least stable phase. Thermal stabilities of the β- and γ-modifications are similar. Solvent and therefore migration fastness of these products were measured to be in the order $\gamma > \beta > \alpha$. Similar observation was also made with respect to their light and weather fastness. Many of the substituted quinacridone derivatives are also polymorphic in nature.[16]

The phenylazo-2-naphthol magenta colorant, C.I. Pigment Red 3 has been prepared in the amorphous and three crystalline polymorphic forms,[12] with distinct differences in electronic absorption and emission spectra as well as in their heats of formation. A yellowish–red α-modification and a bluish–red β-modification have also been reported for perylene diimide.[25]

C.I. Pigment Yellow 17, a dichlorobenzidine disazo compound, has been synthesized to yield two polymorphs of different colour and crystal morphology. Ref. 3 lists a number of distinct differences in colouristic, fastness, and application properties of polymorphic forms of pigments belonging, for example, to the azo chromophore class.

4.2.2.3 Isomorphism. Isomorphism is a somewhat less common but equally useful property of pigments. Pigments of different chemical constitution, but featuring similar spatial requirements are called isomorphs. In other words, isomorphic pigments show similar molecular and crystal lattice geometry and as such are characterized by very similar XRD patterns. Isomorphs generally belong to the same chromophore class, with only minor differences in their substitution pattern. In an isolated number of cases, they may also belong to different classes of chemical compounds.

Isomorphic pigments, particularly when belonging to the same chromophore class, are frequently more uniform in their colouristic, fastness, and other application properties than is true for polymorphous pigments. Some isomorphs of monoazoacetanilide pigments have been reported,[7-9,28,29] the monobromo and dibromo analogues of C.I. Pigment Yellow 3 being examples. Other examples are C.I. Pigment Blues 15 and 15:1 mentioned earlier.

More recently a similar observation was made in the disazo diarylide series of pigments for C.I. Pigment Yellow 14 and C.I. Pigment Yellow 63.[30] The corresponding XRD patterns were essentially similar, with only slight differences in intensities. Furthermore, colour measurement data of C.I. Pigment Yellows 14 and 63 reveal that their colour properties are very similar. Our own investigations have further shown that C.I. Pigment Red 144 and C.I. Pigment Red 214 behave in exactly the same way, featuring very similar XRD patterns and colour properties.

A special case of isomorphism arises when two or more structurally different pigments with distinctly different XRD patterns are mixed by methods other than simple mechanical mixing, to yield a product mixture which shows the same XRD pattern as that of one of the components alone. Such cases are termed solid solutions: either a major or a minor component assumes the role of solvent (host) and solute (guest), respectively. The different molecules (or ions) occupy the lattice sites at random, and this effect occurs over a range of compositions.

In certain cases a specific combination of pigments gives a completely different performance and XRD pattern compared to that of the component molecules. It is customary to denote such compositions as *mixed crystals*.[31] Crystallographically, a mixed crystal is thus different from any of its component molecules. A synergistic effect is common to all such combinations of pigments. While the resultant XRD pattern and colouristic and fastness properties of a mechanical mixture of component molecules are predictable as being in some way additives of the individual compounds, a solid solution or mixed crystal of the same component molecules shows both a non-additive diffraction pattern and unpredictable colouristic and fastness properties.[21]

The phenomena of solid solutions and mixed crystals can be advantageously exploited in practice to extend an existing range of pigments, by widening their colouristic scope and eventually improving fastness and other application properties. Examples of solid solutions involving acetoacetanilide, as well as diacetoacetanilide azo pigments have been reported recently.[30-32] Thus, 'solute' molecules of C.I. Pigment Yellows 12, 13 and 63 can be introduced in substantial proportions into the 'solvent' lattice of C.I. Pigment Yellow 14 without changing the lattice structure. The resulting solid solutions are claimed to display different colouristic properties, dispersibility, aggregation, and particle sizes, as compared with the individual molecules or mechanical mixtures of the same.

A variety of binary, ternary, and quaternary solid solutions have been prepared containing quinacridone derivatives[16,33] and constitute red, orange, gold, and maroon colorants of different pigmentary performance levels. Such solid solution formation can be followed easily by powder XRD.

While component molecules of a solid solution or mixed crystal generally possess closely related chromophore structures, solid solutions with components belonging to completely different classes of compounds have also been reported.[34]

4.2.2.4 Crystallinity. Immediately after formation by grinding or precipitation, organic pigment particles tend to show poor crystallinity, with many internal and external defects. They are prone to difficultly reversible aggregation unless finishing treatments are applied to remove or minimize these defects and improve crystallinity. This improvement leads in general to improved application properties. The lower lattice energy of the more crystalline pigments enhances their fastness properties, and their lower susceptibility to aggregation makes them easier to disperse. Crystallinity can also affect optical properties. For example, the absorption maximum and fluorescence of the untreated amorphous phase of C.I. Pigment Red 31 is shifted to longer wavelengths and the maximum extinction coefficient increased upon thermal conversion to the crystalline phase.[12]

Crystallinity is usually determined by XRD techniques, which can be applied to samples taken at various stages in the preparation of pigments. Poor crystallinity gives a diffuse diffraction pattern; well developed crystals give a sharp pattern. High resolution electron microscopy can also be used for more specialized investigations.[35] Poorly crystalline species may be either very small crystals of size near the resolution limit of XRD (0.01–0.02 μm) or larger particles with internal domains of order (crystallites) with size below or around the resolution limit, and bounded by defects. These larger particles are difficult to define, but could, for example, be nuclei that have aggregated and have begun to sinter or grow together. They need to be annealed.

4.2.3 Crystal habit

Since they are used as particulate solids, organic pigments need to be characterized in terms of their morphology (particle size, shape, and state of aggregation) and crystal lattice properties (modification, crystallinity). The important role of crystal lattice characteristics of pigments in determining their technical performance has already been discussed, but particle morphology also has a very significant influence on various optical, fastness, and other application properties of the pigment.

In pigment technology, a standard terminology has been put forward to describe pigment morphology.[36] Accordingly, pigment particles, depending on their environment, may actually occur as single or primary particles, aggregates, agglomerates, flocculates or a combination of these. The smallest units detected by XRD are termed *primary particles*, which may appear as cubes, platelets, needles, bars, or rods, including various irregularly shaped species which are single crystals with typical lattice disorders or combinations of several lattice structures.

Primary particles, growing together at their surfaces, lead to the formation of *aggregates*, the latter consequently featuring a lower total surface area than the sum of the surfaces of the individual primary particles, and *agglomerates*, groups of single crystals and/or aggregates, joined at their corners and edges only. Unlike aggregates, agglomerates can be separated by a dispersion process, and their total surface area roughly corresponds to the sum of the surfaces of the individual particles. Moreover, the surfaces of the individual particles of an agglomerate are more easily available to adsorption of gases and certain surface additives.

Flocculates are assemblies of wetted particles (primary crystals and/or aggregates or smaller agglomerates) which tend to form after the pigment has been dispersed in a liquid medium. However, being mechanically more labile than agglomerates, flocculates can usually be broken up by weak shearing, such as stirring.

Unfortunately, the terms 'aggregate' and 'agglomerate' are often transposed in surface and colloid science. Therefore, in this chapter the term 'aggregate' is used, in the general sense recommended by IUPAC,[37] to mean a group of particles held together in any way, and the strength of cohesion is indicated where appropriate. In this way the term 'agglomerate' is avoided.

4.2.3.1 Crystal size and state of aggregation. Whereas the potential ability of a pigment to impart colour is dependent ultimately on its chemical constitution, the intensity of colour achieved in practice is highly dependent upon the level of dispersion. Colour strength increases as mean particle size decreases. This is a fundamental principle of organic pigment technology and has been firmly established by theory,[38-41] experiment,[40,42-44] and usage.

The highest dependence occurs at the smallest particle sizes, around 0.1 μm and less. Where particles are elongated the minor dimension has the most influence.[19,45] This also holds in practice. In effect, molecules, many molecular layers below the surfaces of thick particles, are not reached by light and the absorption efficiency is reduced. In thin rods or plates, the molecules are more accessible to light, irrespective of length or breadth. Light scattering theory suggests that, with pigment dispersions, colour strength should pass through a maximum on decreasing particle size at very small sizes. This, however, has not been unequivocally demonstrated in practice. The reliability of theory is limited by multiple scattering at practical pigment concentrations; there are problems with quantitative assessment of particle size, especially with such small sizes. Moreover, a lower limit to achievable particle size is likely to be imposed by the high interfacial energy of highly dispersed systems.

The shade of dispersions can be affected to a small, but significant extent by a decrease in particle size. For example, C.I. Pigment Yellow 13 pigments become greener and C.I. Pigment Red 57:1 pigments become yellower (although the use of bluish shading agents complicates matters in practice). Copper phthalocyanine blues become greener,[3,19] although β-form pigments tend to show an opposite effect; their dichroic character (which makes crystal shape important) may complicate the situation, as may residual α-form crystals having a redder shade. Generally a wide particle size distribution results in dullness or dirtiness of shade. This is usually associated in practice with incomplete dispersion.

The opacity/transparency of a dispersion is also affected by particle size distribution of the pigment. The opacity of a dispersion is determined by the ability of the particles to scatter light, and this requires a significant difference in refractive index between the dispersion medium and the particles. Scattering is optimal at particle sizes around half the wavelength of light. Thus, in a given medium, there is a maximum in opacity at mean particle size in the range from 0.2 to 0.5 μm;[46] opacity at a given mean size is greater the narrower the size distribution. Dispersions having mean particle sizes above and below this region are more transparent.

Fortuitously, for multicolour process printing, high colour strength and transparency go hand in hand at very small particle sizes. Here, successive thin layers (typically 1–3 μm thick) of differently coloured inks are superimposed, and light has to penetrate the outermost layers to reach those underneath. A good example is given by a transparent yellow ink. When printed on a black substrate it is invisible; when printed on a white substrate it is intensely yellow. The light penetrates through the ink to be totally absorbed by the black substrate, so that no yellow colour is visible unless the presence of some large enough particles gives some backscattering of light from the ink layer. On the other hand a white substrate reflects the light back through the ink layer to be perceived (minus the absorbed wavelengths) as yellow.

A high degree of transparency is also an essential requirement of paints for special effects (metallic, pearlescent, etc.). It enables the organic pigments to provide the colour without detriment to the unique light-reflecting and light-scattering effects of the underlying aluminium and mica platelets. In plastics applications, highly transparent pigments often find use as shading agents. There, in combination with highly opaque, dull pigments, they yield more saturated shades. The use of transparent pigments in the colouration of synthetic fibres, result in brilliance and gloss of the intrinsically opaque fibres.

Another consequence of very small particle size is increased strength of structure formation and thixotropy in pigment dispersions, and a consequent dramatic reduction in their ease of handling. This relationship means that desirable improvement of colour strength by reduction in particle size tends to lead to undesirable rheological effects, which in turn can lead to other problems, such as poor gloss of coated films.

A major impact of crystal size is its effect on the available surface area. In addition to crystal size and shape, lattice (modification and crystallinity) and crystal surface characteristics (polarity, pH, ionic charge, lipophilicity or hydrophilicity, etc.) determine the energetics of such a surface. The crystal surface energetics can influence one or more of the following physical parameters of the pigment:

crystal growth, crystallinity, and crystal phase transformation
aggregation, aggregate structures in powders,
interaction with liquid media (wetting, etc.),
solubility,
flocculation and rheology of dispersions,
heat, light and chemical stability.

All of the above parameters obviously have, in turn, a bearing on the various colouristic, fastness, and other application properties of the pigment. The surface area per unit weight of pigment is inversely proportional to particle size; therefore the very small particle sizes of organic pigments surface area is very large. Furthermore, the surface energy of such small particles is enhanced, due to the higher proportion of edges and corners and they have a tendency to interact in any way which reduces the overall energy level. They may aggregate or adsorb other chemical species (solvent, surfactant/resin/diverse additives) from solution. Specific surface areas of $80\,m^2/g$, as determined by nitrogen gas adsorption, are quite common for high strength ink pigments. Primary particle surface areas as high as $200\,m^2/g$ can be calculated from XRD data or electron micrograph analysis.

Ideally organic pigments should be completely dispersed in application media, so that the particles are individual crystals. A few cases appear to come close to this ideal, for example C.I. Pigment Yellow 13-type pigments[47,48] and some calcium 4B toners (C.I. Pigment Red 57:1)[49] extensively three-roll-milled in highly viscous lithographic inks. In these cases, there is quite good correlation between colour strength (corrected for the colourless additive content of the pigments) and mean crystal thickness or breadth determined from XRD line broadening. However, such correlation with crystal size is atypical. Pigment particles usually tend to be a mixture of individual crystals, small aggregates of a few crystals, and a few larger aggregates, sometimes detectable by optical microscopy; this has been established by electron microscopy of ultrathin sections of dried films of liquid inks, gravure inks, and paints. Thus, control of particle size in dispersion requires control not only of crystal size, but also the state of aggregation of crystals.

Decreasing pigment crystal size increases susceptibility to rapid and strongly coherent aggregation, for reasons given above. Moreover, on isolation from the preparative medium by filtration and drying, the tiny pigment crystals (which are in the size range of particles in smokes) need to be aggregated in a controlled way such that grains large enough to give a powder or granules are formed and also in such a way that the crystals can readily be redispersed in the application medium. Processing must be such as to avoid difficultly reversible aggregation caused by compaction on filter presses, by capillary attraction during drying, and by compaction during subsequent blending and handling processes. If not, the potential benefits of small crystal size will not be achieved. The commonly used term 'dispersibility' incorporates both ease and completeness of dispersion.

The aggregate structure of a pigment has a major influence on dispersibility in an application medium. Wettability by the medium is of course essential for good dispersibility, and is comparatively easy to achieve, but it is not by itself sufficient to give satisfactory performance.[50] In reality, the strength of cohesion of the aggregate structure tends to be the determining factor, and has to be minimized. It depends on two factors, namely the area of contact between crystals and the cohesive strength per unit area. The mode of packing of the crystals in the aggregate structure determines the area of the contact, greater for face-to-face packing, less for edge and point contact, other factors being equal. In this context, deformability (softness) of the crystals can be important, because the area of contact may be increased by plastic deformation when crystals are forced together. The other factor, the cohesive strength per unit area of contact increases with an increase in surface polarity of the crystals.

An assessment of the openness of the aggregate structure of a pigment is given by the ratio S_{BET}/S, where S is the specific surface area (area per unit weight) of the crystals (determined from crystal dimensions) and S_{BET} is the specific surface area determined by the BET method applied to nitrogen gas adsorption data.[49,51] Nitrogen molecules do not penetrate between crystals in face-to-face contact under the conditions used

($-196\,°C$, pressure less than 0.4 times the saturation pressure of the gas).[49,52] Therefore, $S_{BET}/S \ll 1$ indicates predominant face-to-face packing; $S_{BET}/S \to 1$ indicates an open structure with little face-to-face packing.

Just how strong the cohesive forces of aggregation can be is well demonstrated by the case of C.I. Pigment Yellow 24, which has been precipitated from solution in sulphuric acid by high turbulency drowning in water, filtered, washed and dried at $100\,°C$. An extremely low S_{BET} value of $2.1\,m^2/g$ indicated dense, compact aggregation, affirmed by TEM of tylose replica (Fig. 4.12a). Not surprisingly, the pigment could not be dispersed by normal application techniques. Instead, it had to be soaked for three days in diacetone alcohol before it could be dispersed by sonication, which set free tiny rod-shaped crystals (Figure 4.12b) having an estimated value of S of about $80\,m^2/g$. Such severely aggregated and poorly dispersible pigments are unacceptable in practice and measures must be taken to improve their dispersibility. The following two examples illustrate how this can be achieved.

The first example illustrates minimization of the area of contact between crystals.[50,53] Mixtures of the α- and β-forms of copper phthalocyanine, prepared by grinding 'crude' material (as synthesized) in the absence of inorganic salt, consist of compact aggregates ($S_{BET}/S < 0.05$) of tiny brick-shaped crystals that are difficult to disperse in lithographic inks. Heating the mixture, however, in an organic liquid produces rod-shaped β-form crystals (by a complex Ostwald ripening process[26]). This material has an aggregate structure with little face-to-face contact of crystals ($S_{BET} \to S$) and markedly improved dispersibility in lithographic inks. The improvement in dispersibility is predominantly due to opening of the aggregate structure. Rod-shaped crystals have lower surface polarity than brick-shaped crystals,[54] as in the starting material, so lowering of surface polarity may be a contributory factor.

The second example illustrates a case where the effect of lowering surface polarity predominates.[49,50] The rectangular plate-shaped crystals of C.I. Pigment Red 57:1 pigments form an open structure naturally ($S_{BET}/S = 0.7-0.8$). The crystal surfaces, however, are polar and despite its openness, the aggregate structure is strongly coherent. Consequently, untreated pigments cannot be dispersed satisfactorily. Both the surface polarity and cohesiveness can be reduced by treatment with abietyl resin to give pigments that disperse well in lithographic inks. The resin (usually predominantly as the calcium salt) adsorbs on the crystal surfaces. This is shown by its effectiveness in inhibiting crystal growth during pigment preparation, thus giving much thinner crystals than otherwise.[55] Consequently despite the fact that pigments contain up to 20% resin per gram of powder, the enhancement of colour strength due to the smaller crystal size offsets the diluent effect of the colourless resin.

CONTRIBUTION OF CRYSTAL FORM/HABIT TO COLORANT PROPERTIES   131

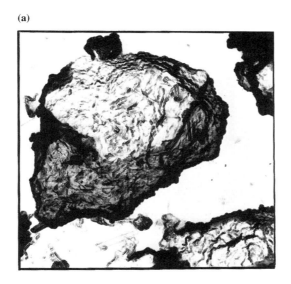

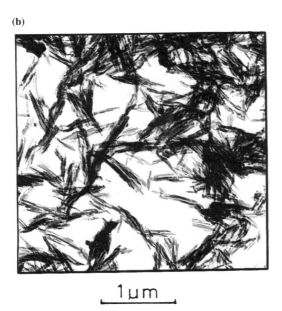

Figure 4.12 TEM pictures of C.I. Pigment Yellow 24; (a) tylose replica of highly aggregated crystals and (b) rod-like crystals subsequent to soaking and sonication of the same dispersion in diacetone alcohol. Reproduced from Mizuguchi and Rihs.[13] with permission.

4.2.3.2 Crystal shape. The tiny crystals of organic pigments are in many cases amenable to examination by electron microscopy and have been seen to vary from brick to rod to plate. Figure 4.13 shows some examples. Crystal shape depends primarily on chemical constitution and lattice structure, but in some important cases it can be altered by modification of manufacturing conditions or by using additives that can inhibit growth by adsorption to crystal faces. The minor dimensions (thickness) of crystals range from around 0.015 to 0.025 μm in pigments for transparent printing inks to around 0.2 to 0.4 μm in pigments for opaque paints.

Crystal dimensions can be determined from transmission electron micrographs in suitable cases. For example, dimensions of brick- to rod-shaped crystals of β-form copper phthalocyanine[51] and various metal-free monoazo pigments having a large crystal size (e.g. C.I. Pigment Yellows 1 and 74, C.I. Pigment Reds 1, 3 and 112[56] and many others[3]) have been determined, assuming breadth to be equal to thickness. The areas of the main faces of plate-shaped crystals of C.I. Pigment Red 57:1 pigments have also been determined.[49] Metal-free disazo pigments with very small crystal size (e.g. C.I. Pigment Yellow 13, breadth < 0.025 μm) are prone to serious damage in the electron beam, and so other methods have had to be used. Powder XRD has featured in this context, average values of specific crystal dimensions being obtained from line breadth in the diffraction line profile. Its value depends on knowing how these dimensions relate to crystal length, breadth, etc. When crystal dimensions are very small, crystallites and crystals are one and the same. The average breadth of C.I. Pigment Yellow 13-type crystals of around 0.02 μm (25 molecular breadths) have been determined in this way.[57,58] So too have the mean thickness of plates of C.I. Pigment Red 57:1 pigments (0.02 μm corresponding to only 11 molecular layers).[49]

As mentioned earlier, the particles of organic pigments in practical dispersions are rarely predominantly individual crystals. Most commonly they are mainly aggregates of a few crystals, sufficiently strongly coherent to have survived the dispersion process. The constituent crystals are packed face-to-face, usually such that the shape of the aggregate ressembles that of the constituent crystals. Thus, rod-shaped crystals tend to give rod-shaped aggregates, etc.

The colouristic properties of organic pigment dispersions are strongly influenced by particle shape. Because colour strength is more dependent on the minor dimensions of the particles, it follows that rod- or plate-shaped particles tend to have higher strength than cube- or brick-shaped particles of the same volume. The hue of dichroic crystals is dependent on crystal shape; for example, rod-shaped crystals of C.I. Pigment Blue 15:3 are redder than cube-shaped crystals of the same volume.[19]

CONTRIBUTION OF CRYSTAL FORM/HABIT TO COLORANT PROPERTIES 133

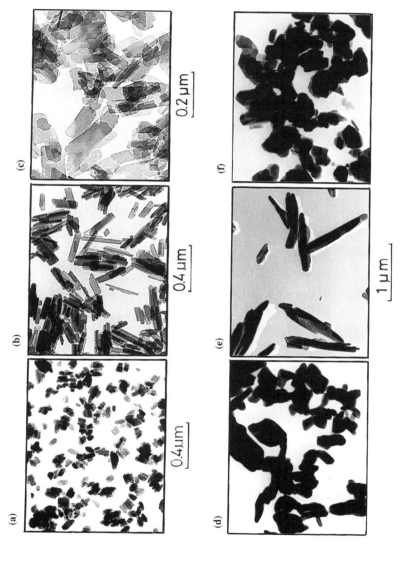

Figure 4.13 Transmission electron micrographs of organic pigments showing brick-, rod-, and plate-shaped crystals. (a) and (b) C.I. Pigment Blue 15:3; (c) C.I. Pigment Red 57:1; (d) C.I. Pigment Orange 36; (e) C.I. Pigment Red 144; (f) C.I. Pigment Red 254.

As a further consequence of the optical anisotropy of organic pigment particles[5,59] their shape may lead to macroscopic effects of preferential orientation in one direction. Thus, needle- or rod-shaped particles may lead to undesirable colouristic performance in all those cases where the application method causes orientation of the crystals. This occurs primarily when pigmented systems are prepared or applied by calendering, moulding or brushing methods, giving rise to directed shear forces. The acicular pigment particles try to minimize the area opposed to these forces by orientating their longer axis in the direction of the applied forces. Thus the intensity of light scattered by a low-density polyethylene film pigmented with C.I. Pigment Orange 43 was experimentally shown to vary strongly as a function of both viewing angle and the wavelength of incident light.[60] The latter dependence, in particular, is responsible for the unmistakable colour change perceived upon rotation of the film. Inspection of the pigmented film by light microscope clearly revealed the presence of oriented rod-like particles in the polymer matrix.

The rheological properties of organic pigment dispersions are significantly influenced by crystal shape. The more elongated the pigment particles, the greater the amount of liquid phase immobilized by flocculation in dispersion under low shear conditions or at rest. This effect is partly geometric since, in randomly packed assemblies of rods that stick on first collision, the void space fraction increases with axial ratio of the rods.[61] Rod-shaped β-form copper phthalocyanine crystals dispersed in phenolic/toluene gravure inks are an example, being especially susceptible to gel formation at pigment concentrations as low as 5% w/w. Polar interaction occurs between ends (electron-rich) and the sides (electron-deficient) of the neighbouring crystals.[62] Likewise, it is well known that β-form copper phthalocyanine pigments having long rod-shaped crystals tend to give very strongly structured lithographic inks.

Another example of the influence of crystal shape on rheology of dispersions is provided by metal-free monoazo pigments with large crystals in decorative paint millbases (long-oil alkyd resin in white spirit).[56,63] The pigments are dispersed substantially as individual crystals. The dependency of strength of flocculate structure on pigment concentration becomes markedly greater above a critical concentration, which varies from pigment to pigment. This critical concentration corresponds to the situation where the spheres of rotation of the crystals begin to overlap. Below the critical concentration only isolated units of structure can form; above the critical concentration a coherent gel structure can form. The more elongated the crystals are, then the larger are their spheres of rotation and the lower is the critical concentration. C.I. Pigment Yellows 1 and 3 and C.I. Pigment Red 3 have much higher critical concentrations than C.I. Pigment Red 112 pigments with relatively thin plate-shaped crystals. This explains why C.I. Pigment Red 112

CONTRIBUTION OF CRYSTAL FORM/HABIT TO COLORANT PROPERTIES 135

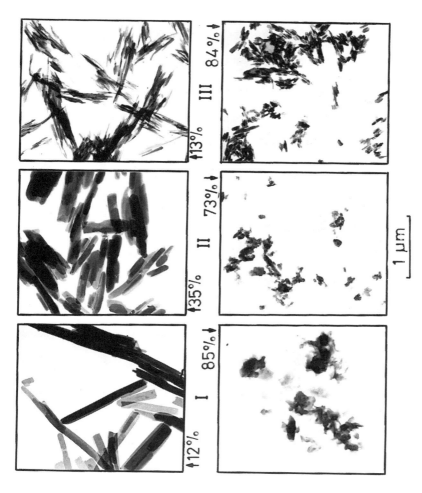

Figure 4.14 TEM pictures of pigments 1–3 before (top row) and after (bottom row) mechanical comminution, and gloss readings of the corresponding alkyd melamine paint films.

pigments cannot be milled at such high pigment loadings as the other pigments and gives lower throughput in paint manufacture. It is clear from these examples that attention to crystal shape is important to obtaining desirable rheological behaviour of dispersion millbases.

Closely related to the general flow properties of pigmented coating systems is the levelling out behaviour of sprayed wet paint films to give smooth surfaces with a high degree of surface gloss. It has been shown that the crystal shape of pigment particles exerts a strong influence on the surface gloss of pigmented paint films; pigments with pronounced acicular crystal shape often prevent the formation of a sufficiently smooth film surface, and this is almost independent of absolute particle size in the range of about 0.02–1 μm.[64] It is not the size of needles or rods that influences the surface structure of sprayed paint films, but the ratio of length to width of those particles. This can be illustrated by three different pigments of pronounced acicular shape. Figure 4.14 shows in the top row the TEM-pictures of pigments **1**, **2** and **3** as obtained after synthesis (two red bisazo pigments and one yellow metal complex pigment). Their average length to width ratios correspond quite well with the gloss readings at the surface of alkyd melamine paint films (pigment **1**: 10:1 ↔ 12%, pigment **2**: 5:1 ↔ 35%, and pigment **3**: 10:1 ↔ 13%), the gloss readings showing no correlation with the average particle size of each pigment. In the bottom row, the same pigments are shown after comminution of the needles, preferentially perpendicularly to their length, by an effective wet milling technique. The almost isometrical particles of similar size now exhibit considerably higher gloss, the gloss readings being of the same magnitude for all three pigments (**1**: 85%, **2**: 73%, and **3**: 84%).

4.3 Control of crystal form/habit

With any given chemical type of pigment, proper control of the various solid state properties is essential to obtain the desired pigmentary performance profile. There are basically two approaches: in-synthesis and post-synthesis control. The former entails controlling the pigment crystals as they precipitate from the chemical synthesis reaction (control of rate of reaction, crystal nucleation/growth and impurity formation). Post-synthesis control entails treatment of either the slurry after precipitation has been completed (e.g. ripening and/or additive treatment) or the dry powder of 'crude' material isolated from the slurry (e.g. grinding, kneading in viscous liquid media, dissolution and reprecipitation, and/or treatment with organic liquid).

Some pigment syntheses are conducted over periods of many hours, for example, those of copper phthalocyanine and quinacridone. In these cases, it is difficult to control crystal growth effectively, due to the elevated temperatures, powerful crystallizing liquids used and timescales involved. It is necessary to isolate the crude non-pigmentary material produced (which enables removal of undesirable impurities) and apply post-synthesis treatment(s) to reduce crystal size and optimize performance. Other pigment syntheses are rapid, with the pigment forming and precipitating virtually instantly on turbulent admixture of solutions or suspensions of reactants, for example, those of classical azo pigments. Such pigments are generally more amenable to in-synthesis control to produce pigmentary-sized material directly. Often they may also require some post-synthesis treatment to modify solid-state character and optimize performance.

4.3.1 In-synthesis control

Crystal size can be controlled by control of reaction rate (dependent, for example, on rate of addition and concentration of reactants, particle size of disperse reactants, pH, temperature), efficiency of mixing (dependent, for example, on stirrer speed and design, vessel size) and the presence of additives that inhibit growth. In this regard, the presence of abietyl resin during the precipitation of C.I. Pigment Red 57:1 pigments controls crystal growth.[55]

With some C.I. Pigment Yellow 13-type pigments, minor proportions of compounds of slightly different chemical composition from the pigment are incorporated into the crystal lattice by co-synthesis. This procedure (known as mixed coupling) can result in pigments having smaller crystals that are resistant to growth. C.I. Pigment Yellow 83 is especially effective as the minor component in this context. The composite products have the crystal lattice structure of C.I. Pigment Yellow 13, even though they are sufficiently different chemically to be given their own C.I. numbers. C.I. Pigment Yellows 127, 174, 176 and 188 are the designations of C.I. Pigment Yellow 13 + 63, 14, 83 and 12, respectively. Recent studies on co-synthesized mixtures of C.I. Pigment Yellows 12 + 14, 13 + 14 and 14 + 63 suggest solid solution as a factor controlling crystal size.[7] Likewise, solid solution formation has been suggested as the mechanism of crystal size control in monoazo pigment mixtures of C.I. Pigment Yellow 1 co-synthesized with C.I. Pigment Yellow 5 or 6.[65] Bluer shade C.I. Pigment Red 57:1 pigments are produced by co-precipitation with minor proportion of Tobias-acid/BON acid coupling.[55] Mixed synthesis is also used to produce diketopyrrolopyrrole pigments having a small particle size.[66]

4.3.2 Post-synthesis control

Further control of crystal size can be achieved by post-synthesis treatment. In some cases, this simply entails *heating in aqueous media*. For example, the aqueous slurry formed in the synthesis stage of C.I. Pigment Yellow 13 contains extremely small, and compactly aggregated crystals having a dull yellow colour (so-called amorphous form). The desired bright yellow colour is produced by heating to over 90 °C. Crystal growth is induced in two stages: annealing of crystals within aggregates followed by sintering of the annealed crystals.[58] The annealing is accompanied by a small change in the XRD pattern (d spacing of 3.42 Å reducing to 3.28 Å),[57] which provided a means of monitoring the effectiveness of influencing factors.

The presence of abietyl resin (derivatives of wood rosin or tall oil) accelerates annealing, possibly by scavenging impurities from crystal surfaces inside the aggregates.[58] The presence of mixed-coupled components retards annealing, as do minor proportions ($< 5\%$ w/w) of sulphonated disazo derivates (which adsorb to the pigment crystal surfaces). However, the influence of resin on annealing tends to dominate and overwhelm the other factors. When resin levels used are higher than 10–15% of the final pigment powder weight, the annealing still proceeds as rapidly as that observed at lower concentrations, but the subsequent sintering stage is retarded. Resin on the annealed crystal surfaces acts as an inhibitor, although not as effectively at this state as sulphonated disazo compounds or certain mixed-coupled species. Suitable resination thus yields well-annealed but still small crystals of C.I. Pigment Yellow 13 type pigments that give high transparency in lithographic inks. Sulphonated disazo and/or certain mixed-coupled species give resistance to crystal sintering, and hence retention of colour strength and transparency, during dispersion in the inks by bead milling in which temperatures $> 80\,°C$ are generated.

In other cases where the crude pigment particles resulting from synthesis are too amorphous and/or have undesirable shape (platelets or needles), the pigment can be converted into the desired pigmentary form either by a simple organic *solvent treatment*, or by first grinding and then treating with solvent (Ostwald ripening). Figure 4.15 illustrates this for a DPP Red pigment. If, however, a pigment is synthesized as large non-pigmentary crystals, they have to be subsequently reduced in size. This can be achieved either by dissolution/precipitation in suitable media, notably concentrated sulphuric acid, or by grinding/kneading. Such post-synthetic measures may also be successfully employed to control crystal lattice properties in polymorphic pigments, the best known examples of which are copper phthalocyanine and quinacridone pigments. The crystals of copper phthalocyanine as synthesized (so-called crude blue) are rod-shaped, β-form, and, depending on the method of synthesis have a length ranging from about 0.5 μm to 100 μm. A variety of processes can be used to produce pigmentary size crystals of α- or β-form.

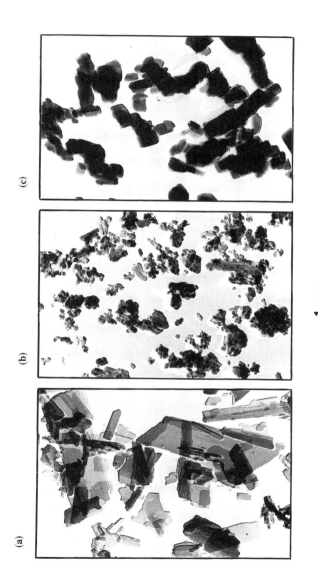

Figure 4.15 TEM pictures of a DPP red pigment (a) as synthesis crude, (b) after pebble milling for 24 h and (c) after treatment of the milled product in N-methyl pyrrolidone for 1 h at 80 °C.

Pigmentary α-form crystals can be obtained as a fine precipitate by slowly drowning out a solution of crude blue in concentrated sulphuric acid into an excess of water (acid pasting). Mixtures of α-, β-forms having small brick-shaped crystals (breadth around 0.02 μm) can be obtained by milling dry crude blue powder with steel balls with or without inorganic salt as a grinding aid. The relative α- and β-form contents depend on the conditions of milling, with α-form contents as high as 98–99% being obtainable. Brick-shaped crystals of 100% β-form can be obtained directly by similar milling of crude blue powder, but with a trace of a powerful crystallizing liquid present (still a dry process). Brick-shaped crystals of 100% β-form can also be obtained by grinding or kneading crude blue with an excess of mildly crystallizing liquid. Processes and mechanisms have been reviewed elsewhere.[26,35]

As mentioned earlier in the context of control of aggregation, heating of milled α-, β-form mixtures (> 50% α-form) in a suitable organic liquid yields pigmentary rod-shaped β-form crystals. Optimization of the process yields > 95% β-form crystals having a length to width ratio in the range from 4 to 10, depending on conditions; crystal breadth, however, is not much greater than that in the starting material. Prolonging treatment after completion of the α- to β-form conversion induces side-by-side fusion of the β-form rods, thereby increasing crystal breadth markedly.[51] This is detrimental to the colour strength of the pigment in subsequent application, so the process has to be carefully optimized. Stabilized α-form blues (C.I. Pigment Blue 15:1) are normally prepared either by dry milling or acid pasting; α-form, stabilized α-form, or β-form pigments prepared by milling processes tend to have brick-shaped crystals. Broadly similar types of processing are also used to produce the various polymorphic forms of quinacridone.

In all above processes where poorly crystalline pigments are subjected to Ostwald ripening, the nature of the solvent used can have a profound influence on the resulting crystal habit. A qualitative method of classifying such effects of solvent properties has been investigated and is called 'Solvent Mapping'.[67] The method considers pigment solubility, a key factor in Ostwald ripening, in terms of solubility parameters of solvents. Although it is important to take into account all (dispersion, polar, and hydrogen bonding) interactions between solute and solvent, the dispersion forces alone tend to give a constant contribution, due to the narrow range of the dispersion parameters of the solvents. For practical and comparative purposes therefore, it is suitable to restrict attention to the polar (δ_P^L) and hydrogen bonding (δ_H^L) contributions. The results of experiments using various solvents can then be represented in a two-dimensional diagram (Figure 4.16), in which every solvent is represented by a point (δ_P^L, δ_H^L).

By making a suitable selection of solvents it is, therefore, possible to

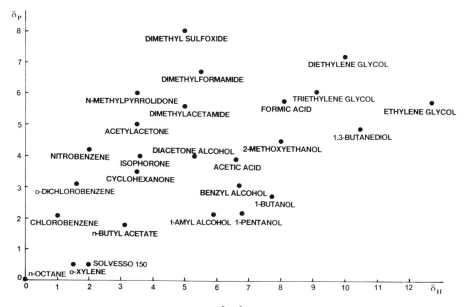

Figure 4.16 Two-dimensional (δ_P^L, δ_H^L)-diagram of common solvents.

span the whole (δ_P^L, δ_H^L)-space and to predict the solvent influence of other solvents, provided their δ_P^L and δ_H^L values are known. It is also possible to detect special effects of certain types of solvents on pigment crystallization. This includes aromaticity, order–disorder, or protonation effects. Use is here made of TEM to show the influence of solvent on C.I. Pigment Orange 65, representative sections of micrographs being arranged in a (δ_P^L, δ_H^L)-diagram (Figure 4.17). As is evident, such mappings allow a simple visualization of the influence of solvents in the Ostwald ripening process on crystal habit without making any of the assumptions required by more sophisticated methods. Although the solvent mapping technique is essentially a post-synthesis measure of morphology control, knowledge accruing from such investigations can often be advantageously employed towards the design of an adequate solvent for the synthesis of a pigment in the desired habit.

4.4 Concluding remarks

Organic coloured pigments find broad general use in many surface coating and plastic industries, including the colouration of paint, varnish,

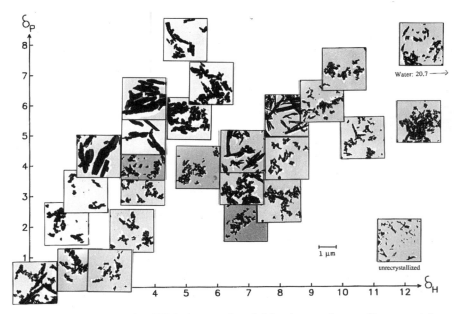

Figure 4.17 Representative TEM photographs of C.I. pigment Orange 65 arranged in a (δ_P^L, δ_H^L)-diagram. Reproduced from Mizuguchi and Wooden[20] with permission.

enamel, printing ink, linoleum rubber, wax, synthetic fibres and bulk plastics. Consequently, they represent a business of high value.

In this chapter, the authors have endeavoured to illustrate that the particulate state of the colorant plays as dominant a role in determining its pigmentary performance as does its chemical constitution. The form and habit of such pigment crystals, however, must derive ultimately from molecular and intermolecular forces prevalent in the crystal lattice. Empirical attempts at product development are hence, increasingly being supplemented by more fundamental approaches towards developing a better understanding of underlying principles governing pigment properties.

Ideally, as sophisticated analytical techniques help to unravel complex structure–performance relationships in pigment crystals, currently evolving powerful computational techniques will also allow reliable prediction of pigment crystal parameters such as crystal structure, habit, and light absorption characteristics on the basis of otherwise easily accessible molecular and powder diffraction data. In combination with this evolving information, a knowledge of better *in-situ* and post-synthetic pigment elaboration techniques allowing control of crystal form and habit will eventually become an even more powerful process for the design of pigments having selective and superior performance.

References

1. R. B. McKay, A. Iqbal and B. Medinger, in *Technological Applications of Dispersions* (R. B. McKay, ed), Marcel Dekker, Inc., New York, Basel and Hong Kong, Chap. 4, pp. 143–176 (1994).
2. Society of Dyers and Colourists, *Colour Index*, 3rd ed. (1982).
3. W. Herbst and K. Hunger, *Industrial Organic Pigments*, VCH, Weinheim, 1993.
4. H. Zollinger, *Color Chemistry: Synthesis, Properties and Applications of Organic Dyes and Pigments*, 2nd rev. ed., VCH, Weinheim, 1991 (English version, 1993).
5. A. Whitaker, *J. Soc. Dyers Colour.*, **94**, 431 (1978).
6. A. Whitaker, *J. Soc. Dyers Colour.*, **104**, 294 (1988).
7. S. Zheng, D. Liu and S. Ren, *Dyes and Pigments*, **18**, 137–149 (1992).
8. M. Klessinger, *Chem. Unserer Zeit*, **12**, 1–11 (1978); further references ibid. S. Dähne, *Z. Chem.*, **10**, 133, 168 (1978); *Science*, **199**, 4334, 1163–1167 (1978).
9. J. Fabian and H. Hartmann, *Light Absorption of Organic Colorants*, Springer-Verlag (1980).
10. J. Griffiths, *Rev. Prog. Color.*, **11**, 37–57 (1981).
11. J. Griffiths, *Colour and Constitution of Organic Molecules*, Academic Press, London (1976).
12. C. H. Griffiths and A. R. Monahan, *Mol. Cryst. Liq. Cryst.*, **33**, 175–187 (1976).
13. J. Mizuguchi and G. Rihs, *Ber. Bunsenges. Phys. Chem.*, **96**, (4), 597–606 (1992).
14. F. H. Chung and R. W. Scott, *J. Appl. Cryst.*, **4**, 506–511 (1971).
15. A. Iqbal, M. Jost, R. Kirchmayr, J. Pfenninger, A. C. Rochat, and O. Wallquist, *Bull. Soc. Chim. Belg.*, **97**, 615 (1988).
16. E. E. Jaffe, *J. Oil Colour Chem. Assoc.*, **1**, 24–31 (1992).
17. A. Thomas and P. M. Ghode, *Paintindia*, **39**, (6), 25–28, 30 (1989).
18. G. Klebe, F. Graser, E. Haedicke and J. Berndt, *Acta Cryst.*, **B45**, 69–77 (1989).
19. P. Hauser, D. Horn, R. Sappok, *FATIPEC Congr. XII*, 191–197 (1974).
20. J. Mizuguchi and G. Wooden, *Ber. Bunsenges. Phys. Chem.*, **95 (10)**, 1264–1276 (1991).
21. S. S. Labana and L. L. Labana, *Chem. Rev.* **67**, 1 (1967).
22. J. B. Conant and P. D. Bartlett, *J. Am. Chem. Soc.*, **54**, 2893 (1932).
23. N. L. Allan, A. L. Rohl, D. H. Gay, C. R. A. Catlow, R. J. Davey, and W. C. Mackrodt, *Faraday Discuss. Chem. Soc.*, **95**, Preprint (1993).
24. W. Herbst and K. Merkle, *Dtsch. Farben-Z.*, **24 (8)**, 365–377 (1970).
25. L. Chromy and E. Kaminska, *Progr. Org. Coat.*, **6 (1)**, 31–48 (1978).
26. B. Honigmann and B. Horn, in *Particle Growth in Suspensions* (A. L. Smith, ed) Academic Press, London and New York, Chap. 18, p. 283 (1973).
27. W. S. Struve, *US Patent 2.844.485* (1959); C. W. Manger and W. S. Struve, *US Patent 2.844.581* (1958).
28. S. J. Chapman and A. J. Whitaker, *J. Soc. Dyers Colour*, **87**, 120 (1971).
29. A. Whitaker, *Z. Kristallogr.*, **171**, 17 (1985).
30. S. Zheng, D. Lui and S. Ren, *Dyes Pigments*, **18**, 137–149 (1992).
31. A. Whitaker, *J. Soc. Dyers Colour*, **103 (12)**, 442–446 (1987).
32. A. Whitaker, *J. Soc. Dyers Colour*, **102**, 66 (1986).
33. F. F. Ehrich, *Kirk-Othmer Encyclopedia of Chemical Technology*, 2nd Edition, Volume **15**, John Wiley, New York, pp. 581–582.
34. E. E. Jaffe and F. Baebler, *US Patent 4.810.304* (1989).
35. J. R. Fryer, R. R. Mather, R. B. McKay, and K. S. W. Sing, *J. Chem. Technol. Biotechnol.*, **31**, 371 (1981).
36. DIN 53206, Teil 1: Teilchengrössenanalyse, Grundbegriffe.
37. D. H. Everett, *Pure Appl. Chem.*, **31**, 579 (1972).
38. A. R. Hanke, Report of a lecture, *J. Opt. Soc. Am.*, **56**, 713 (1966).
39. M. A. Maikowski, *Ber. Bunsenges. Phys. Chem.*, **71**, 313 (1967).
40. P. Hauser, M. Herrmann and B. Honigmann, *Farbe Lack*, **76**, 545 (1970).
41. P. Hauser, M. Herrmann and B. Honigmann, *Farbe Lack*, **77**, 1097 (1971).
42. M. A. Maikowski, *Prog. Colloid Polym. Sci.*, **59**, 70 (1976).
43. W. Carr, *J. Oil Colour Chem. Assoc.*, **54**, 1093 (1971).
44. M. A. Maikowski, *Dtsch Farben-Z.*, **25 (8)**, 386–389 (1971).

45. B. Felder, *Helv. Chim. Acta*, **51**, 1224–1234 (1968); *J. Color Appearance*, **1**, 9 (1971).
46. O. Hafner, *J. Paint Technol.*, **47**, 64 (1975).
47. R. B. McKay, *XX FATIPEC Congr.*, 139 (1990).
48. R. B. McKay, *Farbe Lack*, **97**, 576 (1991).
49. R. B. McKay, *XVIII FATIPEC Congr.*, **2**, 405 (1986).
50. R. B. McKay, *J. Oil Colour Chem. Assoc.*, **71**, 7 (1988).
51. R. B. McKay and R. R. Mather, *Colloids Surf.*, **27**, 175 (1987).
52. R. R. Mather and K. S. W. Sing, *J. Colloid Interface Sci.*, **60**, 60 (1977).
53. R. B. McKay, *J. Oil Colour Chem. Assoc.*, **72**, 89 (1989).
54. R. Sappok and B. Honigmann, in *Characterization of Powder Surfaces* (G. D. Parfitt and K. S. W. Sing, eds), Academic Press, New York, 1875, Chap. 6, p. 231.
55. W. Kurtz, *Am. Ink Maker*, Dec. p. 20, (1986).
56. R. B. McKay, *Surf. Coat. Int., J. Oil Colour Chem. Assoc.*, **75**, 177 (1992).
57. R. B. McKay, *Farbe Lack*, **95**, 471 (1989).
58. R. B. McKay, *Farbe Lack*, **96**, 336 (1990); *Surf. Coat. Int., J. Oil Colour Chem. Assoc.*, **74**, 169 (1991).
59. D. Horn and B. Honigmann, *XII FATIPEC Congr.*, 191 (1978).
60. F. Glaeser, *Dtsch. Farben-Z.*, **32**, (9), 338–342 (1978).
61. M. J. Vold, *J. Phys. Chem.*, **63**, 1608 (1959).
62. W. Ditter and D. Horn, *Fourth Int. Conf. Surf. Coat. Sci. Technol.*, Athens, 251 (1978).
63. R. B. McKay, *XVIII. Int. Conf. Org. Coat. Sci. Technol.*, 249 (1992); *Progr. Org. Coat.*, **22**, 211 (1993).
64. B. Medinger, *XIV FATIPEC Congr.*, 439–443 (1978); *Farbe Lack*, **84 (12)**, 994 (1978).
65. D. Liu and S. Ren, *Dyes Pigm.*, **18**, 69 (1992).
66. A. Iqbal, L. Cassar, A. C. Rochat, J. Pfenninger, and O. Wallquist, *J. Coat. Technol.*, **60**, Mar., 37 (1988).
67. A. Grubenmann, *Dyes Pigments*, (manuscript in preparation).

5 Dye solubility
G. L. BAUGHMAN, S. BANERJEE and
T. A. PERENICH

5.1 Introduction

Dye solubility is a very important parameter to the dye chemist and a number of practical measurement methods of it have been evaluated. However, such methods are usually applied to commercial dye formulations under conditions used for textile dyeing. While the results of these may be of considerable utility, they have little fundamental basis because the solubilities are a function both of the dye formulation and of the experimental methods used.

The solubility of a pure colorant is a fundamental property that is broadly applicable and finds uses in many areas beyond basic science. Thus, knowledge of dye solubilities is useful for problems related to polymers, electronics, solvents, environmental exposure assessment, and for many other products and applications. Also, such fundamental physical constants are increasingly used in the development and testing of methods for predicting properties of diverse classes of colorants.

Although much work has been carried out with commercial dyes, it is surprising that such a fundamental thermodynamic property as solubility has been so seldom reported. Furthermore, there are even less published solubility data on the more modern dyes that are widely used in today's textile industry.

The types of dyes most extensively studied in water are the disperse dyes and biological stains. In both cases, the solubility measurements were motivated by practical considerations. For the disperse dyes, there was a need to understand and improve the dyeing of synthetic fibers and, in the case of stains, it was necessary to standardize materials that were widely used in generally accepted procedures.

5.1.1 Scope of the review

It was originally intended to limit this review to water solubility. However, the available water solubility data are so limited that the review was expanded to include other solvents. As a result, we have compiled data and presented the methodologies that provide realistic measures or estimates of thermodynamic solubility for more than 150 dyes. The references were obtained by searching *Chemical Abstracts*[1] from 1926 through July 1992 and *Chemisches Zentralblatt*[2] from 1897 through 1906.

It is well known to the specialist that commercially used textile dyes generally (1) are impure, (2) can be very difficult to purify, and (3) include structures that are not in the public domain. Therefore, data have been included in this review only if the original source states that the dyes studied were colorants containing no diluents, i.e. that they were not commercial formulations. Also, data were eliminated that could not be related to a known structure with reasonable confidence.

We have also chosen not to include discussion of the thermodynamics of dissolution and aggregation since extensive reviews of these topics are widely available. However, it must be noted that much of the available data on dye solubility, especially that for disperse dyes, were originally developed for use in thermodynamic studies.

Many of the commerical dye names and structures reported in this review are given in the *Colour Index*.[3] Additionally, because much of the data reported are from the Russian literature, we have given the transliterated Russian name, structure, and CAS number whenever possible; many of the structures of Russian dyes were taken from the book by Stepanov.[4] In virtually all cases where data are given, the original references have been consulted; *Chemical Abstracts* references are used only for a few cases when conclusions from difficult-to-obtain foreign publications are reported.

5.2 Solubility measurement

5.2.1 Difficulties in solubility measurement

Although solubility measurements have been a long established procedure in chemical studies, many reported solubilities often differ by orders of magnitude from one worker to another. This is especially true when solubilities are very low, as is usually the case for non-ionic and/or non-polar organic compounds in polar solvents. This situation has recently been covered by Yalkowsky and Banerjee[5] in their comprehensive treatment of water solubility measurement. Because their emphasis was on highly insoluble compounds in water, in this update we only briefly address factors that are most likely to complicate work with dyes when they are quite insoluble.

By definition, dye solubility is the concentration of a dye in equilibrium with its pure solid or liquid phase at a given temperature.

$$(DYE)_{Solid} \rightleftarrows (DYE)_{Dissolved} \tag{5.1}$$

Therefore, all solubility measurements require separation of the saturated solution from the solid dye that may be present at equilibrium. Also, this must be done without disturbing the equilibrium. Further, the solid must be removed to a level such that any remainder contributes only negligibly to the measurement of dissolved dye. Large errors may occur if the solubility is low and if significant amounts of very small or colloidal particles remain in suspension. This can be difficult or impossible to overcome for highly insoluble dyes since for quantitation such compounds are usually extracted into a solvent in which they are much more soluble. The presence of residual solid material is less of a problem when solubility is high, i.e. greater than a few milligrams per liter.

It is also important to recognize that highly insoluble (hydrophobic) solutes always have a strong driving force to leave the solution phase for any phase that is more like the solute. Under these conditions, cleanliness of glassware, composition of filter material, use of tubing, etc. are all factors that must be carefully considered.

Impurities in dyes can also contribute significantly to inaccuracies in solubility measurements, especially since dyes are often difficult to purify. For reasonably pure dyes, major impurities are likely to be structurally related, similarly colored compounds. In this case, if equilibration is performed below the melting point, and if the dye and the impurity do not interact, the actual solubility remains unchanged.[5] Even in this case however, errors will result if the same dye is used both for equilibration and for the quantitation standard. This occurs due to gross error in the dye mass as a consequence of the mass of impurities or diluents; among the latter is water of hydration, which is present in most, if not all, sulfonic acid dye salts.

Errors due to impurities may also be compounded by the use of non-specific quantitation methods such as spectrophotometry. This is probably the major source of error in many dye solubility studies. The problem is heightened if the impurities are much more soluble than the dye, and thus contribute a disproportionately large share to the total absorbance. This problem can be eliminated or minimized by (1) using compound-specific quantitation methods such as gas or high performance liquid chromatography, (2) reducing the amount of excess dye used for equilibration, and (3) using higher purity dye.

Methods for the purification of water-soluble dyes have included adsorption on cotton,[6] high performance liquid chromatography (HPLC),[7] and other techniques reviewed by Giles and Greczek.[8] Hydrophobic dyes are purified much like other organic compounds. The potential magnitude of impurities in commercial dyes is readily apparent on examination of analytical data which have been reported for commercial thiazines[9] and rhodamines.[10]

Several other factors should also be noted with regard to the solubility of any compound, but especially that of large multi-functional molecules such as dyes. Some dyes may not have a measurable solubility and chemists have known for many years that certain dyes aggregate (especially in water) at high concentrations to form colloidal suspensions, liquid crystal-like arrays, dimers, etc. (see Duff and Giles[11] for a summary). Thus, it was reported by Ostwald and Walter in 1936[12] that benzopurpurine solubility was a function of the amount of dye used. This observation was later shown to be valid not only for benzopurpurine[13] and chrysophenine,[13] but also for metal complex dyes of the disperse type[14] in water and other solvents. Under these conditions, equation (5.1) is violated and measurement of solubility is not possible. Further, these solutions may become so viscous that the solid phase cannot be separated.

Another area of considerable misunderstanding concerns the influence of particle size and crystal form on solubility. Both influences are very important to practical dyeing, especially with disperse dyes. Braun[15] has discussed the former and Biederman,[16] Shenai and Sadhu[17] and Odvarka *et al*.[18] have shown the effect that the latter can have on dye activity in a dyebath. From the thermodynamic point of view, it is important to remember that, while both influences are real, they are transient, i.e. they do not influence the solubility of the thermodynamically stable form.

Under conditions used to determine the solubility of purified dyes, particle size is unlikely to have a significant effect. Similarly, solubility differences between the amorphous and the most stable crystalline form are probably small, i.e. less than an order of magnitude. Also, prolonged equilibration times and elevated temperatures will further reduce any contribution from these effects. Nevertheless, it is important to be fully aware of the contribution that these factors may have on the measured solubility.

Finally, there are some potentially serious pitfalls for those unfamiliar with the literature on dyes and the textile dyeing field. For example, some structures in the *Colour Index*[3] are not totally correct, particularly in the case of certain ionic dyes. In the absence of other confirming information, it is unreliable to assume that the non-dye ion (e.g. Na^+ or K^+) is the same as that shown in the *Colour Index*.[3] An incorrectly known ion can result in a measured solubility that is correct, but ascribed to the wrong compound. Similarly, water of hydration is seldom mentioned or taken into account, although this factor is undoubtedly important for most ionic dyes.

It is also common to have difficulties in associating correct names and even *Chemical Abstracts* Registry numbers with the correct structure. This is particularly true with many older dyes and with commercial names. Such problems are especially likely when structures differ with respect to only one functional group, e.g. a chloro group in place of iodo, or the orientation of the group.

Most of these difficulties can be obviated, or reliably surmounted, by taking nothing for granted and testing everything. At the minimum, physical properties like melting points, infrared or mass spectra, nuclear magnetic resonance, etc. need to be used to confirm structures, unless the data are known with absolute certainty.

5.2.2 Measurement methods

All methods for determining solubility comprise three functions: equilibration, phase separation and quantitation. In many cases, these processes are performed by straightforward methods. For example, equilibration is simply a matter of stirring for sufficient time to reach a constant solute concentration (usually at a controlled temperature). Therefore, we will only discuss methods that differ from common practice in some useful or interesting manner.

Any adequately sensitive analytical method can be used for quantitation, but spectrophotometry is, by far, the most widely used. When dye solubilities are high, they have sometimes been measured by weight.[10] Radioactivity has been employed in only one case for dyes,[19] although this method has often been applied to other compounds. More recently, solubilities of disperse and solvent dyes have been quantified by HPLC.[20,21]

Solubility methods are usually differentiated by the manner in which the phases are separated. On this basis, the most widely used method is filtration. Choice of filter materials is determined by type of dye, particle size to be removed, solubility, filtration rate required, solvent, amount of filtrate required, and temperature. The filter must be inert and adsorb a negligible amount of the dye. Filters have been made of sintered glass, wool, cotton, paper, fluorinated polymer, and metal. In general, filtration is simple if adequate phase separation can be achieved without shifting the equilibrium, i.e. without excessive solvent loss or cooling.

Some workers[18,22,23] have used a metal apparatus specially fabricated for solubility studies at temperatures above ambient. All of these devices were designed for measuring water solubilities of disperse dyes at elevated pressures typical of dye bath conditions (pressure filtration method). Using such an apparatus, Odvarka et al.[18] compared results for purified C.I. Disperse Red 121 (at 130 °C) with data obtained by other methods and with several filters. They found that the filtration method gave comparable results for the purified dyes, but concluded that filtration methods are unreliable for use with commercial formulations of disperse dyes.

Georgieva and Mostoslavskii[24] have described an all-glass apparatus for working with small quantities of dyes. Their apparatus is most suitable for studies using organic solvents and dyes that are at least moderately soluble, since filtration is through sintered glass.

Other methods have been used that do not depend on filtration. Přikryl et al.[25] developed a polymer cell method based on diffusion through polymer films in water at temperatures of 40 to 140 °C. They also found a favorable comparison between their results for purified dyes and results from the pressure filtration method. The polymer cell method, although attractive, was found in our laboratories to be unusable at ambient temperatures for C.I. Disperse Blue 79 because of the slow diffusion rate. Possibly, a better choice of film would eliminate this difficulty.

In another study, Sah[26] measured dye solubility in small amounts of liquid crystals. Because of the small sample sizes, the phase separation was made by centrifugation. This is the only use of centrifugation that we found for dyes, although it has been frequently used with other compounds when solubility was not prohibitively low.

The generator column method has been used in measurement of the water solubilities of disperse and solvent dyes at room temperature.[21,22] Unlike all of the static methods mentioned above, the generator column method is dynamic. In this procedure, the water (or other solvent) flows, at a known rate, through a stainless steel HPLC column that is packed with a finely divided sorbent and excess solid dye. The effluent is collected on a sorbent and eluted for analysis (in this case by HPLC). Although the dye and water may not reach true equilibrium, variations in flow rate can show whether the system is at a steady state that is detectably different from equilibrium.

For example, using a sensitive HPLC technique for analysis, Yen et al.[20] measured a disperse dye solubility of 10^{-7} g/l with a standard deviation of 5×10^{-8} g/l. At higher solubilities, the precision was usually significantly better. A variant of this method utilized mixed solvents to estimate the water solubility of C.I. Solvent Green 3 as 10^{-10} g/l.[21] The method is suitable for use only with compounds that have extremely low solubility.

An unusual method was used by Prusov et al.[27] to measure the solubility of two reactive dyes, Reactive Bright Red 6S and Reactive Violet 4K. The method is based on the Dubinina–Astahova sorbtion isotherm[28] (equation (5.2)):

$$\mathrm{Ln}\, C^s = \mathrm{Ln}\, C^s_\infty - \tan \alpha (\mathrm{Ln}\, C_\infty / C_p)^n \qquad (5.2)$$

where C^s and C_p are the equilibrium concentrations of dye on the adsorbent, cotton, and in solution; C^s_∞ is the limiting adsorbed concentration at the solubility of the adsorbate, C_∞; n is an integer; and $\tan \alpha$ is the slope. Solving the equation for C_∞ provides the solubility based on the other measured parameters. This method seems difficult to apply and subject to serious question regarding the general validity of equation (5.2). However, it may provide solubilities that are unattainable by other methods.

5.3 Measured solubilities

Most of the solubility data in the literature were obtained as a part of general thermodynamic studies. This is particularly true of disperse dyes. In such cases, we have reported data for the lowest temperatures nearest ambient. If data were available at other temperatures, the temperature range covered in the original study is also given. All of the solubility data and dye structures are reported in Tables 5.1A and 5.1B. Two major sources of data on disperse dyes and biological stains are not incorporated in these tables. Each of these will be discussed briefly below.

In 1988, Baughman and Perenich[29] compiled room-temperature (25 °C) water solubilities and vapor pressures (within the temperature range of study) for most of the hydrophobic dyes measured prior to that time. This included the comprehensive data for 43 dyes given by Bird[30] in 1954, who reported that the water solubilities of these dyes ranged from about 10^{-7} to 10^{-6} M. Unfortunately, few of the dyes are widely used today and many of the values were extrapolated from higher temperatures. The solubility values are not reported here since they are readily available elsewhere.[29,30]

Many dyes of several different classes are used as biological stains, and in 1927, Holmes[31] pointed out the need for data on their solubility. In 1928[32] and 1929,[33] Holmes published solubility data for about 80 stains in both water and 95% alcohol. He has the distinction of being the only person to show the effect on solubility of different metal salts for several anionic days.[33] More recently, the data of Holmes, Gurr[34] and others have been summarized and compiled in *H. J. Conn's Biological Stains*[35] to give the largest single collection of dye solubilities. This book also presents the solubilities of many dyes in cellosolve, ethylene glycol and xylene.

Data from *H. J. Conn's Biological Stains* are excluded because the compilation is readily available and also because of questions concerning reliability. Most of stains are ionic dyes, the solubility of which is strongly a function of the counter ion. Unfortunately, the stain literature relies heavily on common (trivial) names and *Color Index*[3] constitution numbers for dye identification. This frequently precludes determination of an exact formula for the dye actually used. Further, there is often little or no information on purity or methodology.

5.3.1 *Solubility in water*

Many of the papers from which data were taken for the compilation of dye solubilities (Table 5.1) will not be discussed. However, some studies have features that merit note. First, Přikryl *et al.*[25,36] measured the solubilities of 37 disperse dyes at 125 to 130 °C. Some were commercial

Table 5.1A Dye solubilities

Dye	Name	Solvent	Solubility (g/l)	T (°C)	$T_r^\dagger$ (°C)	CAS No.	C.I. No.
1^{46}	1-Aminoanthraquinone	PER	23.0	121		82-45-1	
1^{52}	1-Aminoanthraquinone	EG	7.45×10^{-4d}	25	10–40		
2^{46}	2-Aminoanthraquinone	PER	1.05	121			
3^{41}	C.I. Basic Blue 22 Cationic Blue 1Z*(CNS$^-$)	Water	3	25	5–15	14254-18-3	61512
4^{46}	1,4,5,8-Tetraaminoanthraquinone	PER	0.319	121			
5^{37}	C.I. Disperse Blue 3	Water	4.9×10^{-4e}	Room		2475-46-9	61505
5^{49}		EtoH	0.3	Room			
		ACE	2	Room			
		DCE	34	Room			
		BENZ	27	Room			
		DBP	9	Room			
	Disperse Blue K*						
6^{37}	C.I. Disperse Blue 7	Water	1.1×10^{-3c}	23		3179-90-6	62500
	C.I. Solvent Blue 69						
6^{39}	C.I. Disperse Blue 7 Disperse Blue-Green*	Water	0.041	40	40, 60	1379-90-6	62500
7^{37}	C.I. Disperse Blue 14	Water	1.9×10^{-5c}	≈25		2475-45-8	65500
8^{48}	C.I. Disperse Blue 26	PER	0.58	23	23–70	3860-63-7	63305
	Duranol Blue G*	TRIC	1.13	23	23–70		
		DIOX	6.4	30	30–70		
9^{37}	C.I. Disperse Blue 56 Disperse Blue PE*	Water	5.9×10^{-6c}	≈25		31810-89-6	
		PER	0.47	23	23–70		
		TRIC	0.84	23	23–70		
9^{48}		DIOX	16.1	23	23–70		
		DCE	1.71	23	–		
		MC	2.13	23	–		
10^{25}	C.I. Disperse Blue 152	Water	0.0335	125			

DYE SOLUBILITY

#	Dye	Solvent	Solubility	T (°C)	Range	CAS	CI
11[48]	C.I. Disperse Orange 11	PER	2.75	23	23–70	82-28-0	60700
	Disperse Orange*	TRIC	12.1	23	23–70		
		DIOX	7.2	30	30–70		
12[21]	C.I. Disperse Red 9	Water	1.2×10^{-4}	25		82-38-2	60505
13[21]	C.I. Solvent Red 111	Water	4.8×10^{-4}	25			
13[48]	C.I. Disperse Red 11	PER	0.26	23	23–70	2872-48-2	62015
	C.I. Solvent Violet 26	TRIC	0.65	23	23–70		
	Disperse Bright Rose*	DIOX	5.40	23	23–70		
		DCE	3.45	23	—		
		MC	5.30	23	—		
		EG	2.64	23	—		
14[39]	C.I. Disperse Red 15	Water	$1.3 \times 10^{-3\ddagger}$	40	40, 60	116-85-8	60710
	Disperse Red 2S*						
14[46]	1-amino-4-hydroxy-AQ	PER	34.2	121			
14[48]	C.I. Disperse Red 15	PER	2.10	23	23–70		
	Disperse Red 2S*	TRIC	4.74	23	23–70		
		DIOX	20.1	30	30–70		
15[21]	Disperse Red 60	Water	6.4×10^{-7}	25		17418-58-5	60756
15[25]	C.I. Disperse Red 60	Water	0.0074	130		17418-58-5	60756
16[25]	C.I. Disperse Red 121	Water	0.0088	130			
		Water	3.3×10^{-4}	25			
17[21]	1,4-Diaminoanthraquinone	PER	4.19	121		128-95-0	61100
17[46]	C.I. Disperse Violet 1	PER	0.81	23	23–70		
17[48]	C.I. Solvent Violet 11	TRIC	1.70	23	23–70		
17[52]	Disperse Violet K*	DIOX	7.2	30	30–70		
		EG	6.27×10^{-2d}	25	10–40		
18[37]	C.I. Disperse Violet 8	Water	7.5×10^{-3c}	25		82-33-7	62030
21[21]	C.I. Solvent Green 3	Water	$9.2 \times 10^{-11\ddagger}$	121		128-80-3	61565
22[46]	1,5-Diaminoanthraquinone	PER	0.400	121			
23[46]	1,4-Dihydroxy-AQ	PER	153	25			
23[52]	C.I. Pigment Violet 12	EG	7.53×10^{-2d}	25	10–40	81-64-1	58050
	Quinizarin						
24[52]	1-Hydroxyanthraquinone	EG	8.5×10^{-3d}	25	10–40	129-43-1	
26[21]		Water	1.0×10^{-7}	25			

Table 5.1A continued

Dye	Name	Solvent	Solubility (g/l)	T (°C)	$T_r^\dagger$ (°C)	CAS No.	C.I. No.
27[67]		Water	0.001	21			
		TEA	0.25	21			
		HEP	0.001	21			
		CT	1.68	21			
		EBA	0.50	21			
		EB	0.03	21			
		PER	0.04	21			
		MEK	0.94	21			
		CB	0.34	21			
		ACE	1.25	21			
		CHEX	2.13	21			
		DCB	0.44	21			
		CPN	5.00	21			
		PYD	4.37	21			
		BuOH	0.001	21			
		DMF	10.00	21			
		EtOH	0.15	21			
		DMSO	0.12	21			
		MeOH	0.10	21			
		EG	0.13	21			
28[41]	C.I. Basic Blue 3 Cationic Turquoise-2Z* (CNS⁻)	Water	16	25	5–15	55840-82-9	51004
29[42]	C.I. Basic Blue 9 Methylene Blue (Cl⁻)	Water	43.6	20–25		61-73-4	52015
31[42]	C.I. Basic Blue 12 Nile Blue A (HSO₄⁻)	Water	31	20–25		3625-57-8	51180
31[42]	C.I. Basic Blue 17 Toluidine Blue 0 (Cl⁻)	Water	38.2	20–25		92-31-9	52040
32[42]	C.I. Basic Blue 24 New Methylene Blue N (ZnCl₂)	Water	39	20–25		1934-16-3	52030

DYE SOLUBILITY

#	Dye	Solvent	Solubility	T (°C)	Conc. range	CAS	C.I.
34[42]	C.I. Basic Green 5	Water	16	20–25			52020
35[42]	Methylene Green (Cl⁻)	Water	12	20–25		8188-9	45170
36[42]	Rhodamine B (Cl⁻) C.I. Basic Red 1	Water	54	20–25		989-38-8	45160
37[42]	Rhodamine 6G (HCl) C.I. Basic Red 2	Water	29	20–25		477-73-6	50240
38[42]	Safranin T (Cl⁻) Brilliant Cresyl Blue (HCl)	Water	55	20–25		10127-36-3	51010
39[43]	Fluorescein (Yellow)	Water $I = 0.1$	3.8×10^{-4} M^b	23		2321-07-5	45350:1
39[43]	Fluorescein (Red)	Water	1.45×10^{-4} M^b	23		2321-07-5	45350:1
41[42]	Pyronine Y (Cl⁻)	Water	89.6	20–25		92-32-0	45005
42[42]	Thionine (HCl)	Water	2	20–25		581-64-6	52000
42[42]	Janus Green (Cl⁻)	Water	32	20–25		2869-83-2	11050
43[52]	p-Aminoazobenzene	EG	19.5^d	25	10–40	60-09-3	11000
44[52]	Azobenzene Diazene	EG	0.54^d	25	10–40	103-33-3	
45[41]	Cationic Red 5 Zh*	Water	365	25	5–25		
	HCOO⁻		274	25	5–25		
	CH₃COO⁻		10.5	25	5–25		
	Cl⁻		0.09	25	5–25		
	CNS⁻		0.011	25	10–40		
	ClO⁻₄						
46[52]	4-(Phenylazo)azobenzene	EG	1.68×10^{-4} ^{d}	25	5–15	15000-59-6	11052
52[41]	C.I. Basic Blue 54	Water	0.14	25			
53[41]	Cationic Blue 2K(CNS⁻) C.I. Basic Yellow 28 (CH₃SO⁻₄) Golden Yellow 2K*	Water				054060-92-3	48054
	CH₃COO⁻		280	25			
	C₂H₅COO⁻		196	25			
	Cl⁻		7.5	25			
	CNS⁻		0.5	25			
	I⁻		0.26	25			
	CH₃SO⁻₄		2.4	25			
55[21]	C.I. Disperse Blue 79	Water	5.2×10^{-6}	25		3618-72-2	11345:1

Table 5.1A continued

Dye	Name	Solvent	Solubility (g/l)	T (°C)	$T_r^\dagger$ (°C)	CAS No.	C.I. No.
56^{37}	C.I. Disperse Blue 165	Water	3.0×10^{-6c}			41642-51-7	11077
57^{37}	C.I. Disperse Brown 1	Water	6.0×10^{-5c}	≈ 25		23355-64-8	11152
58^{37}	C.I. Disperse Orange 25	Water	3.6×10^{-5c}	≈ 25		31482-56-1	11227
59^{37}	C.I. Disperse Orange 29	Water	3.7×10^{-6c}	≈ 25		19800-42-1	26077
60^{37}	C.I. Disperse Orange 30	Water	4.3×10^{-5c}	≈ 25		5261-31-4	11119
61^{21}	C.I. Disperse Red 1	Water	1.6×10^{-4}	25		2872-52-8	11110
	Disperse Scarlet Zh*						
61^{39}	C.I. Disperse Red 1	Water	8×10^{-4}	40	40, 60	2872-52-8	11110
	Disperse Scarlet Zh*						
61^{48}		PER	1.10	23	23–70		
		TRIC	3.27	23	23–70		
		DIOX	7.80	23	23–70		
61^{49}		EtoH	2.00	Room			
		ACE	17	Room			
		DCE	14	Room			
		BENZ	4	Room			
		DBP	10	Room			
62^{21}	C.I. Disperse Red 5	Water	1.0×10^{-4}	25		3769-57-1	11215
63^{25}	C.I. Solvent Red 117	Water	0.0345	125			
64^{21}	C.I. Disperse Red 166	Water	7.9×10^{-6}	25			
65^{21}	C.I. Disperse Red 274	Water	4.5×10^{-6}	25		83929-87-7	11855
66^{37}	C.I. Disperse Yellow 3	Water	2.8×10^{-5c}			2823408	
	C.I. Solvent Yellow 77						
67^{37}	C.I. Disperse Yellow 23	Water	5.2×10^{-7c}			6250-23-3	26070
68^{25}	C.I. Disperse Yellow 68	Water	0.0166	125		21811-64-3	21005
69^{21}	C.I. Solvent Red 1	Water	3.3×10^{-7}	25		1229-55-6	12150
70^{36}	Oil Soluble Yellow 2Zh	Water	0.772	130			
			$T > MP$				
70^{52}	C.I. Solvent Yellow 2	EG	1.03^d	25	10–40	60-11-7	11020
71^{52}	C.I. Solvent Yellow 7	EG	1.96^d	25	10–40	1689-82-3	11800
72^{36}		Water	0.937	130			

DYE SOLUBILITY

Ref	Solvent	Solubility	T	T range	CAS
73[36]	Water	1.76	130		
74[36]	Water	0.593	130		
75[36]	Water	0.772	130		
76[36]	Water	0.276	130		
77[36]	Water	1.23	130		
78[36]	Water	0.304	130		
79[36]	Water	0.528	130		
80[36]	Water	1.30	130		
81[36]	Water	0.978	130		
82[36]	Water	1.56	130		
83[36]	Water	1.03	130		
84[36]	Water	0.446	130		
85[36]	Water	0.676	130		
86[36]	Water	0.081	130		
87[36]	Water	0.328	130		
88[36]	Water	1.03	130		
89[36]	Water	0.291	130		
90[36]	Water	0.316	130		
91[36]	Water	0.690	130		
92[36]	Water	0.346	130		
93[21]	Water	$T > MP$	25		
94[21]	Water	5.9×10^{-7}	25		
95[21]	Water	1.6×10^{-5}	25		
96[21]	Water	2.8×10^{-4}	25		
97[47]	PER	6.9×10^{-7}	30	30–90	72828-64-9
98[47]	PER	16	30	30–90	71617-28-2
99[47]	PER	20	30	30–60	6657-33-6
100[47]	PER	30	30	20–45	68877-63-4
101[47]	PER	90	30	30–90	
102[47]	PER	0.83	30	30–90	
103[47]	PER	0.5	30	30–90	
104[47]	PER	0.3	30	30–90	
105[47]	PER	0.5	30	30–60	
	PER	2.5	30		
106[36]	Water	0.054	130		
107[36]	Water	0.090	130		

Table 5.1A continued

Dye	Name	Solvent	Solubility (g/l)	T (°C)	$T_r^\dagger$ (°C)	CAS No.	C.I. No.
108[36]		Water	0.046	130			
109[36]		Water	0.087	130			
110[36]		Water	0.062	130			
111[36]		Water	0.082	130			
112[36]		Water	0.037	130			
113[36]		Water	0.049	130			
114[67]		Water	0.004	21			
		TEA	0.60	21			
		HEP	0.003	21			
		CT	0.005	21			
		EBA	0.12	21			
		EB	0.06	21			
		PER	0.01	21			
		MEK	2.40	21			
		CB	0.26	21			
		ACET	1.20	21			
		CHEX	2.0	21			
		DCB	0.31	21			
		CPN	1.50	21			
		PYD	6.00	21			
		BuOH	0.37	21			
		DMF	20.00	21			
		EtOH	0.90	21			
		DMSO	30.30	21			
		MeOH	3.5	21			
		EG	0.17	21			
115[67]		Water	0.007	21			
		TEA	0.032	21			
		HEP	0.005	21			
		CT	0.05	21			
		EBA	0.014	21			
		EB	0.21	21			

DYE SOLUBILITY

116[67]

PER	0.06	21
MEK	0.75	21
CB	0.11	21
ACET	0.94	21
CHEX	0.002	21
DCB	0.005	21
CPN	0.75	21
PYD	1.00	21
BuOH	0.41	21
DMF	0.75	21
EtOH	0.05	21
DMSO	0.04	21
MeOH	0.03	21
EG	0.02	21
Water	0.002	21
TEA	0.25	21
HEP	0.001	21
CT	0.037	21
EBA	0.32	21
EB	0.057	21
PER	0.023	21
MEK	0.87	21
CB	0.22	21
ACE	5.34	21
CHEX	3.62	21
DCB	0.52	21
CPN	8.25	21
PYD	10.25	21
BuOH	0.09	21

Table 5.1A continued

Dye	Name	Solvent	Solubility (g/l)	T (°C)	$T_r^\dagger$ (°C)	CAS No.	C.I. No.
117[67]		DMF	13.33	21			
		EtOH	0.75	21			
		DMSO	7.18	21			
		MeOH	1.37	21			
		EG	0.83	21			
		Water	0.008	21			
		TEA	1.37	21			
		HEP	0.05	21			
		CT	0.444	21			
		EBA	1.25	21			
		EB	0.55	21			
		PER	0.56	21			
		MEK	6.75	21			
		CB	2.12	21			
		ACE	5.00	21			
		CHEX	0.81	21			
		DCB	0.63	21			
		CPN	20.00	21			
		PYD	8.50	21			
		BuOH	0.68	21			
		DMF	9.50	21			
		EtOH	0.25	21			
		DMSO	0.17	21			
		MeOH	0.09	21			
		EG	1.25	21			
118[21]		Water	7.4×10^{-5}	25			
119[49]	Sudan yellow G	EtOH	0.3	Room		70528-90-4	
		ACE	2	Room			
		DCE	34	Room			
		BENZ	27	Room			
		DBP	9	Room			
122[37]	C.I. Disperse Blue 354	Water	5.1×10^{-9c}			74239-96-6	

DYE SOLUBILITY

	Dye	Solvent	Solubility	Temp	CAS	CI
123[21]	C.I. Disperse Yellow 54	Water	7.4×10^{-7}	25	7576-65-0	47020
123[25]	C.I. Disperse Yellow 54	Water	0.0033	125	7576-65-0	47020
124[37]	C.I. Disperse Yellow 64	Water	2.0×10^{-7c}		10319-14-9	47023
125[21]	C.I. Solvent Yellow 33	Water	1.7×10^{-4}	25	8003-22-3	47000
126[39]	C.I. Disperse Yellow 13	Water	$4.5 \times 10^{-4\ddagger}$	40	3688-79-7	58900
127[39]	Disperse Yellow 6Z*	Water	0.019	40	119-15-3	10345
	C.I. Disperse Yellow 1			40, 60		
	Disperse Yellow Fast 2K*			40, 60		
127[48]	C.I. Disperse Yellow 1	PER	0.68	23	119-15-3	10345
	Disperse Fast Yellow 2K*	TRIC	1.73	23		
		DIOX	39.0	23		
		DCE	14.0	23		
		MC	14.8	23		
		EG	10.6	23		
				23–70		
				23–70		
				23–70		
				—		
				—		
128[21]	C.I. Disperse Yellow 42	Water	2.0×10^{-4}	25	5124-25-4	10338
128[25]		Water	0.239	130		
130[42]	C.I. Basic Blue 26	Water	65	20–25	2580-56-6	44045
131[42]	Victoria Blue B (Cl⁻)	Water	97	20–25	633-03-4	42040
	C.I. Basic Green 1					
	Brilliant Green (HSO⁻₄)					
132[42]	C.I. Basic Green 4	Water	40	20–25	569-64-2	42000
	Malachite Green (Cl⁻)					
133[42]	C.I. Basic Violet 1	Water	33	20–25	8004-87-3	42535
	Methyl Violet (Cl⁻)					
134[42]	C.I. Basic Violet 3	Water	1	20–25	548-62-9	42555
	Crystal Violet (Cl⁻)					
135[42]	C.I. Basic Violet 4	Water	2.65	20–25	632-99-5	42510
	Fuchsine (HCl)					
136[42]	Methyl Green (Cl⁻, Br⁻)	Water	42	20–25	14855-76-6	42590
137[37]	C.I. Vat Blue 1	Water	1.4×10^{-4c}		482-89-3	73000
	C.I. Pigment Blue 66					
	Indigo					
138[50]	C.I. Vat Blue 4	DMF	<0.01	30	81-77-6	69800
	C.I. Food Blue 4					
	Vat Clear Blue 0*					

Table 5.1A *continued*

Dye	Name	Solvent	Solubility (g/l)	T (°C)	$T_r^\dagger$ (°C)	CAS No.	C.I. No.
139[50]	C.I. Vat Blue 6	DMF	<0.01	30		130-20-1	69825
	C.I. Pigment Blue 64						
	Vat Azure K*						
140[50]	C.I. Vat Blue 20	DMF	<0.01	30		116-71-2	59800
	Vat Dark Blue O*						
141[50]	C.I. Vat Blue 33	DMF	0.16	30		4215-99-0	67915
	Vat Turquoise ZHa*						
142[50]	C.I. Vat Brown 1	DMF	<0.01	30		2475-33-4	70800
	Vat Cinnamon SK*						
143[50]	C.I. Vat Brown 5	DMF	<0.01	30		3989-75-1	73410
	C.I. Pigment Brown 27						
	Thioindigo Red-Cinnamon Zh						
144[50]	C.I. Vat Green 1	DMF	0.16	30		128-58-5	59825
	Vat Bright Green S*	DEA	0.011				
		BENZ	0.012				
		ETCE	0.020				
		CYAM	0.056				
		DIOX	0.072				
		MEM	0.065				
145[24]	C.I. Vat Red 41	CB	0.32	50		522-75-8	73300
	Thioindigo Red S*						
145[50]		DMF	0.38	30			
146[50]	C.I. Vat Orange 1	DMF	0.09	30			59105
	Vat Golden Yellow KHa*						
147[50]	C.I. Vat Red 47	DMF	<0.01	30		6492-68-8	73305
	Thioindigo Pink 2S*						
148[50]	C.I. Vat Yellow 4	DMF	0.01	30		128-66-5	59100
	Vat Golden Yellow ZhHa*						
149[50]	Vat Black S*	DMF	0.15	30		67446-03-1	
150[50]	Vat Bright Green Zh*	DMF	0.03	30			
151[50]	Vat Yellow ZhHa*	DMF	0.10	30			
152[50]	Indanthrene Bright	Water	0.18	30			

153[50]	Violet VVK*	DMF	0.12	30
154[24]	Thioindigo Orange K*	CB	0.70	50
155[24]	Thioindigo	CB	0.01	50
156[24]		CB	0.61	50
157[24]		CB	0.96	50
158[24]		CB	0.64	50
159[24]		CB	0.092	50
160[24]		Water	0.27	50
161[24]		Water	0.73	50
162[24]		Water	0.67	50
163[24]		Water	0.44	50
164[24]		Water	2.40	50
165[24]		Water	2.75	50
166[24]		Water	3.62	50
167[24]		Water	4.61	50
168[24]		Water	4.35	50
169[24]		EtOH	0.2	Room
174[49]		ACE	1	Room
		DCE	10	Room
		BENZ	7	Room
		DBP	4	Room

Digits in dye designation are reference nos.
* Russian name.
† Temperature range of study.
‡ Estimated, see text.
a As the salt.
b Intrinsic solubility, see also Table 5.2.
c Estimated from eq. (5.3), ref. 37.
d Calculated from sub-cooled liquid solubility and ΔH_f.

ACE, acetone; AQ, anthraquinone; BENZ, benzene; BuOH, n-butanol; CB, chlorobenzene; CHEX, cyclohexanone; CPN, cyclopentanone; CT, carbon tetrachloride; CYAM, cyclohexylamine; DCB, o-dichlorobenzene; DCE, 1,1-dichloroethane; DEA, diethylamine; DIOX, dioxane; DMF, dimethylformamide; DMSO, dimethylsulphoxide; EB, ethylbenzene; EBA, ethylbenzoate; EG, ethylene glycol; EOAM, ethanolamine; ETCE, ethyl cellosolve; EtOH, ethanol; HEP, n-heptane; MC, methylene chloride; MEK, methylethylketone; MEM, monomethylamine; MeOH, methanol; PEG, polyethylene glycol; PER, perchloroethylene; PYD, pyridine; TEA, triethylamine; TRIC, trichloroethylene, DBP, dibutylphthalate.

Table 5.1B Dye structures

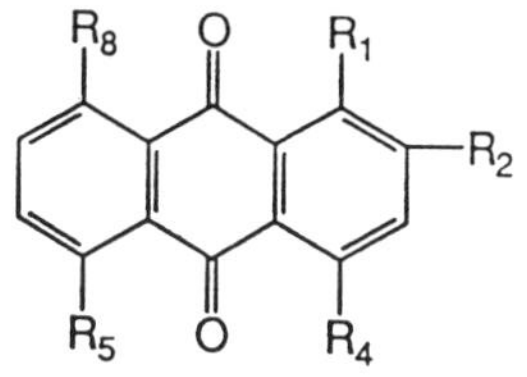

Dye	R_1	R_2	R_4	R_5	R_8
1	NH_2				
2		NH_2			
3	$NH-CH_3$		$NH(CH_2)_3N^+(CH_3)_3$		
4	NH_2		NH_2	NH_2	NH_2
5	$NHCH_3$		$NH-C_2H_4OH$		
6	$NH-C_2H_4OH$		$NH-C_2H_4OH$	OH	OH
7	$NHCH_3$		$NHCH_3$		
8	$NHCH_3$		OH	$NH-CH_3$	OH
9	NH_2	Br	OH	NH_2	OH
10	NH_2	Br	NH_2	NO_2	
11	NH_2	CH_3			
12	$NHCH_3$				
13	NH_2	OCH_3	NH_2		
14	NH_2		OH		
15	NH_2	OC_6H_5	OH		
17	NH_2		NH_2	NO_2	
18	NH_2		NH_2		
20	$NH-CH_3$		$NH-\phi-CH_3$		
21	$NH-C_6H_4CH_3-4'$		$NHC_6H_4CH_3-4'$	NH_2	
22	NH_2		$NHC_6H_4CH_3-4'$		
23	OH		OH		
24	OH			OH	
25	$NHC_6H_4CH_3-4'$		$NHC_6H_4CH_3-4'$	OH	OH

DYE SOLUBILITY

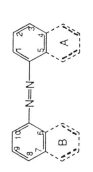

Dye	R_1	R_2	R_3	R_4	R_5	R_6	R_7	R_8	R_9	R_{10}
43			NH_2							
44										
45			$N(CH_3)(C_2H_4N^+(CH_3)_3)$							
46		OCH_3	$N=N \cdot C_6H_5$							
55			$N(C_2H_4OCOCH_3)_2$		$NHCOCH_3$	Cl		NO_2		
56	$NHCOCH_3$		$N(C_2H_5)_2$			NO_2		NO_2		Br
57	Cl		$N(C_2H_4OH)_2$			CN		NO_2		CN
58			$N(C_2H_5)(C_2H_4CN)$			Cl		NO_2		Cl
59			$N=N \cdot C_6H_4 \cdot OH-4'$	OCH_3				NO_2		
60			$N^-(C_2H_4CN)(C_2H_4OCOCH_3)$					NO_2		
61			$N(C_2H_5)(C_2H_4OH)$			Cl		NO_2		Cl
62	CH_3		$N(C_2H_4OH)_2$					NO_2		
63			$N(C_2H_4OH)(CH_2CBrCH_2)$					NO_2		Cl
65			$N(C_2H_4COOCH_3)_2$	CH_3	$NHCOCH_3$	CN		NO_2		
66	OH							$NHCOCH_3$		OCH_3
67			$N=N \cdot C_6H_5$				$NHCOCH_3$	OH		
68			$N=N \cdot C_6H_4 \cdot OH-4'$				$N(CH_3)COCH_3$	OH		
70			$N(CH_3)_2$				$NCOCH_3$			
71			OH				$N(CH_3)COCH_3$			
72			$N(CH_3)_2$							
73			$N(CH_3)_2$							
74			$N(C_2H_5)_2$							
75			$N(C_2H_5)_2$							
76			$N(CH_3)_2$					$NHCOCH_3$		
77			$N(CH_3)_2$					$N(CH_3)COCH_3$		
78			$N(C_2H_5)_2$					$NHCOCH_3$		
79			$N(C_2H_5)_2$					$N(CH_3)COCH_3$		

Table 5.1B continued

Dye	R_1	R_2	R_3	R_4	R_5	R_6	R_7	R_8	R_9	R_{10}
80			$N(CH_3)_2$				$CONH_2$			
81			$N(CH_3)_2$				$CONHCH_3$			
82			$N(CH_3)_2$				$CON(CH_3)_2$			
83			$N(C_2H_5)_2$				$CONH_2$			
84			$N(C_2H_5)_2$				$CONHCH_3$			
85			$N(C_2H_5)_2$				$CON(CH_3)_2$			
86			$N(CH_3)_2$					$CONH_2$		
87			$N(CH_3)_2$					$CONHCH_3$		
88			$N(CH_3)_2$					$CON(CH_3)_2$		
89			$N(C_2H_5)_2$					$CONH_2$		
90			$N(C_2H_5)_2$					$CONHCH_3$		
91			$N(C_2H_5)_2$					$CON(CH_3)_2$		
92			$N(C_2H_5)_2$							
93	CH_3		$N(C_4H_9)(C_2H_4OCOCH_3)$					NO_2		CN
94		Cl	$NHCH_2CHOHCH_3$					NO_2		Cl
95			$N(C_2H_4CN)(C_2H_4OH)$		$NHCOCH_3$	CN		NO_2		Cl
96		OCH_3	$N(C_2H_4CN)(CH_2CHCH_2)$		$NHCOCH_3$	Br		NO_2		NO_2

Dye	R_1	R_2	R_3	R_4	R_5	R_6	R_7	R_8	A	B
49	OH							SO_3Na	+	
	NHC_2H_5							$N\!=\!N\cdot C_6H_5$	+	+
69	OH								+	
97			$NHCOCH_3$							
98			$NHCOC_4H_9$							
99			$NHCOC_9H_{19}$			OCH_3				
100			$NHCOC_{17}H_{35}$							
101			$NHCOCH_3$						+	
102			$NHCOC_4H_9$						+	
103			$NHCOCH_3$						+ + +	+ + +
104			$NHCOC_4H_9$						+ +	
105			$NHCOC_{17}H_{35}$						+	

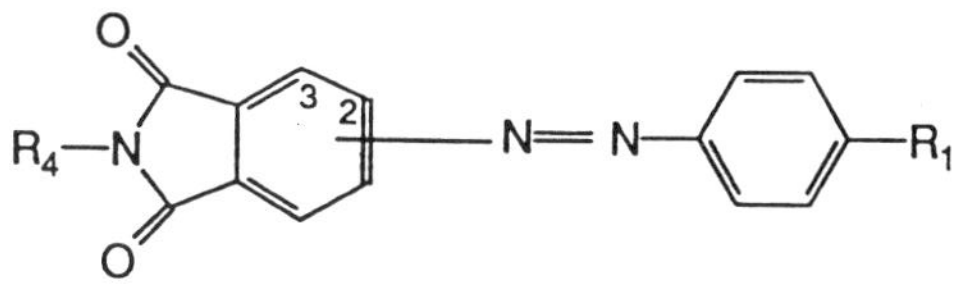

DYE SOLUBILITY

Dye	Position of azo group	R_1	R_2
106	3	$N(CH_3)_2$	H
107	3	$N(CH_3)_2$	CH_3
108	3	$N(C_2H_5)_2$	H
109	3	$N(C_2H_5)_2$	CH_3
110	2	$N(CH_3)_2$	H
111	2	$N(CH_3)_2$	CH_3
112	2	$N(C_2H_5)_2$	H
113	2	$N(C_2H_5)_2$	CH_3

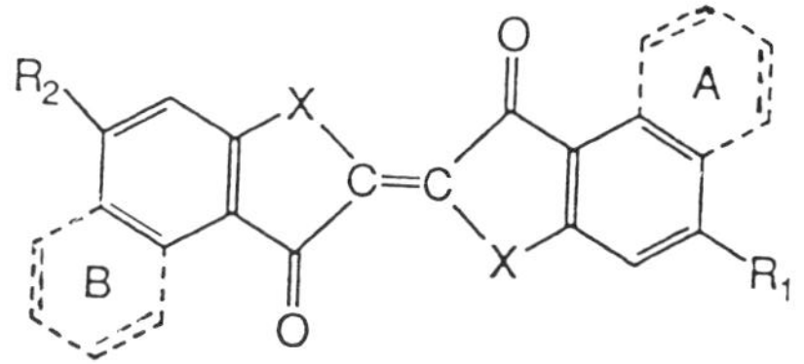

Dye	R_1	R_2	A	B	X
137	H	H	+		NH
143	H	H		+	S
146					S
147	Cl	Cl			S
154	H	H			

168 PHYSICO-CHEMICAL PRINCIPLES OF COLOR CHEMISTRY

Table 5.1B *continued*

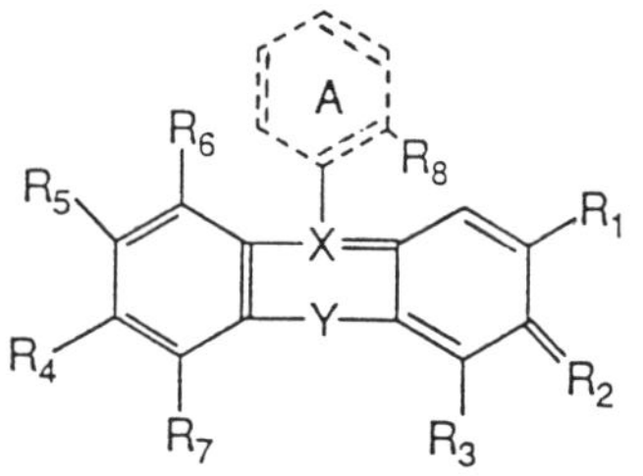

Dye	R_1	R_2	R_3	R_4	R_5	R_6	R_7	A	R_8	X	Y
28		$=N^+(C_2H_5)_2$		$N(C_2H_5)_2$						N	O
29		$=N^+(CH_3)_2$		$N(CH_3)_2$						N	S
31		$=N^+H_2$		$N(CH_3)_2$						N	S
32	CH_3	$=N^+HC_2H_5$	CH_3	NHC_2H_5						N	S
34		$=N^+(CH_3)_2$	NO_2	$N(CH_3)_2$			CH_3			N	S
35		$=N^+(C_2H_5)_2$		$N(C_2H_5)_2$				+	COOH	C	O
36	CH_3	$=N^+HC_2H_5$		NHC_2H_5	CH_3			+	$COOC_2H_5$	C	O
37	NH_2	CH_3		CH_3	NH_2					N	N
38	CH_3	$=N^+(C_2H_5)_2$		N^+H_2		NH_2		+	COOH	N	O
39		$=O$		OH						C	O
41		$=N^+(CH_3)_2$		$N(CH_3)_2$						C	O
42		$=N^+H_2$		NH_2						N	S

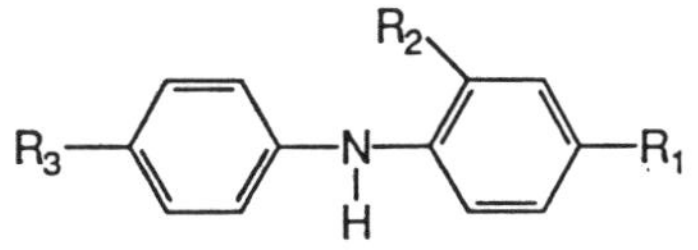

Dye	R_1	R_2	R_3
127	NO_2	NO_2	OH
128	$SO_2NHC_6H_5$	NO_2	

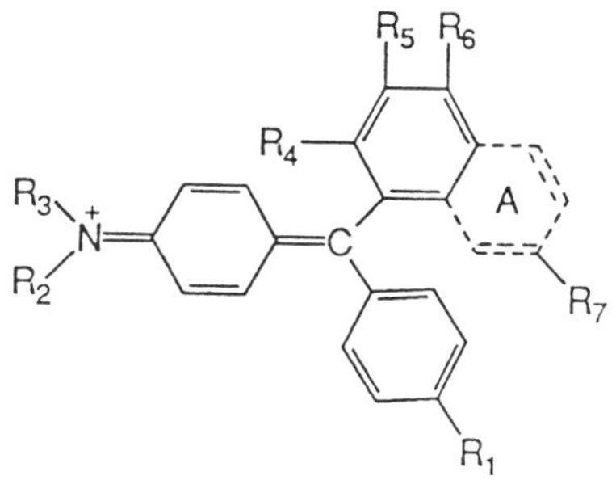

DYE SOLUBILITY

Dye	R_1	R_2	R_3	R_4	R_5	R_6	A	R_7
129	$N(C_2H_5)_2$	C_2H_5	C_2H_5	SO_3^-				
130	$N(CH_3)_2$	CH_3	CH_3			NHC_6H_5	+	SO_3^-
131	$N(C_2H_5)_2$	C_2H_5	C_2H_5				+	H
132	$N(CH_3)_2$	CH_3	CH_3					
133	$N(CH_3)_2$	CH_3	CH_3			$N(CH_3)_2$		
134	$N(CH_3)_2$	CH_3	H		CH_3	$N(CH_3)_2$		
135	NH_2	H	H			NH_2		
136	$N(CH_3)_2$	CH_3	CH_3			$N^+(C_2H_5)(CH_3)_2$		

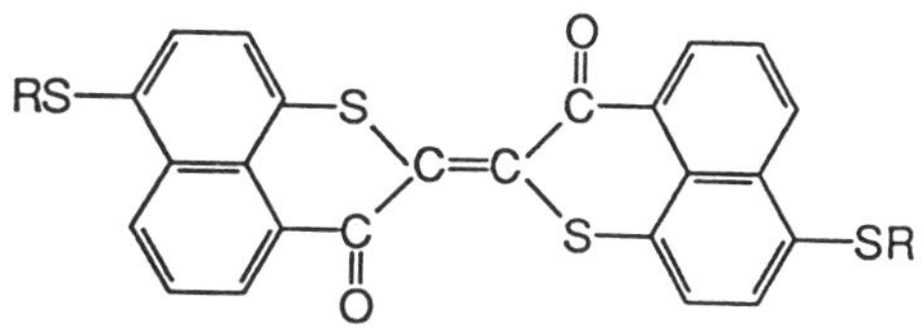

Dye	R
155	H

Table 5.1B *continued*

Dye	R
156	CH_3
157	C_4H_9
158	*iso* C_4H_9
159	C_5H_{11}
160	$C_{12}H_{25}$
161	$C_{18}H_{37}$
162	$SO_2N(C_4H_9)_2$
163	SR=H
164	CH_3
165	C_4H_9
166	C_5H_{11}
167	$C_{12}H_{25}$
168	$C_{18}H_{37}$
169	$SO_2N(C_4H_9)_2$

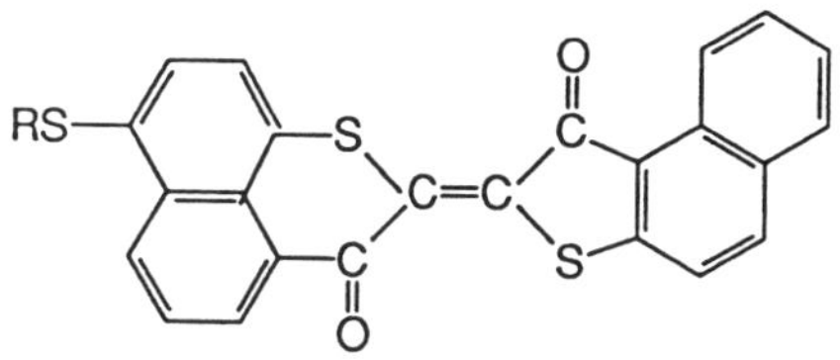

Dye	R_1	R_2
114	H	H
115	NO_2	H
116	Cl	H
117	Cl	Cl

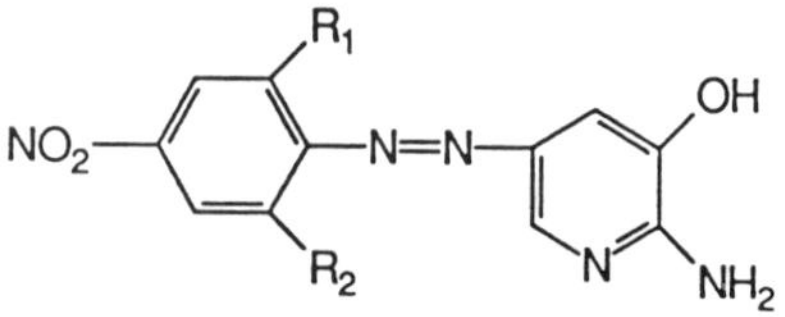

	R_1	R_2
123		OH
124	Br	OH
125		

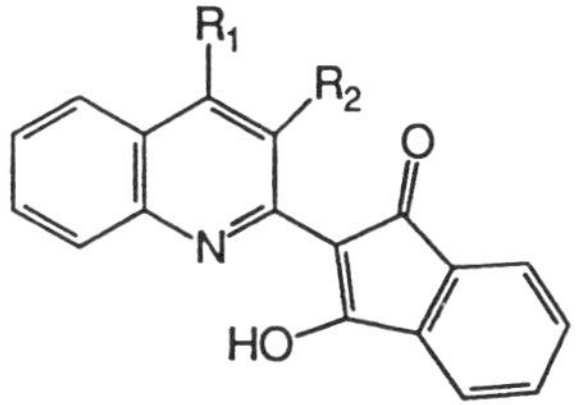

16

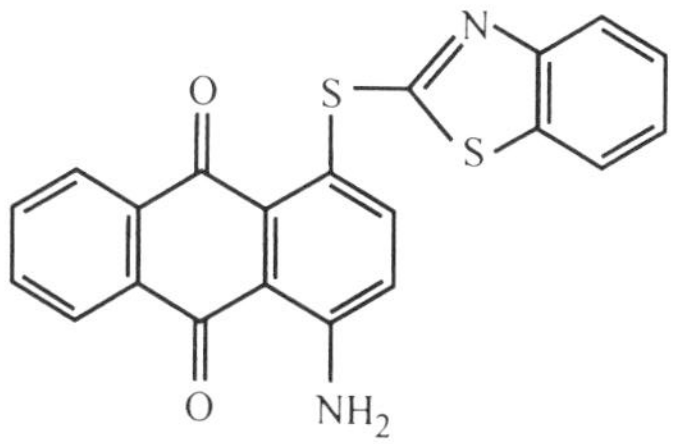

26

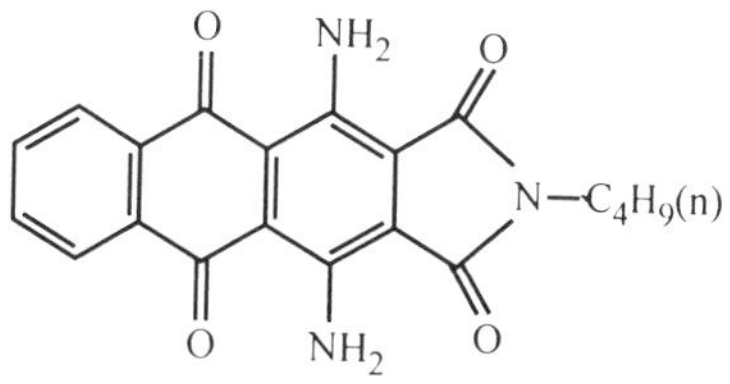

Table 5.1B *continued*

Dye	
31	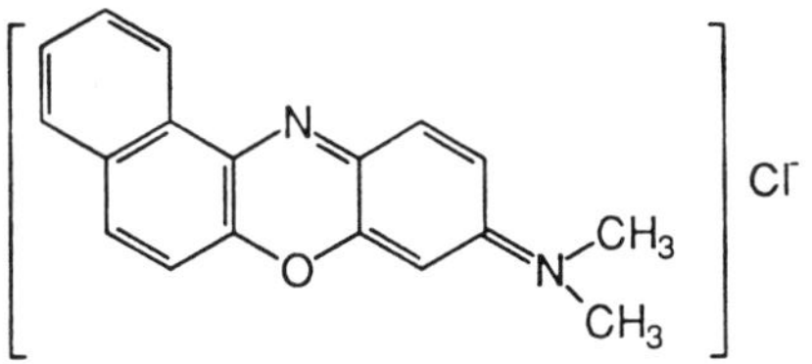 (structure: benzo-fused phenoxazine with =N(CH$_3$)$_2$ group, Cl$^-$ counterion)
42	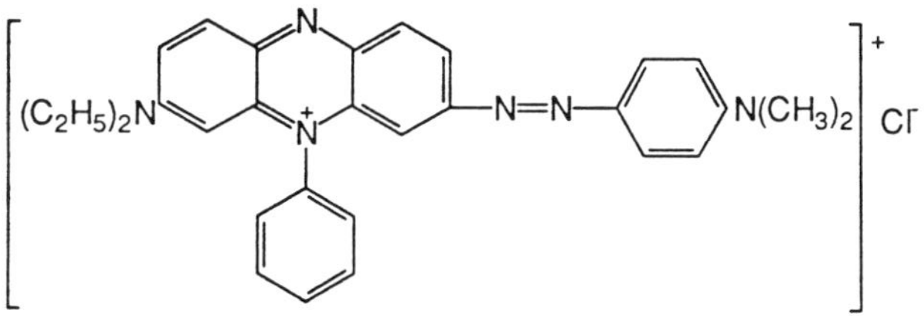 (structure: phenyl-substituted phenazinium cation with –N=N–C$_6$H$_4$–N(CH$_3$)$_2$ and (C$_2$H$_5$)$_2$N substituents, Cl$^-$ counterion)
52	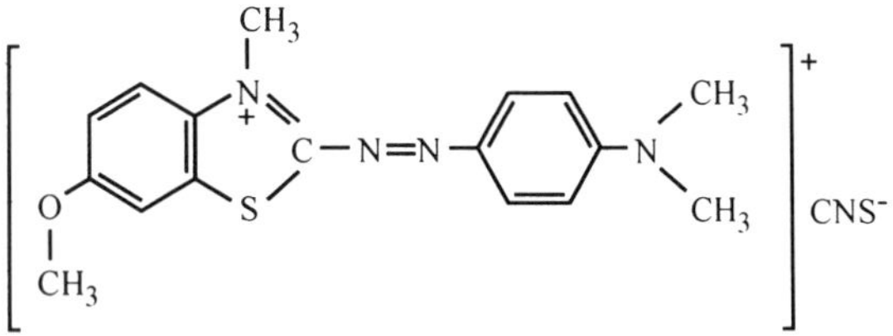 (structure: methoxy-substituted benzothiazolium with –C=N–N=C$_6$H$_4$–N(CH$_3$)$_2$, CNS$^-$ counterion)

53 64 118 119

Table 5.1B *continued*

Dye	
120	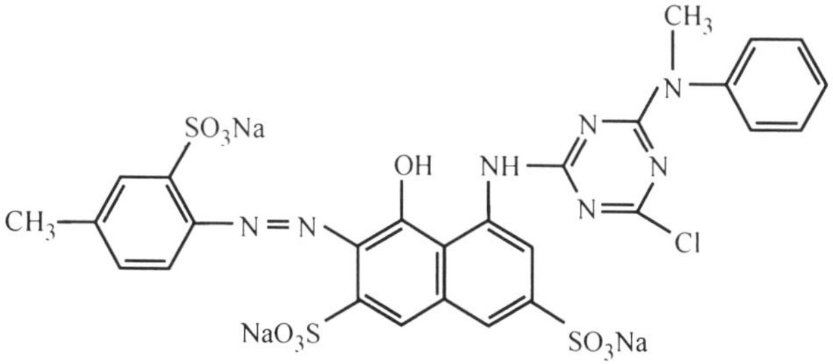 (structure)
121	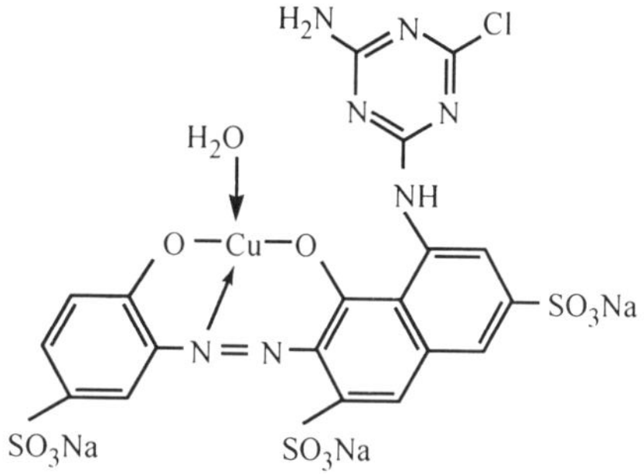 (structure)
122	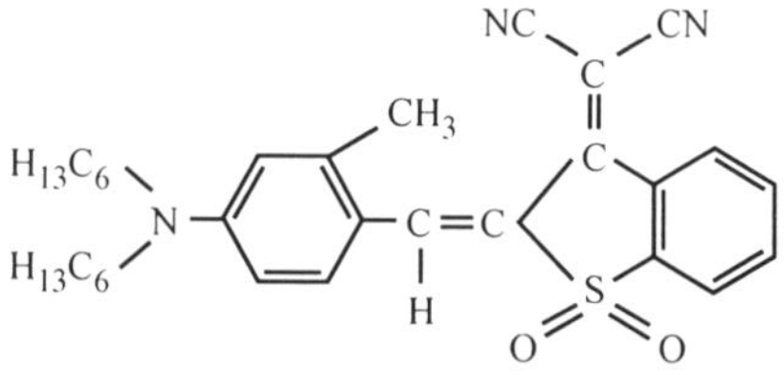 (structure)

126

138

139

Table 5.1B *continued*

Dye

140

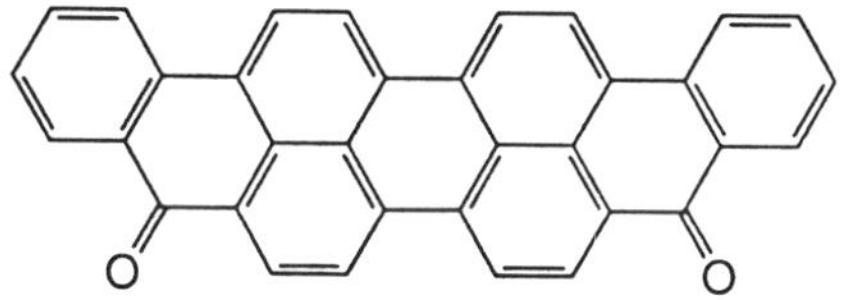

141

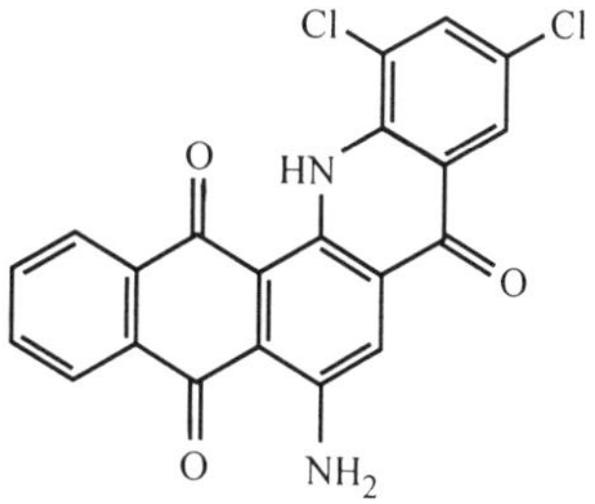

142

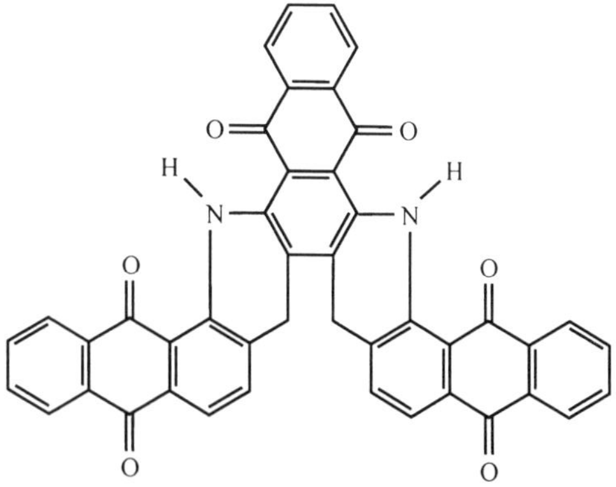

149

144

148

Table 5.1B *continued*

Dye
146 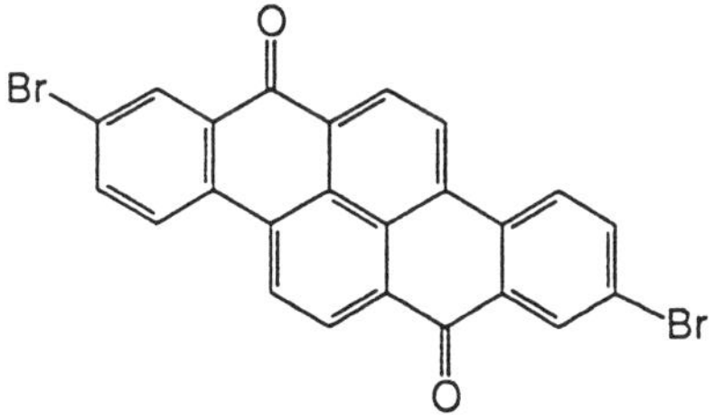
27 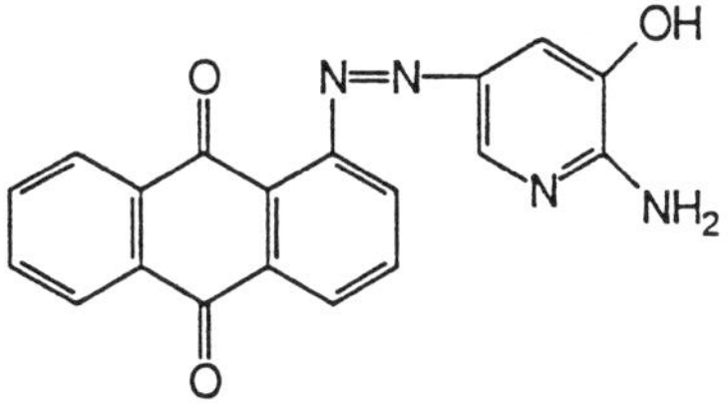

Identification numbers refer to data in Table 5.1A
[a] SCN⁻ salt
[b] Cl⁻ salt
[c] ZnCl₂ salt
[d] Hydrochloride salt
[e] HSO₄⁻ salt
[f] chloro ethobromide salt
[g] Na⁺ salt

dyes,[25] but most were synthesized N,N-dialkylaminoazobenzenes containing N-acetylamino, carbamido or imido substituents.[36] In the latter work,[36] the Nernst distribution coefficient, K_N, between the dyebath and polyester fibers was calculated. Our calculation of K_N from their saturated fiber concentrations and measured distilled water solubilities[36] permits the following observations. First, the values of K_N based on dyebath concentrations are smaller than the values based on water solubility (i.e. saturation) for all eight imino compounds, but for only one of the other 22 dyes. Second, the same nine compounds are the only ones having water solubilities that are less than 100 g/l. Third, for all dyes except eight of the nine imino compounds, the two K_Ns differ by less than 100% from the value of K_N based on dyebath concentration. For the imino compounds, the difference is from 300 to 4000%. For the other 20 compounds, agreement to better than a factor of 2 shows that the solubility is strongly predictive of the dyebath equilibria.

More recently, Baughman and Weber[21] reported the room-temperature solubilities of 20 purified disperse and solvent dyes. The solubilities were shown to have the expected linear log–log relationship with partition coefficient (for n-octanol/water). Hou et al.[37] have also estimated solubilities for 14 other dyes, including indigo, by the chromatographic method vide infra. Most of the dyes in these two studies[21,37] are presently in commerce and several are typical of important modern dye structures.

Several workers have examined the influence of non-detergent solutes on solubility. Among these is a study by Popov et al.[22] They report that the presence of 50 g/l of caprolactam increases the solubility of C.I. Disperse Orange 30 and C.I. Disperse Blue 73 ten-fold, but that ethylene carbonate results in a negligible increases. They also report the heat of solution to be unaffected. Unfortunately, they only give solubility data in graphical form.

In a recent study, Prusova et al.[38] examined the effect of low concentrations (< 0.1 M) of ammonia and ammonium acetate on the solubilities of C.I. Disperse Violet 1 and C.I. Disperse Red 11. They reported that solubility increases to a maximum and then decreases over the concentration range used. Although the maximum increase was only about 20%, they give no basis for determining whether the differences are statistically significant. In any event, the effects, as would be expected, are not very large.

Primak et al.[39] looked at the influence of acetonitrile concentration (up to 50 g/kg) on the solubility of five disperse dyes at 40 and 60 °C. The data show the approximate log linear dependence of solubility with volume fraction of solvent (see Estimation methods vide infra).

Finally, Starikovich et al.[14,40] examined the solubilities of several metal complex disperse dyes in water, trichloroethylene, and tetrachloroethylene. The studies are interesting because such dyes are apparently not widely used. The dyes are more soluble in water than in the other solvents and have solubilities of 0.05 to 3.4 g/l at 40 °C. Unfortunately, we could not reliably determine the structures of the compounds and, as noted earlier, some, apparently, have variable solubility.

As noted earlier, water solubilities for the relatively easy-to-measure ionic dyes are much more sparse than for the hydrophobic compounds. Except for the stains, we found only five papers that contained such data.

Sazanova et al.[41] published an important study of cationic dyes in which the anion was systematically varied for two of five dyes. For those two dyes, they report that the solubility increases with decreasing radius of the anion in the following order: $ClO_4^- < I^- < CNS^- < Cl^- <$ propionate $<$ acetate $<$ formate. They also noted that solubility was greatest for the cation with the most electronegative atoms and the smallest size.

Cationic dyes are used in the colorimetric analysis of boron as the tetrafluoroborate salt, a species which can be extracted into an organic solvent. To aid in such analyses, Buldini[42] measured the solubilities of 19 dyes, at 20–25 °C, and the stability constants of their tetrafluoroborate salts. None of these dyes is currently used for textiles.

In an elegant and detailed study of fluorescein, Diehl and Markuszewski[43] determined the intrinsic solubility (i.e. solubility of the neutral acid) of the red and yellow forms. They then measured the pH dependence of the solubilities to determine the protonation constants of each form.

Except for stains, few solubilities have been reported for the metal salts of anionic dyes, and in most cases, they are for the alkali metal salts. These are the most common dye salts because of their high solubilities. In the case of other metals, only three reports have been found. The first of these is Holmes' work on stains, as previously discussed.[33] Ogawa and Takase[44] reported graphical data on the solubility and activity coefficients for the sodium and barium salts of C.I. Acid Red 88. Their results show, as did the data of Holmes[33] in Table 5.2, the expected lower solubility of the Ba salt.

To assess the likelihood of precipitation in natural waters, Hou and Baughman[45] examined the solubility of the calcium salts of 52 acid and direct dyes. They also measured the solubility of five of the salts (four acid and one direct) by two independent methods. Both of the latter two works[44,45] deal with solubility as the solubility product constant. All of the dye salt solubilities obtained in our review are presented in Table 5.2.

5.3.2 Solubility in organic solvents

Many dyes are more soluble in organic solvents than in water, which should make their solubility easier to measure. However, the solubilities of relatively few dyes have been reported in organic solvents. Not surprisingly, most of the dyes studied are disperse or solvent dyes, i.e. the most soluble non-ionic dyes. These studies have been prompted primarily by interest in solvent dyeing and fastness to dry cleaning.

There is also a large body of data on the solubility of stains. Holmes measured many solubilities in 95% ethanol (EtOH)[32,33] and his data are included in the compilation edited by Lillie.[35] The latter also contains data with respect to a number of other solvents. For the reasons noted above, in the introduction to the section on Water solubilities, most of these data are not included here.

Belkin and Karpov[46] measured the solubility of a series of amino and aminohydroxy anthraquinones in perchloroethylene (PER) at 121 °C. Their data illustrate the importance of intramolecular hydrogen bonding in increasing the solubility. They also conclude that the dyes behave as ideal solutions in PER at 121 °C, since there is a log–linear relationship between solubility and the reciprocal of absolute temperature of the melting point.

Baranova and Kuznetsova[47] measured PER solubilities of alkanoylamino azobenzene and alkanoylamino azonaphthyl dyes at 30–90 °C. Their results show that the presence of a naphthalene ring in place of a benzene ring decreases solubility, but that the presence of longer alkyl chains increases it.

A number of workers have measured the solubilities of dyes in several solvents. Berezin et al.[48] measured the solubility of eight disperse dyes (six anthraquinone, one amino and one azo) at 23–70 °C. In all cases, dioxane (DIOX) was the best solvent. The other solvents used were PER, trichlorethylene (TRIC), 1,1-dichloroethane (DCE), methylene chloride (MC), and ethylene glycol (EG). They found TRIC to be a better solvent than PER, but all of the other solvents used were better than either of these.

Veller and Porai-Koshits[49] measured the 'room temperature' solubility of several disperse or solvent dyes and one acid dye. The measurements were made spectrophotometrically in ethanol (EtOH), acetone (ACE), DCE, benzene (BZ) and dibutylphthalate (DBP). In view of the lack of temperature control, their results must be viewed as approximate.

The water solubility study of metal complex disperse dyes[14] noted earlier was also accompanied by measurements in PER and TRIC. Graphical data for these systems also show that some of the dyes have solubilities which varied with the amount of dye added. The authors found that TRIC was a better solvent than PER, but that water was the best solvent.

Except for some thioindigo and related compounds in chlorobenzene,[24] Lunyka and Artim[50] have provided the only other data on vat dye solubilities. They measured the solubility of C.I. Vat Green 1 in diethylamine (DEA), BZ, ethyl cellosolve (ETCE), monoethylamine (MEM), cyclohexylamine (CHAM), ethanolamine (EtAM), DIOX, and

Table 5.2 Anionic dye salt solubilities

Dye	Name	Salt	Solubility (g/kg)	Solvent	T (°C)	C.I. No.	CAS No.
39[33]	Fluorescein (Acid)	DH$_2$	0.3	H$_2$O	26	45350:1	2321-07-5
	C.I. Solvent Yellow 94		22.1	EtOH*	26		
	C.I. Acid Yellow 73	DNa$_2$	502	H$_2$O	26	45350	518-47-8
	Uranin		71.9	EtOH*	26		
		DMg	45.1	H$_2$O	26		
			3.5	EtOH*	26		
		DCa	11.3	H$_2$O	26		
			4.1	EtOH*	26		
		DBa	65.4	H$_2$O	26		
			5.6	EtOH*	26		
47[45]	C.I. Acid Blue 113	DCa	0.14	H$_2$O	25	26360	
48[45]	C.I. Acid Orange 8	D$_2$Ca	0.24	H$_2$O	25	15575	
49[44]	C.I. Acid Red 88	DNa	≈ 4	H$_2$O	35	15620	1658-56-6
		D$_2$Ba	≈ 0.005	H$_2$O	35		
50[45]	C.I. Acid Red 114	DCa	0.015	H$_2$O	25	23635	
51[45]	C.I. Acid Red 151	D$_2$Ca	0.0053	H$_2$O	25	26900	
54[45]	C.I. Direct Yellow 28	DCa	0.21	H$_2$O	25	19555	
125[27]	Reactive Bright Red 6S*	DNa$_3$	123	H$_2$O	25		8005-72-9
120[27]	Reactive Violet 4K*	DNa$_3$	102	H$_2$O	25		70210-46-7
121[49]	Acid Bright Green Zh*	DNa	105	EtOH			56646-12-9
			1.9	ACE			
			0.4	DCE			
170[33]	Eosin (sodium)	DNa$_2$	442	H$_2$O	26	45380	17372-87-1
	C.I. Acid Red 87		21.8	EtOH*	26		
	Eosin Y	DMg	14.3	H$_2$O	26		
			2.8	EtOH*	26		
		DCa	2.4	H$_2$O	26		
			0.9	EtOH*	26		
		DBa	1.8	H$_2$O	26		
			0.6	EtOH*	26		

DYE SOLUBILITY

171[33]	Erythrosin (sodium) C.I. Acid Red 51 C.I. Food Red 14 Erythrosin B	DNa$_2$ DMg DCa DBa	111 18.7 3.8 5.2 1.5 3.5 1.7 0.4	H$_2$O EtOH* H$_2$O EtOH* H$_2$O EtOH* H$_2$O EtOH*	26 26	45430	16423-68-0
172[33]	Phloxine (sodium) C.I. Acid Red 98	DNa$_2$ DMg DCa DBa	509+ 90.2 208.4 291 35.7 4.5 60.1 11.7	H$_2$O EtOH* H$_2$O EtOH* H$_2$O EtOH* H$_2$O EtOH*	26 26 26 26 26 26 26 26	45405	6441-77-6
173[33]	Rose Bengal (sodium) C.I. Acid Red 94	DNa$_2$ DMg DCa DBa	362.5 75.3 4.8 15.9 2.0 0.7 1.7 0.5	H$_2$O EtOH* H$_2$O EtOH* H$_2$O EtOH* H$_2$O EtOH*	26 26 26 26 26 26 26 26	45440	632-69-9

* 95% ethanol.
ACE, acetone; DCE, dichloroethane.

dimethylformamide (DMF). They report that the dye was highly insoluble in methanol, isobutanol, EG, and carbon tetrachloride. It was most soluble in DMF and the majority of their solubilities were measured in DMF at 30 °C.

Two other studies have focused particularly on glycols. Szadowski and Niewiadomski[51] examined the solubilities of a series of mono and disazo disperse dyes in polyethylene glycol (PEG). Nango et al.[52] measured the solubilities of five azo and four anthraquinone compounds in EG at 10–40 °C. These authors also measured the enthalpies of fusion for the dyes. Entropies of fusion calculated from their data are in good agreement with theory.

5.4 Estimation methods

5.4.1 Computational

The UNIFAC procedure[53] is a useful blend of theory and empiricism, and was designed principally to provide activity coefficients of mixtures. However, the procedure also works quite well for determining pure-component solubilities. The UNIFAC (UNIQUAC Functional-group Activity Coefficient) model is derived from the UNIQUAC (Universal Quasi Chemical) equation. UNIFAC is a group contribution method; it considers the solute–solvent mixture as a solution of groups. For example, a solution of benzene in water is considered as a mixture of ACH (aromatic carbon) and H_2O groups. Activity coefficients are computed for the groups, and are then combined to give the activity coefficient of the solute. Solubility (expressed in mole fractions) is the reciprocal of the infinite dilution activity coefficient. In the UNIFAC equation:

$$\ln \gamma_i = \ln \gamma_i^C + \ln \gamma_i^R \qquad (5.3)$$

the activity coefficient γ_i, of a solute is expressed as a sum of a combinatorial part (C) that reflects differences in the size and shape of the molecules in the system, and a residual (R) part that derives from intermolecular interactions. The combinatorial term models the cavity forming process in water and contains terms in solute area and volume. The residual reflects enthalpic effects and requires interaction coefficients for the various groups in the system. The most recent compilation of these coefficients is that of Hansen et al.[54]

The advantage of the group concept is its versatility, since several compounds are accessible using only a few groups. A related advantage is that mixtures of solutes can be handled almost as easily as pure solutes, since calculations are based on the groups rather than on the pure compounds. A limitation is that the group approach ignores the differing molecular environments in which the groups reside. As a result, distinction among isomers is often lost. For example, UNIFAC gives the same result for *o*-dibromo- and *p*-dibromobenzene.

Activity coefficients are easily converted to solubility in molar units by

$$S_w = 55.5 a_{org}/\gamma_w \tag{5.4}$$

where a_{org} and γ_w are the activity and activity coefficient, respectively, of the organic compound in water. In situations where the solubility of water in the organic phase is small, as is the case for hydrophobic compounds, a_{org} is approximately 1.

In their original paper on UNIFAC, Fredenslund *et al*.[53] showed that the calculated water solubilities of C_4–C_8 alkanes, olefins and dienes compared well with measured values. More recently, Arbuckle[55] used UNIFAC to calculate the solubilities of a number of 'priority' pollutants and found that UNIFAC underestimated the solubilities of the larger and more hydrophobic members of the series. This is not really surprising, since the reliability of group contribution techniques tends to decrease as the number of groups increases, and deviations from simple additivity are compounded. Furthermore, available UNIFAC parameters were not derived from solubility, and calculated activity coefficients in the infinite dilution region are particularly sensitive to variations in the interaction parameters.

In order to assess the degree to which UNIFAC underestimates water solubility, Banerjee[56] calculated the activity coefficients for 50 structurally diverse compounds varying in solubility by eight orders of magnitude. In accord with Arbuckle's findings,[55] Banerjee observed that the solubility of compounds having $\log S_w < -2$ is underestimated by UNIFAC. The deviations, however, are systematic, and can be corrected easily by inclusion of a term with the melting point (MP). Regression of the data led to

$$\log S_w = 1.2 + 0.78 \log a_{org}/\gamma_{aq} - 0.01 \text{ (MP-25)} \tag{5.5}$$

$$n = 50; r = 0.98$$

For most of the compounds in the series, a_{org} is approximately 1, and in essence, the coefficient 0.78 corrects for a systematic and cumulative error in γ_w. This type of error is quite common in group additivity studies, where the properties of molecular fragments are usually less than additive.

A present limitation of UNIFAC is that interaction parameters are available for relatively few functional groups. The method was designed for use in chemical engineering, and parameters for industrial chemicals are more likely to be found than those for drugs, pesticides, dyes, etc.

5.4.1.1 Application to dyes. Three values are necessary for a UNIFAC water solubility calculation, i.e. group surface area, group surface volume, and the interaction parameters for the various groups with water. In order to extend UNIFAC to azo dyes, we defined a new group, ($-N{=}N-$). Volume and area parameters of this group were obtained from standard bond lengths and bond angles.[57] Interaction parameters for the azo groups with water were obtained by Monte Carlo methods from the water solubilities of azobenzene and its 4-nitro, 4-amino, 4-hydroxy and 2,4-hydroxy derivatives. These compounds were chosen since apparently reliable solubilities were available.[58] The UNIFAC parameters calculated for the groups are R: 0.510, C: 0.431; interaction parameters: $AZO-H_2O$ (1037), H_2O-AZO (709).

Water solubilities calculated with these parameters and equation (5.5) are compared to experimental values in Table 5.3. The solubilities are for the super-cooled 'liquid' form of the dyes. The starred values in Table 5.3 are for situations where the measured and experimental values deviate by more than a factor of 25. Notably, for all these dyes, the calculated values are higher than the measured solubilities. No obvious explanation is available for the differences for C.I. Solvent Yellow 2 and Disperse Dye. However, for Disperse A5 and Disperse A7, the solubilities were measured at temperatures above 70 °C and extrapolated to 25 °C, a procedure likely to generate substantial error.

Table 5.3 Comparison of UNIFAC and experimental solubility values of supercooled liquid dyes (experimental data from Ref. 29)

	MP (°C)	Solubility (M)	
		meas.	calc.
C.I. Disperse Orange 3	217	3.4E-4	7.4E-4
C.I. Disperse Red 1	167	5.3E-5	1.7E-4
C.I. Disperse Red 5	192	5.2E-5	8.4E-5
C.I. Disperse Red 7	190	1.3E-4	2.5E-4
C.I. Disperse Red 17	160	1.2E-4	4.0E-4
C.I. Disperse Red 19	211	1.9E-4	1.2E-3
C.I. Solvent Yellow 1	125	1.5E-3	1.3E-3
C.I. Solvent Yellow 2	123	2.3E-5	9.1E-4*
C.I. Solvent Yellow 58	134	4.0E-3	1.3E-3
Azobenzene	68	1.2E-4	1.1E-4
DIS A.1	190	1.0E-4	4.8E-5
4-Nitroazobenzene	135	7.8E-5	9.6E-5
4-Chloroazobenzene	88	1.6E-5	2.3E-5
C.I. Disperse Orange 5	127	4.5E-7	9.9E-6
C.I. Disperse Dye	156	1.9E-7	1.0E-5*
C.I. Disperse A2	111	7.6E-5	6.3E-5
C.I. Disperse A5	152	4E-11	7.5E-7*
C.I. Disperse A6	118	2E-8	4.7E-7
C.I. Disperse A7	236	2E-9	5.7E-9*

*Calculated value differs from measured value by more than a factor of 25.

These deviations notwithstanding, the UNIFAC method is promising, since its accuracy compares favorably with other techniques. Also, the interaction coefficients involved were obtained from simple azobenzenes rather than from dyes, and the accuracy of the method should not deteriorate much when applied to other dyes. A negative aspect is that UNIFAC interaction parameters are unavailable for many dye functionalities. In other words, UNIFAC should work quite well where it can be applied, but the range of structures that are presently covered is not extensive.

5.4.2 Solubility parameters

Scatchard[59] and Hildebrand et al.[60] independently developed what is now called regular solution theory. Its basic premise is that the entropy of mixing of the solute and solvent is ideal, but that the heat of mixing is not. The heat of mixing ΔH_{mix} is calculated from a consideration of the energy required to isolate one mole of solute from its pure phase, to create a hole in the solvent large enough to accommodate it, and to insert the solute into the hole. It is related to the heats of solute–solute (H_{2-2}), solvent–solvent (H_{1-1}), and the solute–solvent (H_{1-2}) interactions by

$$\Delta H_{mix} = -H_{1-1} - H_{2-2} + 2H_{1-2} \tag{5.6}$$

If there is no volume change on mixing, equation (5.6) can be written as

$$\Delta E_{mix} = -E_{1-1} - E_{2-2} + 2E_{1-2} \tag{5.7}$$

The cohesive energy of the pure components can be envisaged as the product of the internal pressure, P, and the volume of the hole. Since the latter equals the molecular volume of the solute, V_2, it follows that

$$-E_{1-1} = P_1 V_2 \tag{5.8}$$

$$-E_{2-2} = P_2 V_2 \tag{5.9}$$

The internal pressure, which is frequently called the cohesive energy density, is defined as

$$P_i = \frac{\Delta E_{i,vap}}{V_i} \tag{5.10}$$

where $\Delta E_{i,vap}$ is the energy of vaporization, and V_i is the molecular volume. E_{1-2} cannot be determined experimentally, since the internal pressure and the heat of vaporization can only be measured for a pure compound. For substances that interact only through van der Waals forces, this term can be approximated by the geometric mean rule, i.e.

$$E_{1-2} \approx \sqrt{(E_{1-1} \cdot E_{2-2})} \tag{5.11}$$

or
$$P_{1-2} \approx \sqrt{(P_1 \cdot P_2)} \tag{5.12}$$

Substitution into equation (5.7) gives
$$\Delta E_{mix} \approx (P_1 + P_2 - 2P_{1-2})V_2 \tag{5.13}$$

or
$$\Delta E_{mix} \approx [P_1 + P_2 - 2\sqrt{(P_1 \cdot P_2)}]V_2 \tag{5.14}$$

which can be rearranged to give
$$\Delta E_{mix} \approx (\sqrt{P_1} - \sqrt{P_2})^2 V_2 \tag{5.15}$$

The square root of the cohesive energy density is called the solubility parameter δ, i.e.
$$\delta_i = \sqrt{P_i} \tag{5.16}$$

Combining equations (5.15) and (5.16) gives
$$\Delta E_{mix} \approx (\delta_1 - \delta_2)^2 V_2 \tag{5.17}$$

where δ_1 and δ_2 are the solubility parameters of solvent and solute, respectively.

If the entropy of mixing is zero, then the free energy of mixing is equal to the heat of mixing, i.e.
$$\Delta G_2 \approx V_2(\delta_1 - \delta_2)^2 \tag{5.18}$$

Therefore,
$$-RT \ln x_2 \approx V_2(\delta_1 - \delta_2)^2 \tag{5.19}$$

Since the geometric mean assumption that forms the basis of equation (5.19) does not hold in situations where hydrogen bonding occurs, the equation does not apply to solubility in water. However, there have been many attempts to extend solubility parameter theory to polar and hydrogen bonding systems. The most notable of these is the approach of Hansen,[61] who proposed that the solubility parameter be partitioned into three components reflecting dispersion, dipolar effects, and hydrogen bonding interactions. The relationship between δ_2 and these components is given by
$$\delta_2 = \delta_2^{disp} + \delta_2^{dip} + \delta_2^{hb} \tag{5.20}$$

where the superscripts disp, dip, and hb represent dispersion, dipolar effects, and hydrogen bonding, respectively. The individual components of the solubility parameter are then treated separately in equation (5.19) and the results added.

Hansen's approach offers no theoretical improvement over the original method of Scatchard and Hildebrand because the hydrogen bonding portion does not follow the geometric mean assumption. For example, chloroform and acetone strongly hydrogen bond with each other, but each would be assigned a δ^{hb} value of zero. Also, the strength of the hydrogen bond between water and hexane is much less than the average of the large δ^{hb} value for water and the zero value for hexane.

5.4.2.1 Applications. Solubility parameters have been used extensively in industry, more for rationalizing the transfer of dye to fabric than for estimating solubility *per se*. Transfer improves as the δ value for the dye approaches that of the fabric, since decreasing $\Delta\delta$ (on an absolute basis) signals increasing miscibility or penetration of the dye into the fabric. For example, Chavan and Jain[64] studied the transfer of a number of disperse dyes to polyester during sublimation transfer printing, and found that transfer was best when the δ value for the dye was similar to that of polyester. In another study,[65] the same authors found that dye transfer to ethylene-glycol-treated cotton increased as $\Delta\delta$ between the dye and the glycol decreased.

Solubility parameters for dyes have usually been estimated from structure as discussed earlier, although direct determinations have been made on occasion. For example, Belkin and Karpov[66] obtained δ values for a related family of disperse dyes by thin layer chromatography on silica, after calibration with dyes of known δ. These results are surprising, since solubility parameters do not reflect differences in hydrogen bonding. However, this approach will only be viable for closely related structures, and then only when interpolated within a good calibration curve.

Both the advantages and limitations of solubility parameters can be appreciated from the data of Siddiqui,[67] Tables 5.4 and 5.5, which relate solubility to δ. Solubility is expected to increase as the solubility parameters of solute and solvent converge. A rank-order trend is clear in Table 5.4. Equally clear is that the trend is not quantitative; the solubility of dye **116** in butanol is poor despite the similarity in δ for butanol and the dye. This inconsistency is not surprising, since a low $\Delta\delta$ implies that

Table 5.4 Solubility of dye **116** ($\delta = 12.25$) in various solvents at 21 °C[67]

	δ	Solubility (g/l)
Triethylamine	7.41	0.25
Ethylbenzene	8.80	0.057
Acetone	9.84	5.34
Cyclohexanone	10.0	3.62
Butanol	11.46	0.09
Methanol	14.5	1.37
Water	23.47	0.02

Table 5.5 Solubility of various dyes in water ($\delta = 23.47$) at 21 °C[67]

Dye	δ	Solubility (g/l)
27	12.38	0.001
114	13.19	0.004
115	10.9	0.007
116	12.25	0.002
117	10.70	0.008

the heat of mixing is small and that other factors come into play. The effect of other factors has been emphasized by Biedermann and Datyner.[68] However, these discrepancies are not so much a breakdown in solubility parameter theory as situations where δ cannot be expected to relate to solubility. Since a small $\Delta\delta$ implies a low heat of mixing, solubility will lose its dependence on the heat of mixing and will be governed by other parameters that fall outside the scope of solubility parameter theory.

Table 5.5 lists water solubilities of several dyes and their δ. Here, there is no apparent correlation (for reasons previously discussed) other than the fact that the solubilities are all low since $\Delta\delta$ is large. A similar trend has been observed by Biedermann and Datyner.[68]

In conclusion, solubility parameters are useful for generally classifying solvents for their performance with regard to a given dye. Solubility parameters are relatively easy to compute for dyes, and are known for most industrially important solvents. On the other hand, the approach is of little use for water or for situations where small differences in solubility need to be estimated.

5.4.3 Regression analysis

Hansch et al.[69] first recognized the relationship between aqueous solubility and the mole-fractional octanol-water partition coefficient, $(K(X)_{ow})$. $K(X)_{ow}$ can be expressed as

$$K(X)_{ow} = \frac{\gamma_w^{i,ms}}{\gamma_o^{i,ms}} \qquad (5.21)$$

where the superscripts i and ms indicate infinite dilution and mutually saturated phases, respectively. If the activity coefficient ratio is (a) relatively insensitive to changes in solute concentration, and (b) approximately the same whether the octanol and water phases are pure or mutually saturated, then

$$K(X)_{ow} = \frac{X_o}{X_w} \qquad (5.22)$$

or

$$\log x_w = -\log K(X)_{ow} + \log x_o \qquad (5.23)$$

Since, at room temperature, octanol is miscible with most organic liquids (i.e. $x_o \approx 1.0$), we get

$$\log x_w = -\log K(X)_{ow} \qquad (5.24)$$

which on conversion to molar units gives

$$\log S_w = -\log K_{ow} + 0.8 \qquad (5.25)$$

For crystalline solutes, an additional term with the melting point (MP in °C) is required to compensate for the entropy of fusion. For most mid-molecular weight rigid organic compounds, the entropy of fusion (ΔS_f) can be approximated as 56.5 J mol^{-1} K^{-1}, upon which

$$\log S_w = -0.01 (\text{MP} - 25) - \log K_{ow} + 0.8 \qquad (5.26)$$

results.

Equations (5.25) and (5.26) have been applied successfully to several hundred solutes.[5] Baughman and Weber[21] obtained the following equation from regression analysis of a number of hydrophobic disperse dyes

$$\log S_w = 0.34 - 1.02 \log K_{ow} \ (r = 0.89) \qquad (5.27)$$

Inclusion of a term for MP did not materially improve the relationship.

The regression coefficient is lower than those obtained for other solutes[5] where r typically exceeds 0.96. A possible reason is that equation (5.26) assumes ideal solute behavior in octanol, which is probably erroneous for complex dyes. Also, uncertainty in measuring dye solubilities and partition coefficients is much greater, and dyes can associate or even assume different tautomeric forms, in octanol and in water.[21] As a case in point, we attempted to generate UNIFAC coefficients for the carbonyl group in anthraquinones using the 1-hydroxy, 1-amino, and 1,4-dihydroxy derivatives for which reliable solubility and entropy of fusion data were available. To our surprise, we were unable to generate consistent parameters even for these three simple compounds, and we suspect that they tautomerize to different degrees. If so, then the anthraquinone dyes represent a range of structures varying not only in functionality, but also in the degree to which the carbonyl group enters into tautomerism.

The assumption of a constant ΔS_f also does not apply to dyes. For example, the rigid anthraquinone dyes have much lower ΔS_f values than the azo dyes, which are much more flexible. Also, the MP term in equation (5.8) tends to be linear with log K_{ow}, and both variables usually increase with molecular size.[5] Hence the dependence on MP is frequently masked, and regression leads to the conclusion that MP is unimportant in solubility.

Baughman and Weber[21] also used experimental ΔS_f values in place of a constant entropy of fusion and obtained

$$\log S_w = -1.22 \log K_{ow} - 1.28 \times 10^{-4} \Delta S_f(\text{MP-25}) - 0.04 \quad (5.28)$$

$$(r^2 = 0.91)$$

which is, perhaps, the best equation available for estimating dye–water solubility. Unfortunately, ΔS_f values are sparse, and as a result, equation (5.28) is of limited practical use. For rigid anthraquinone type structures, an average ΔS_f of 66 J mol^{-1} K^{-1} may be assumed, and for more flexible structures, an average ΔS_f of 78 J mol^{-1} K^{-1} has been suggested.[21] It should be noted that ΔS_f for some of the large azo dyes is greater than 100 J mol^{-1} K^{-1}.

5.4.4 Empirical/estimation

At least two experimental methods are available for estimating the solubility of highly insoluble compounds. The first is based on work of Morris et al.[70] who utilized the fact that the logarithm of solubility varies approximately linearly with the volume fraction of co-solvent. The water solubility is obtained by extrapolation to 100% water. Baughman and Weber[21] used this method to estimate the water solubility of C.I. Solvent Green 3, which was too insoluble for measurement by the generator column method vide ante. Although there is little basis for assessing the accuracy of this method, the data of Primak et al.[39] provide an interesting comparison. Regression of their data for the solubility of five dyes in acetonitrile–water gives the results shown in Table 5.6. In this case, the estimates agree with the measured values to within a factor of two to three.

A second method makes use of the log-linear relationship between solubility and reverse-phase HPLC capacity factors. This method was applied by Hou et al.[37] to the estimation of the solubilities of several commercial dyes. The calibration curve had a regression coefficient of 0.92, but the root mean square deviation was 0.63 log units, or about a

Table 5.6 Water solubilities of disperse dyes estimated from acetonitrile–water solubilities at 40 °C[39]

Dye	Measured (mg/l)	Estimated (mg/l)
6	41	18
14	–	1.3
61	0.8	0.37
126	–	0.45
127	19.0	14

factor of four. Use of the sub-cooled liquid solubility only resulted in a slight improvement.

Although the above methods can be used to estimate the water solubility of dyes, the quality of the estimates is less than desired. Computational methods usually ignore several features important to dyes, whereas the calibration techniques tend to average them across the selected dyes. For example, dyes have variable entropies of fusion, which adds to the uncertainty if ΔS_f is unavailable. Also, dyes are prone to association, polymorphism and tautomerization, all of which are usually neglected, but are expected to greatly influence solubility.[21] Thus, the uncertainty in solubility estimates for dyes is likely to remain greater than that for simpler compounds.

Acknowledgment

The authors acknowledge the partial support of this work under Cooperative Agreement CC812638 from the United States Environmental Protection Agency.

References

1. *Chemical Abstracts*, American Chemical Society, Washington, DC.
2. *Chemisches Zentralblatt*, Akademie, Berlin.
3. *Colour Index*, Society of Dyers and Colourists, 3rd ed., 1976, Yorkshire, UK.
4. Stepanov, B. I., *Introduction to the Chemistry and Technology of Organic Dyes*, 2nd ed., Chimia, Moscow, 1977.
5. Yalkowsky, S. H. and Banerjee, S., *Aqueous Solubility: Methods of Estimation for Organic Compounds*, Dekker, NY, 1992.
6. Goldthwait, C. and Kirby, R., *Am. Dyest. Rep.*, 1966, **55**, 625.
7. Bailey, J., *J. Chromatogr.*, 1984, **314**, 379.
8. Giles, C. and Greczek, J., *Text. Res. J.*, 1962, **32**, 506.
9. Clemens, H. J., Toepfer, K., *Acta Histochem.*, 1968, **31**, 117.
10. Schmidt, T., *Acta Histochem.*, 1970, **38**, 250.
11. Duff, D. G. and Giles, C. H., in *Water: A Comprehensive Treatise*, Vol. 4, Franks, F. ed., Plenum Press, 1974.
12. Ostwald, W. and Walter, R., *Kolloid Z.*, 1936, **76**, 291.
13. Mel'nikov, B. N. and Moryganov, P. V., *Kolloidn. Zh.*, 1955, **17**, 99.
14. Starikovich, E. E. and Kushko, A. A., *Izv. Vyssh. Uchebn. Zaved.*, 1973, **6**, 100.
15. Braun, H., *Rev. Prog. Coloration. Rel. Top.*, 1983, **13**, 62.
16. Biedermann, W., *J. Soc. Dyers Colour.*, 1971, **87**, 105.
17. Shenai, V. A. and Sadhu, M. C., *J. Appl. Polym. Sci.*, 1976, **20**, 3141.
18. Odvárka, J., Schejbalová, H. and Gärtner, F., *J. Soc. Dyers Colour.*, 1980, **96**, 410.
19. Krasnoperova, A. P. and Chernyy, Y. Y., *Izn. Vgssh. Uchebn. Zaved, Khim. Khim. Teknovl.*, 1974, **17**, 298.
20. Yen, C-P. C., Perenich, T. A. and Baughman, G. L., *Environ. Toxicol. Chem.*, 1989, **8**, 981.
21. Baughman, G. L. and Weber, E. J., *Dyes, Pigm.*, 1991, **16**, 261.
22. Popov, R., Decheva, R. and Milkova, A., *(Vyssh. Khim. Tekn. Inst. Sofia, Bulg.) Tekst Prom. Sofia*, 1980, **29(2)**, 70.

23. Datyner, A., *J. Colloid Interface Sci.*, 1978, **65**, 527.
24. Georgieva, E. B. and Mostoslavskii, 1984, Deposited Doc. VINITI 2409-84. (C.A.120:150909S)
25. Přikryl, J., Ružička, J. and Burgert, L., *J. Soc. Dyers Colour.*, 1979, **95**, 349.
26. Sah, R. E. *Mol. Cryst. Liq. Cryst.*, 1985, **129**, 315.
27. Prusov, A. N., Zaharov, A. G. and Kochergin, A. B., *Izv. Vyssh. Uchebn. Zaved., Khim. Khim. Tekhnol.*, 1991, **34**, 66.
28. Dubinin, M. M. and Astahov, V. A., *Izv. Akad. Hauk. SSSR, Ser. Khim.*, 1971, **1**, 5.
29. Baughman, G. L. and Perenich, T. A., *Environ. Toxicol. Chem.*, 1988, **7**, 183.
30. Bird, C. I., *J. Soc. Dyers Colour.*, 1954, **70**, 68.
31. Holmes, W. C., *Stain Technol.*, 1927, **2**, 40.
32. Holmes, W. C., *Stain Technol.*, 1928, **3**, 12.
33. Holmes, W. C., *Stain Technol.*, 1929, **4**, 73.
34. Gurr, E., *Encyclopedia of Microscopic Stains*, Williams and Wilkins Co., Baltimore, 1960.
35. *H. J. Conn's Biological Stains*, 9th ed., R. D. Lillie ed., Williams and Wilkins Co., Baltimore, 1977.
36. Szadowski, J., Přikryl, J. and Wojciechowski, K., *Chem. Stosow.*, 1987, **31**, 415.
37. Hou, M., Baughman, G. L. and Perenich, T. A., *Dyes Pigm.*, 1991, **16**, 291.
38. Prusova, S. M., Proprokova, N. P., Kalinnikov, Yu. A. and Melnikov, B. N., *Izv. Vyssh. Uchebn. Zaved., Khim. Khim. Teknol.*, 1992, **35**, 48.
39. Primak, L. I., Saribekov, G. S. and Androsov, V. F. *Izv. Vyssh. Uchebn. Zaved., Tekhnol. Tekst. Prom.*, 1976, **(2)**, 89.
40. Starikovich, E. E., Androsov, V. F. and Kushko, A. A., *Izv. Vyssh. Uchebn. Zaved.*, 1973, **5**, 97.
41. Sazanova, A. A., Stavinchek, B. G. and Chalik, E. A., *Iziv. Vyssh. Uchebn. Zaved. Khim. Khim. Tekhnol.*, 1982, **25**, 348.
42. Buldini, P. L., *Anal. Chim. Acta*, 1976, **82**, 187.
43. Diehl, H. and Markuszewski, R., *Talanta*, 1985, **32**, 159.
44. Ogawa, T. and Takase, Y., *Seni Gakkaishi.*, 1967, **23**, 182.
45. Hou, M. and Baughman, G. L., *Dyes, Pigm.*, 1992, **18**, 35.
46. Belkin, A. I. and Karpov, V. V., *Zh. Prikl. Khim.*, 1975, **48**, 160.
47. Baranova, T. A. and Kuznetsova, S. S., *Izv. Vyssh. Uchebn. Zaved.*, 1978, **21**, 1517.
48. Berezin, B. D., Shkrobvisheva, V. I. and Melnikov, B. H., *Trudy Ivan. Khim-Tekhnol.*, 1974, **17**, 18.
49. Veller, E. A. and Poraï-Koshits, B. A., *Zh. Prikl. Khim.*, 1955, **27**, 857; *Chemi. Abstr.*, **49**, 16437h.
50. Lunyaka, K. V. and Artim, M. I., *Izn. Vyssh. Uchebn. Zaved., Tekhnol, Tekst. Prom.* 1982, **6**, 60.
51. Szadowski, J. and Niewiadomski, Z., *Przegl. Wlok.*, 1988, **42**, 436; *Chemi. Abstr.*, **110**, 136859J.
52. Nango, M., Katayama, A. and Kuroki, N., *Sen'i Gakkaishi*, 1980, **36**, T228.
53. Fredenslund, A., Jones, R. L. and Prausnitz, J. M., *AIChE J.*, 1975, **21**, 1086.
54. Hansen, H. K., Rasmussen, P., Fredenslund, Aa, Schiller, M. and Gmehling, J., *Ind. Eng. Chem. Res.*, 1991, **30**, 2352.
55. Arbuckle, W. B., *Environ. Sci. Technol.*, 1986, **20**, 1060.
56. Banerjee, S., *Environ. Sci. Technol.*, 1985, **19**, 369.
57. Bondi, A., *Physical Properties of Molecular Crystals, Liquids, and Glasses*, 1968, Wiley, New York.
58. Yalkowsky, S. H., *The AQUASOL Database of Water Solubility*, University of Arizona, 1991, Tucson, AZ.
59. Scatchard, G., *Chem. Rev.*, 1931, **8**, 321.
60. Hildebrand, J. H., Prausnitz, J. M. and Scott, R. L., *Regular and Related Solutions*, 1970, Van Nostrand, New York.
61. Hansen, C. M., *J. Paint Technol.*, 1967, **39**, 104.
62. Hansen, C. M., *Ind. Eng. Chem. Process Des. Dev.*, 1969, **8**, 2.
63. Hansen, C. M., and Beerbower, A., Solubility Parameters in *Kirk Otmer Encyclopedia of Chemical Technology*, Vol. 2, Suppl., Wiley, New York, 1971, 889–910.

64. Chavan, R. B. and Jain, A. K., *J. Soc. Dyers Colour.*, 1989, **105**.
65. Chavan, R. B. and Jain, A. K., *Am. Dyest. Rep.*, 1988, **77**, 29.
66. Belkin, A. I. and Karpov, V. V., *Zh. Prikl. Khim.*, 1982, **55**, 877.
67. Siddiqui, S. A., *Text. Res. J.*, 1981, **51**, 527.
68. Biedermann, W. and Datyner, A., *Text. Res. J.*, 1991, **61**, 637.
69. Hansch, C., Quinlan, J. E. and Lawrence, G. L., *J. Org. Chem.*, 1968, **33**, 347.
70. Morris, K. R., Abramowitz, R., Pinal, R., Davis, P. and Yalkowsky, S. H., *Chemosphere*, 1988, **17**, 285.

6 The photodegradation of synthetic colorants
N. KURAMOTO

6.1 Introduction

The action of light on dyes, whether in solution or on textile materials, involves a complex series of processes. Although it is over 100 years since Bancroft[1] reported that the light-induced fading of dyes proceeds by oxidation at times and by reduction at other times, the photodegradation of synthetic dyes continues to be a commercially important problem and the subject of much research.

Many published papers indicate that the light stability of dyes depends not only on (a) various structural factors but also (b) the physical state of the absorbed dye, including the size and location of associated particles, (c) the nature of the substrate, (d) the extent of penetration of the dye and its concentration, (e) the ease of diffusion of air, moisture, and gasses, (f) the energy of the incident radiation, and (g) the relative ease with which transient reactive species are formed upon absorption of light.[2-4] Earlier reviews[2,3,5-8,10-13,16,17] and books[4,9,14,15] on this subject provide useful background information on the basic principles of the photodegradation of dyes.

The absorption of light by organic dyes, as mentioned above, is followed by various chemical and physical interactions which result from the dye molecules being promoted to an excited state which is more reactive than the ground state. The excited state results from the promotion of an electron from the HOMO (Highest Occupied Molecular Orbital) to the LUMO (Lowest Unoccupied Molecular Orbital) level. The excitation energy is released in a very short time when the promoted electron returns to the ground state. Under certain environments the energy is dissipated as heat because of radiationless transitions between the energy states. There is also the possibility of release of the excitation energy as fluorescence or phosphorescence. The excitation energy can also be converted into chemical energy by dissociation, intramolecular rearrangement, redox processes, or other photochemical reactions. In addition, the absorbed energy may be transferred to another molecule. In such a transfer process, the absorbing molecule acts as a photosensitizer or photocatalyst. The sensitization effect of organic dyes invokes many important photoreactions, e.g. the photographic processes, photosynthesis in plants, visual perception by the human eye, the self-sensitized photofading of dyes, and the dye-sensitized phototendering of textiles. It is for these reasons that the photochemistry of dyes is of great interest and has continued to be a field of intense research activity in the last decade.

This chapter reports studies involving the photodegradation of synthetic dyes and pigments published since the earlier reviews by Giles et al.,[11,15] Egerton et al.,[7,8] and Allen.[13,14]

6.2 Basic photochemical principles

6.2.1 Excited electronic states and photochemistry

In general, a dye molecule can be regarded as existing in either the ground state or an excited state before it reacts chemically. The excited dye, in either the singlet or the triplet state, can undergo several reactions.

When a molecule absorbs a quantum of light, it is raised from a low energy level to a higher level by displacement of an electron. Because of the relationship between this quantized electronic transition and quantized rotational and vibrational motions, the absorbed energy is divided between different degrees of freedom. The magnitude of the energy necessary for displacing electrons from an occupied orbital to a higher orbital is larger than the magnitude of vibrational or rotational energy. Therefore, in general, transitions within a molecule can be illustrated by a simple energy-level diagram (see Figure 6.1) in which only the vibration levels are outlined without any indication of rotational levels. Such diagrams are convenient for describing intramolecular transitions, because higher vibrational and rotational levels of an electronic state are deactivated very rapidly by internal conversion to lowest levels.[18,19]

Since the ground state is the singlet state (S_0), in which the spins of the electrons are paired, for most organic molecules the excited states directly attained from the absorption of a quantum of light (photons) are also singlet states S_1, S_2 .... Molecules excited to a higher state (S_2, S_3 ...) usually fall to S_1 very rapidly (in $\sim 10^{-12}$ s), dissipating the excess energy to the surrounding environment, especially in condensed systems.

In radiationless transitions, which can occur as intra- and intermolecular processes, an electronic state is converted to another state without the absorption or emission of light. Such transitions include the internal conversion mentioned above, in which there is an intramolecular radiationless transition between states of like multiplicity, and also intersystem crossing involving intermolecular transition between the singlet and triplet state. These transitions occur isoenergetically because of the

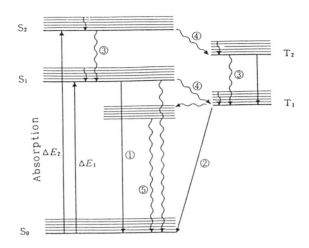

① Fluorescence ② Phosphorescence ③ Internal conversion
④ Intersystem crossing ⑤ Vibrational relaxation

────▶ Radiative transition ∼∼▶ Radiationless transition
≡≡≡ Vibrational level ──── Energy level

Figure 6.1 Excited states and transitions between these states in a typical organic molecule (Jablonski diagram).

necessary crossing of the potential energy surfaces of the different states. This crossing can be illustrated using the potential energy diagram of a diatomic molecule (Figure 6.2). This diagram approximates the multi-dimensional surface which exactly represents the potential energy of a polyatomic molecule as a function of its dynamic configuration. In addition to these intramolecular transitions, intermolecular radiationless energy transitions can be of great relevance in photochemical reactions. In these processes, the energy of an excited state is transferred to another molecule. For example, catalytic fading and sensitized phototendering may be the result of such a reaction.

The transition ($S_0 \to T^*$) between singlet and triplet states is spin-forbidden, and the energy absorption for this transition is too weak to be observed under normal condition. It is, therefore, not considered to be important from the photochemical viewpoint. For photochemical reactions, triplet states produced by intersystem crossing via an indirect excitation process from the singlet states are very important. The order of the rate constant for the intersystem crossing from the lowest excited singlet state to the triplet state greatly influences the photoreactivity of an excited molecule.

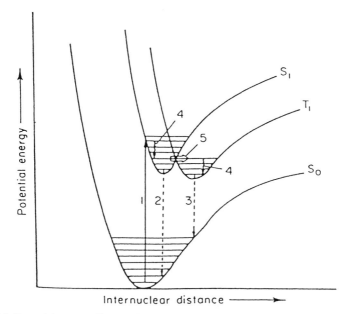

Figure 6.2 Potential energy diagram for a diatomic molecule. 1, excitation; 2, fluorescence; 3, phosphorescence; 4, internal conversion of the vibrational of S_1, T_1; 5, isoenergetic intersystem crossing.

In radiative transitions a molecule emits light on returning to the ground state. Fluorescence is observed by deactivation of an excited singlet state ($S_1 \rightarrow S_0$) and phosphorescence by a radiative triplet–singlet ($T_1 \rightarrow S_0$) transition. Thus, fluorescence is the emission accompanying the transition between levels of the same multiplicity, and phosphorescence is due to the transition from the triplet to the ground state. Because of the difference in the radiative lifetimes between singlet and triplet states, which vary from 10^{-9} to 10^{-6} s for the singlet state and from 10^{-3} to 10 s for the triplet state, fluorescence has a very short lifetime compared with phosphorescence emission. Phosphorescence spectra are observed at longer wavelengths than fluorescence.[19–22]

6.2.2 General photochemical reactions of dyes

The stability of colorants to light in both solution and substrates encompasses a number of problems of technological interest and importance. With more recent developments in photochemistry and the associated instrumentation, it is becoming possible to better understand the nature of the complex chemical processes involved.[2–9,23]

With regard to photosensitization in solutions, Koizumi et al.[21] have summarized the primary processes involved in the most important photo-redox reactions (Scheme 6.1). The light absorbing species is denoted as S. In some cases the singlet excited states participate in the reactions, but they are less important than the triplet states. The D-R_e and D-O_x mechanisms indicate that the reaction is initiated with interaction of the excited dye with a reductant (Re) and an oxidant (Ox), respectively. Processes depicted on the right and left sides in Scheme 6.1 are those in the deaerated and the aerated solutions respectively. Two types of reactions are classified, according to whether the reaction involves radical species (Type I) or not (Type II). Of particular importance in Type II reactions is that such reactions involve the singlet oxygen species produced by energy transfer.

In deaerated solutions, when only a reducing or oxidizing agent is present, either the D-R_e or the D-O_x mechanism is operative. But, in the presence of both reagents, the reaction is initiated according to a D-O_x and/or D-R_e mechanism, depending on the nature of the reagents and also on the experimental conditions (Type I reactions).[2,3,11,13,16,24-31]

On the other hand, in aerated solutions the reaction is initiated by the interaction of the triplet dye and oxygen, provided no reducing agent is present. For many dyes the main reaction is energy transfer from the triplet dye to oxygen, generating singlet oxygen,[2-12,15,21,23,32-39] i.e. Type II reaction $[^3D^* + {}^3O_2 \rightarrow {}^1D + {}^1O_2(^1\Delta g)]$. However the electron transfer reaction $[^3D^* + {}^3O_2 \rightarrow D^{+\cdot} + O_2^{-\cdot}]$ also occurs although to a much lower extent.[2-12,16,21,40-41] When a reducing agent is present, or the solvent acts as a reductant, then the initial process which occurs is a D-R_e of Type I or a D-O of Type II mechanism, depending on the experimental conditions.

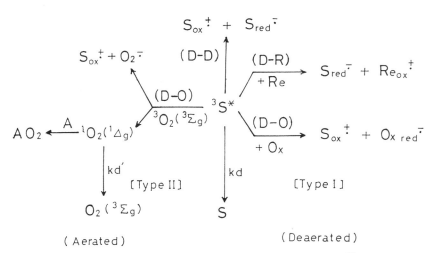

Scheme 6.1 Primary processes in photo-redox reactions.[21]

A D-D mechanism is also possible in aerated solutions when the dye concentration is high.[3,21]

6.2.3 Photosensitized oxidation

The probability of Type II photosensitized oxidation by singlet oxygen needs to be generally taken into account in aerated conditions. This mechanism was for many years not taken into consideration until Kautsky[33,42,43] proposed that singlet oxygen must be involved as an energy carrier in the photosensitized oxidation of a substrate. Egerton[36] also suggested that this oxidizing intermediate was responsible for the photosensitized degradation of dyed textile fibers in a dry oxygen atmosphere. In 1969, Foote et al.[44,45] showed that the concept of singlet oxygen could no longer be ignored in studies of photosensitised oxidations. Foot and Wexer[44,46,48,49] showed that the oxidation of olefinic dienoid and aromatic compounds with singlet oxygen generated by sodium hypochlorite and hydrogen peroxide[44,45,47–49] resulted in the same products as those formed by dye-sensitized photo-oxidation. Corey et al.[50] and others[51–54] made a somewhat similar study to that of Foot et al., using singlet oxygen produced by a radiofrequency discharge method.

$$D + h\nu \rightarrow {}^1D^* \rightarrow {}^3D^* \tag{6.1}$$

$$^3D^* + {}^3O_2 \rightarrow D + {}^1O_2 \tag{6.2}$$

$$^1O_2 + A \rightarrow AO_2 \tag{6.3}$$

Both $^1\Sigma_g^+$ and $^1\Delta_g$ states of oxygen are usually produced, and have been identified by their characteristic emission;[49,55–57] in addition, the $^1\Delta_g$ state gives a characteristic ESR signal in the vapor phase;[57,58] the two metastable singlets differ in the electronic configuration of their degenerate highest occupied (antibonding) orbitals. The $^1\Delta_g$ state has both electrons in one orbital and the other vacant, whereas the $^1\Sigma_g^+$ state has one electron in each orbital.[51,59] This $^1\Delta_g$ state has an energy of 22 kcal above the ground state, a longer lifetime than the $^1\Sigma_g^+$ state, an electrophilic character, and the possibility of two-electron reactions because of a vacant orbital. The $^1\Sigma_g^+$ state (37.5 kcal above the ground state) should resemble the ground state and undergo radical-like reactions (see Table 6.1).[60] The $^1\Delta_g$ state is relatively long-lived and can survive more than 10^8 collisions in the gas phase,[61] whereas the $^1\Sigma_g^+$ state is much shorter-lived.[55,57] Singlet oxygen ($^1\Sigma_g^+$) is more rapidly quenched than $O_2(^1\Delta_g)$ by gases such as H_2O, MeOH, CO_2 and $O_2(^3\Sigma_g^-$; ground state).[57,62,63] The high efficiency of deactivation of the $^1\Sigma_g^+$ state may be attributed to the fact that $^1\Sigma_g^+$ oxygen is deactivated initially to $^1\Delta_g$ oxygen, a spin-allowed process, whereas $^1\Delta_g$ oxygen must undergo a spin-forbidden transition to reach the ground state.[57–59,64]

Table 6.1 Electronic states of the oxygen molecule

Oxygen molecule	Configuration of electrons in highest occupied orbitals	Energy above ground state
Second excited ($^1\Sigma_g^+$)	↑ ↓	37.5 kcal/mol
First excited ($^1\Delta_g$)	↑↓ —	22.5 kcal/mol
Ground state ($^3\Sigma_g^-$)	↑ ↑	

The relative lifetimes of singlet oxygen ($^1\Delta_g O_2$) in solvents are markedly influenced by the solvent employed.[55,57,65,67] Solvents such as carbon tetrachloride, carbon disulfide, chloroform and acetonitrile are particularly favorable.[65–67] Furthermore, the lifetime is thought to be about ten times longer in both D_2O and in a 1:1 mixture of D_2O–CD_3OD than in the corresponding non-deuterated solvents.[54,66,69] It has been suggested that such deuterium effects can provide a simple diagnostic test for singlet oxygen.[54,69]

The dye-sensitized photo-oxidation by the singlet oxygen mechanism is of interest with respect to the following reactions: the synthesis of endo-peroxides (1,4-addition, in a reaction analogous to Diels–Alder addition) from cyclic dienes and conjugated *cis*-dienes, allylic hydroperoxides ('ene' reaction) from olefins having allylic hydrogen, 1,2-dioxetanes from electron rich olefins, and sulfoxides from sulfides.[21,46,55,58] The products of reactions which give several oxidation products have been shown to be the same as those formed in chemical oxidations using hypochlorite and hydrogen peroxide.[45,48] From further detailed kinetic studies of stereoselectivity with different acceptors and measurements, Foot concluded that a reactive intermediate in both dye-photosensitized and chemical oxidation is $O_2(^1\Delta_g)$.[46]

6.3 Physical factors affecting dye lightfastness

It is well established that the photostability of dyed materials is influenced by several external factors, such as light source, humidity, atmosphere, temperature and aggregation, as well as by the chemical structure both of the dye and of the substrate.[2–4,9,11,15,17]

Photofading induced by light of different wavelengths can be anticipated to a certain extent. For example, not only excitation of the dye in the visible region, but also excitation of substances of the fibers by UV light may be effective in fading. McLaren[70,71] reported relationships between the effect of light of different wavelengths and the lightfastness of the dye on the substrate. It is well known that the photofading of Congo Red on cellophane films occurs rapidly with 350 nm light and results from transitions in the naphthalene nucleus.[72] The ISO (blue wool) standard lightfastness test has been shown to give excellent correlations with both Xenon arc lamps and outdoor tests.[73-78] The lightfastness properties of dyes and pigments have also been examined using color difference measurements,[79] and Xenotest irradiation sources have been concluded to give the best results for measuring lightfastness.[17] Xenotest sources are the preferred system for industrial evaluations.

Many investigations on the effect of atmospheric components on the fading process have been carried out.[70,71,80-85] Thus, McLaren et al.[70,71,80,81] showed that the fading rates of many dyes were accelerated by the presence of moisture; for example, azoic and vat dyes show a decrease in fastness with rise in humidity.[83,85] Although the production of H_2O_2 is possible in a moist oxygen environment, the explanation of the humidity effect is not totally understood. Earlier Giles et al.[86] and Kamel et al.[87] showed that monoazo disperse dyes on polypropylene and nylon fade by an oxidative process, but in some cases the fading of the same azo dyes involves reduction, depending on the nature of the substrate.[24,88,89] Other important factors pertinent to the influence of atmospheric components are the influence of singlet oxygen on dye fading[10,38,90-98] and phototendering via a dye-sensitized process;[99,102] these topics are discussed in later sections. In some cases, however, oxygen has an inhibitory effect on fading. For example, Diazacyanine Blue shows a higher fastness in the presence of oxygen than in nitrogen on cellulose or protein fibers.[103]

It is a well-known fact that the lightfastness of a dye on textile fibers increases with the dye concentration, and this has been related to dye aggregation in the fiber.[2,3,9,82,104-107] It has also been shown that, in many cases, the lightfastness of a dye on a textile increases with increase in the depth of shade.[108] Giles et al.[109] first demonstrated that these observations could be used for constructing 'characteristic fastness grade' (CFG) or 'characteristic fading' (CF) curves which give information about the physical state of the dye in dye-substrate systems. The CFG curve is a linear plot of the average lightfastness rating of the dyes and the logarithm of the original dye concentration in the fiber.[109,110] Instead of the lightfastness rating, the time (t_F) required for a given percentage fade can be used to give the CF curve. It is assumed that fading occurs only at that part of the surface of the dye particles that is exposed to air, light and moisture, with the rate of fading dependent on the extent to which this surface is exposed. Usually, the CF or CFG curve has a positive slope with water-soluble dyes, because either the proportion of larger particles increases with rise in concentration, or decreases the accessibility of their surfaces.[86,109,110] However, this is not always the case, and it has been found, for example, that some hydrophobic fluorescent brightening agents on polyamide and polyester films have a negative slope.[111] Measurements of CFG or CF curves for various dyed fibers have been extensively reported.[2,86,109-112]

6.4 Mechanisms of dye photodegradation

The photodegradation of dyes has been extensively studied for several decades,[2,16,82,113–117] but, much of this work has been qualitative in nature and related to the visual behavior of the color stability of various fabrics as a function of experimental parameters such as atmospheric components, humidity and temperature. While certain conclusions about the mechanism of dye photodegradation have been derived from these investigations, a limited use can be made of them. A number of investigations have since been carried out on the mechanism of dye photodegradation based on specific chemical reactions.[10,38,90–95,118–128] The results of such experimentation is important, because the prevention of photodecomposition of dyes in various substrates is difficult to control without some understanding of the degradation mechanisms involved. Data concerning the photodegradation mechanisms of dyes of various chemical classes are now reviewed.

6.4.1 Azo dyes

6.4.1.1 Reductive fading. Apart from the possibility of photoisomerization, azo dyes will either undergo photooxidation or photoreduction reactions[129–131] depending to a large extent on the nature of the substrate and the surrounding atmosphere. Azo dyes in anaerobic solution can be reduced to the corresponding amines by hydrogen donors produced on photo-excitation of suitable simple molecules. The reductive path shown in Scheme 6.2 will vary from one dye type to another, but, essentially, reduction gives a hydrazo compound, and disproportionation eventually gives substituted anilines and the original dye. The reductive process is more predominant in reductive environments typified by proteins such as wool and silk.[132] The photodecomposition of azo dyes involving reductive cleavage of the azo group under anaerobic conditions has been reported by many workers.[24,25,29,30,132,133]

$$Ar_1-N=N-Ar_2 + H\cdot \rightarrow Ar_1-\underset{|}{N}-\dot{N}-Ar_2 \quad (6.4)$$

where the product has H on the first N.

$$2Ar_1-\underset{H}{\overset{|}{N}}-\dot{N}-Ar_2 \rightarrow Ar_1-\underset{H}{\overset{|}{N}}-\underset{H}{\overset{|}{N}}-Ar_2 + Ar_1-N=N-Ar_2 \quad (6.5)$$

$$2Ar_1-\underset{H}{\overset{|}{N}}-\underset{H}{\overset{|}{N}}-Ar_2 \rightarrow Ar_1-N=N-Ar_2 + Ar_1-NH_2 + Ar_2-NH_2 \quad (6.6)$$

Scheme 6.2 Photoreduction mechanism for an azo dye.

The photodecomposition of an azo dye is also possible by excitation of compounds which are present together with the dye molecules in the system. In such photochemical processes the non-dye molecules, (X) react via their excited state with dye molecules according to the following scheme:

$$X \xrightarrow{h\nu} X^* \tag{6.7}$$

$$X^* + \text{Dye} \rightarrow X' + \text{product} \tag{6.8}$$

The reactive intermediates, X^*, which react in this way with dyes are hydrogen donors, and tend to reduce dye bonds. As an example, the conversion of an azo dye into the corresponding amine by mandelic acid is shown in Scheme 6.3. Van Beek et al.[24,25] studied some azo acid dyes, such as **1**, in aqueous and ethanolic solutions using both flash-photolytic and rapid-flow techniques. With D,L-mandelic acid (**2**) as an additive, they showed that the reaction takes place in several distinct stages. Similar reductive fading of azo dyes has been investigated using a variety of conditions.[28,29,134,135]

More recently, Freeman et al.[136] have reported on photodegradation, using a 254-nm light source or a Xenon arc light of C.I. Acid Orange 60 (**3**) in dimethylformamide (DMF) and in nylon 66 fiber. From mass spectrometric and chromatographic analyses of the degradation products, they showed that the photodecomposition takes place via a reductive process involving hydrogen atom abstraction from the solvent and polymer, producing mainly the starting compound, 1-hydroxy-2-aminobenzenesulfonamide, together with a smaller amount of a phenylpyrazolone derivative.

6.4.1.2 Oxidative fading.

In recent years, the oxidative photofading of azo dyes has been attributed to the active participation of singlet oxygen. Griffiths and Hawkins[90,91] proposed a mechanism in which singlet oxygen

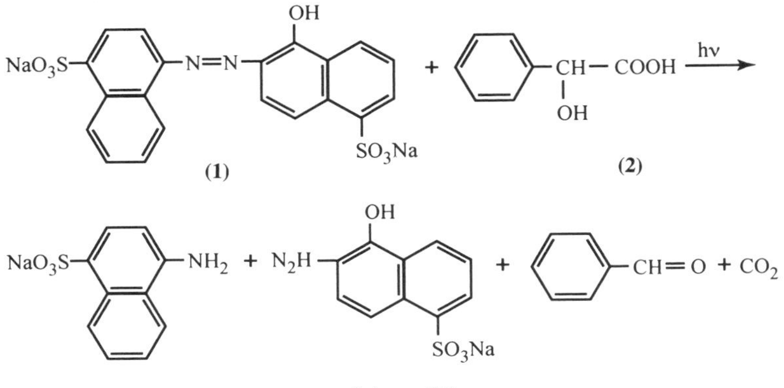

Scheme 6.3

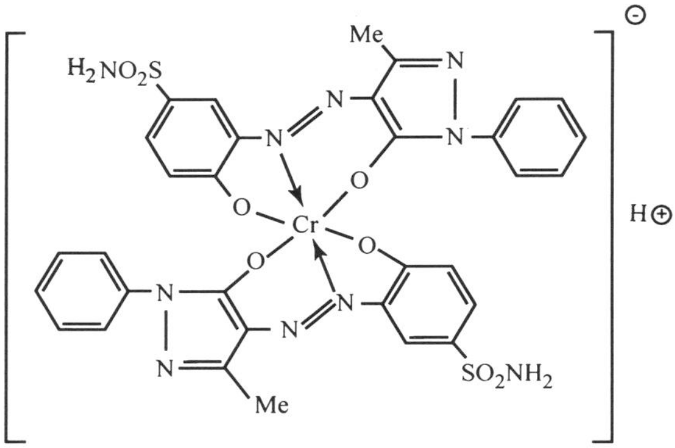

(3) C.I. Acid Orange 60

attacks the hydrazone form of 4-arylazo-1-naphthol dyes (**4**) in solution, to yield the unstable peroxide **6**, which then decomposes to produce the arenediazonium ion and 1,4-naphthoquinone **7**, as shown in Scheme 6.4. They showed that singlet oxygen was a key intermediate in the fading, which occurred via dye-sensitized photooxidation in solution and in polypropylene films.[90,91,137]

Kuramoto and Kitao[38,93,94,138] have also reported on the photofading behavior and products of 1-arylazo-2-naphthols (**8**), about 55–60% of which occurs via the hydrazone form (**9**) in solution. Thus the photo-oxidation of 1-4'-tolylazo-2-naphthol in methanol containing small amounts of acidic catalyst, gave 4-methylanisole (23%), 4-anisic acid (43%),

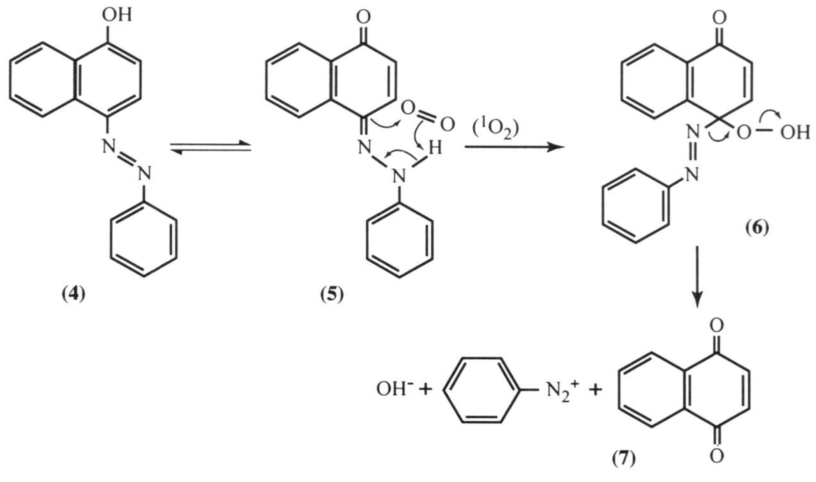

Scheme 6.4

methyl-4-anisate (19%), anisaldehyde (trace), and phthalic acid and its ester as identified products. The action of light on the 4-methylphenyldiazonium salt in methanol gave mainly 4-methylanisole, which was further oxidized to give anisaldehyde or 4-anisic acid.[139] Phthalic acid and its ester were also formed via the secondary reaction of 1,2-naphthoquinone (**12**), which is unstable towards the radiation employed (Scheme 6.5).

The rates of the photooxidation of 1-arylazo-2-naphthols were accelerated in the presence of sensitizers of singlet oxygen, e.g. Methylene Blue or Rose Bengal. On the other hand, the rates were retarded by 1,4-diazabicyclo[2,2,2]octane (DABCO) or nickel dibutyldithiocarbamate, efficient singlet oxygen quenchers, and over 90% of unreacted dye was recovered. Photofading takes place only slowly in the solutions not containing sensitizers. Thus, it appears that 1-arylazo-2-naphthols undergo self-sensitized or dye-sensitized photooxidation in solution, giving a hydroperoxide intermediate (**10**) by an 'ene' reaction involving singlet oxygen. It is also known that these dyes are oxidized by singlet oxygen formed from hydrogen peroxide and sodium hypochloride,[49,55] and that both reactions are suppressed by DABCO.

Substituent effects on the photofading of 1-(4'-substituted phenylazo)-2-naphthols in air-saturated methanol have been examined.[94] Figure 6.3 shows the relation between the logarithm of the percentage of fading ($\log f$), and the Hammett σ constant of the substituent groups in the aryl ring. The results, with the exception of those for the nitro group, suggest that the light fastness of these dyes is dependent on the electron density on the azo nitrogen atoms, and that the relative rate of photofading increases with increase in the electron-releasing nature of the substituents

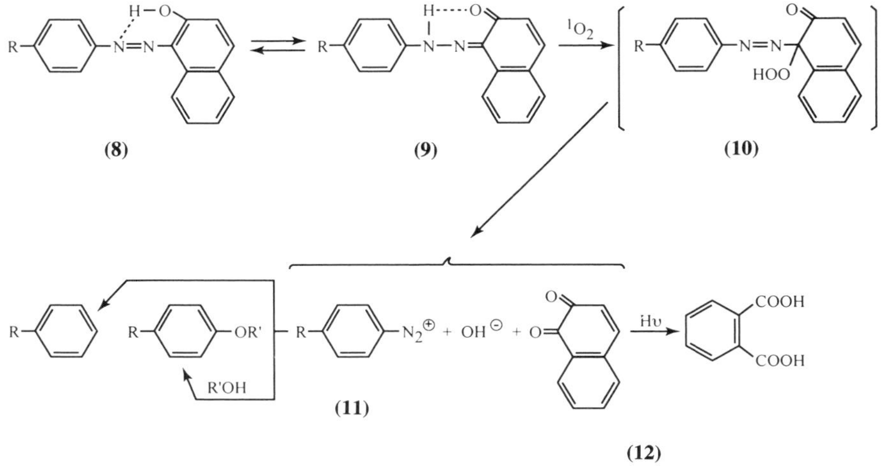

Scheme 6.5

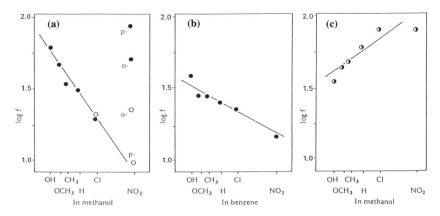

Figure 6.3 Relationships between the logarithm of the relative % fading, $\log f$, and the Hammett σ-constants of substituent groups on the photofading of 1-(4'-substituted phenylazo)-2-naphthols. The solutions containing 4.0×10^{-4} moles of the dyes in 500 ml of methanol or benzene were irradiated for 5 h with a 100 W high-pressure mercury lamp, respectively, —●—, under normal conditions; —○—, under oxygen atmosphere; —◐—, under nitrogen atmosphere. (a) In methanol, (b) in benzene, (c) in methanol.

in the aryl ring. This result is that expected on the basis of an initial electrophilic attack by singlet oxygen on the hydrazone tautomer. In the case of 1-(4-nitrophenylazo)-2-naphthol (**13**) however, anomalous photofading behavior in methanol and in ethanol was observed, but not in oxygen-saturated methanol or in air-saturated benzene. The photofading products of **13** in methanol were found to be nitrobenzene, phthalic acid and its ester (formed by the oxidative process), together with the reduction products 1,4-phenylenediamine, 4-nitroaniline, 1-amino-2-naphthol and 1-(4'-aminophenylazo)-2-naphthol (**19**). In 2-propanol, yields of the products from the photoreductive process increased, whereas in the case of acetone solution, only photooxidative products are formed. Thus, it appears that the anomalous photofading in Figure 6.3a may be due to photoreduction of the nitro and azo groups to amino groups in alcoholic solvents, instead of, or prior to, a self-sensitized photooxidation via a singlet oxygen mechanism.[94] Precedent for such a photoreaction of the nitro group is provided by Naphthol Yellow S (2,4-dinitro-1-naphthol-7-sulphonic acid) which undergoes light induced reduction to give 2-nitro-4-amino-1-naphthol-7-sulphonic acid.[3]

In recent years, Mustroph *et al.*[126,127] have also investigated the Methylene Blue-sensitized singlet oxygen photooxidation of phenylazoacetylacetones (**20**) and phenylazoprazolones (**21**) in air-saturated methanol. An excellent relationship was found between the fading-rate of a series of dyes (**20**) and Hammett σ-constants. On the other hand, an unexpected trend was found for the photofading of dyes (**21**); the

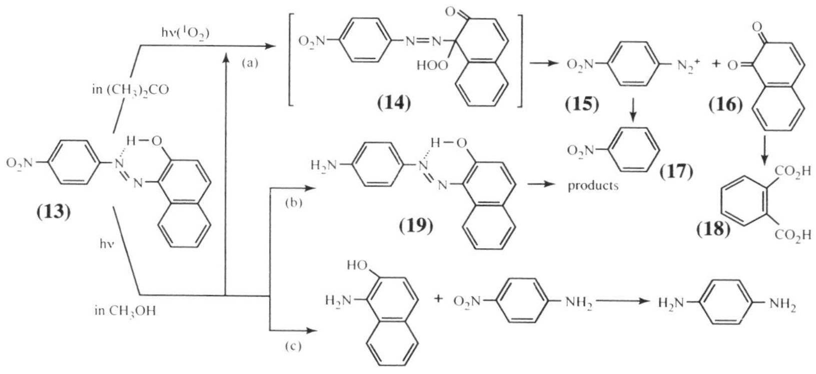

Scheme 6.6

Hammett plots were of two types, one with a negative slope derived from dyes containing electron-releasing substituents, and another with a positive slope derived from electron-withdrawing substituted dyes. The anomalous photofading is due to attack of singlet oxygen on the more reactive anion of the phenylazopyrazolones containing electron-withdrawing substituents. In methanol containing acetic acid, however, **21** exist only as hydrazone tautomers, and the fading rates show a linear relationship with σ-constants ($\rho = -1.43$, $r = -0.97$). It has been suggested that phenylazopyrazolone dyes undergo such oxidation by singlet oxygen, in relation to catalytic fading (discussed in a later section).

Results from several other studies have appeared which demonstrate in one way or another that singlet oxygen is an important intermediate in the self-sensitized or dye-sensitized photofading of azo dyes.[95,96,140-143]

However, Neevel et al.[26,27] reported an oxidation mechanism resultant from the flash-photolysis of biacetyl-sensitized tautomeric azo dyes in aqueous systems. In addition to a singlet oxygen mechanism, they proposed a radical mechanism involving the adduct of oxygen and triplet biacetyl, and the acylperoxy radical, in the sensitized photooxidation process (Scheme 6.7).

6.4.1.3 Anomalous photodecomposition in azo dyes. The influence on lightfastness of a substituent in a position *ortho* to the azo group has been examined using 4-nitro-4'-(*N*-ethyl-*N*-cyanoethyl)-aminoazobenzene (**22**) as reference compound. 2'-Nitro-substituted dyes (**23**), on polyester and on nylon faded in light much more rapidly than the corresponding 4'-nitro derivative (**22**).[144] 4-Nitroazobenzenes are more resistant to fading under anaerobic than under aerobic conditions, whereas the fading of 2-nitroazobenzenes is the same under both conditions. The anomalous decomposition of 2-nitroazobenzenes may be

210   PHYSICO-CHEMICAL PRINCIPLES OF COLOR CHEMISTRY

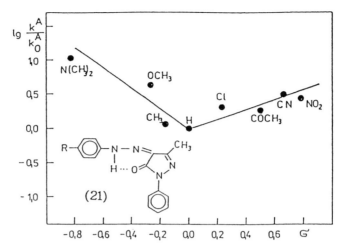

Figure 6.4 Relationship between the fading rate and Hammett σ constants on Methylene Blue-sensitized photooxidation of phenylazopyrazolone (**21**) in methanol.[126]

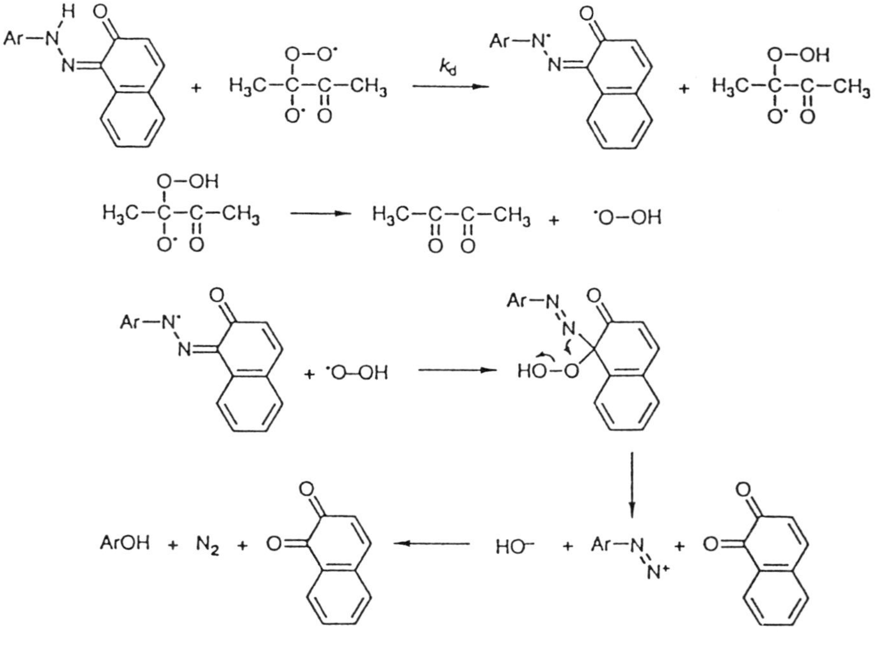

Scheme 6.7

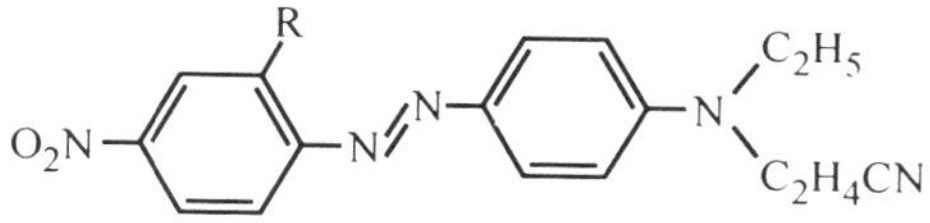

(22) R = H
(23) R = NO$_2$

explained by the assumption that intramolecular photooxidation of the azo group to the azoxy group occurs through transfer of an oxygen atom of the 2-nitro group.[145] On the other hand, in 1-arylazo-2-naphthols, the 2'-nitro compound (24) gives results which contrast with those for nitroazobenzenes. This can be explained on the basis of an intramolecular hydrogen bonded configuration (Scheme 6.8) for the azo-tautomeric form,[146,147] which is not possible in the corresponding 4-nitro compound.

6.4.1.4 Cis ⇌ trans isomerization. The photofading behavior via *cis-trans* isomerization of some azo dyes have been studied kinetically in polymer media by many workers.[148–154] Dubini-Paglia *et al.*[152–154] have reported the thermal *cis-trans* isomerization and photofading of azo dyes XC$_6$H$_4$N=NC$_6$H$_4$N(C$_2$H$_5$)$_2$ (where X is H, OCH$_3$, CN, NO$_2$, and COOC$_2$H$_6$) in poly(ethylene terephthalate) or polypropylene matrices below the glass transition temperature, and in dibutylphthalate solutions. The light stability of the dye in the polymer matrices was affected by several factors external and/or internal to the dye–matrix systems. In particular, polymer properties such as morphology, crystallinity and orientation strongly affected the photofading. Under Xenotest irradiation, a photostationary state was reached as a result of competition between direct *trans-cis* photoisomerization and thermal processes. Consequently, the concentration of the *cis* isomer is determined by the light intensity and by the temperature of the sample; it is much more affected by temperature than is the photochemical reaction and, an

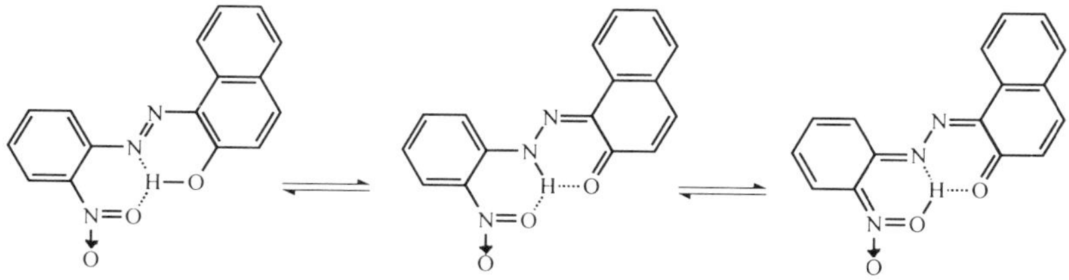

(24)

Scheme 6.8

increase in temperature reduces the *cis* isomer concentration at the photostationary state.

The nature of the dye also appears to be important; the dye in which X=OCH$_3$ was markedly less stable. Electronic factors are likely to play a role, since this dye was the only one containing a strong electron-donating group.[154]

6.4.2 Anthraquinone dyes

Anthraquinone dyes are generally characterized by high fastness to light in comparison with those of many other dye classes.

The photochemical reactions of anthraquinone acid,[113–117,136,156–162] disperse[14,128,163–167] and vat dyes[168–171] have been extensively studied both in solution and in polymer substrates.[165,166,172–174, 178–181] In some cases, a similarity was found between the photochemical reaction of the dye in solution and on the polymer film. From the results of earlier studies,[3,5,6,12–14,16,118,175–181] the photochemistry of *N*-alkylaminoanthraquinone compounds can be summarized as involving the following reactions: (a) *N*-dealkylation, (b) nuclear hydroxylation, (c) substitution of amino by hydroxy groups, (d) introduction of an oxygen function at the C-atom adjacent to the N-atom in *N*-alkylamino groups, and (e) formylation of amino groups. Reactions of types (a)–(c) were observed by Couper,[182] *N*-dealkylation (a) was observed by Giles and Sinclair,[184] and reactions of type (d) have been reported from other laboratories. The formylation process (e) was found to occur on irradiating 1,4-diaminoanthraquinone and its *N*-alkyl derivatives.

Photodecomposition and light stability vary with the environmental conditions during fading and with the nature of the substrate. The fading behavior of *N*-alkylaminoanthraquinone dyes on nylon, cellulose acetate and poly(ethylene terephthalate) (PET) have been extensively studied.[3,118,179,180] The dealkylation reaction occurs in the initial stages of fading and one of the main reasons for reddening of hue which occurs on the fading of these dyes on substrates is the relative ease with which the *N*-alkyl groups can be removed by irradiation. On the basis of reported data,[3,182,183,175,178–180,184] the oxidative aspect of the photodecomposition of *N*-alkylaminoanthraquinone dyes is illustrated by Scheme 6.9.

The photofading products of *N*-alkylaminoanthraquinone dyes (**25**) have been examined in order to account for the difference in their fading characteristics in PET compared with those in nylon and cellulose acetate.[103,175,179] Although similar products were observed in all cases, differences were apparent in their rate of formation. The *N*-dealkylation occurs by cleavage of the C–N or by dehydrogenation to form an imine. However, doubts have been cast on the validity of products **26** and **27**. Wegerle[180] concluded that formylation of the amino group was important,

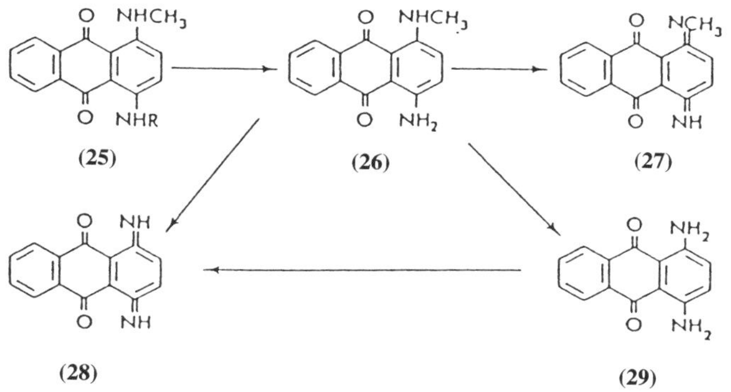

Scheme 6.9

leading to structure **30** and in later investigations,[185] also reported that one of the final products of fading was a 1-amino-4-formimidoanthraquinone of structure **31**.

Allen et al.[14,115,116,162] found that 1,4-bis(ethylamino)- and 1,4-bis(isopropylamino)anthraquinones photodecomposed on polyester and polyamide fibers to give 1-imino-4-formimido-anthraquinone, together with 1,4-diaminoanthraquinone and 1,4-bis-methylaminoanthraquinone. These products were the same as those formed by thermal decomposition of the dyes on silica gel plates.

In order to relate the excited-state properties of anthraquinone dyes to their structure and lightfastness properties, several studies have been made in different substrates. In this respect, Allen et al.[185–190] carried out investigations on the spectroscopic and flash photolysis behavior of some amino- or hydroxy-substituted anthraquinone dyes in solution, and related the resulting data to the lightfastness properties of the dyes in polyamide and poly(ethylene terephthalate). They reported from examinations of the triplet excited state or luminescence properties that low lightfastness in polyamide is accompanied by strong transient absorption in 2-propanol, while high lightfastness in polyester is accompanied by a weak transient absorption. In case of flash photolysis of 1- or 2-substituted anthraquinone dyes, Allen et al.[13,14,185–190] also reported the contribution of a semi-quinone radical, abstracting a hydrogen atom, and of a radical anion produced by the photoexcited dye abstracting an electron from the environment. This reaction would take place more readily in polyamide substrates. Chang and Miller[117] showed that with 1,4-diaminoanthraquinone and 1,4-dihydroxy-anthraquinone in N-ethyl-acetamide (a nylon model) hydrogen abstraction from the solvent to form 'leuco' anthraquinone derivatives occurred in the absence of oxygen. Lishan et al.[159] also showed that the photodegradation of some acid anthraquinone dyes such as C.I. Acid Blue 277 occurs by a reductive process on nylon, involving hydrogen transfer from the nylon polymer to the photoexcited dye molecule. A simplified mechanism of this dye photodegradation is illustrated in Scheme 6.10.

Furthermore, anthraquinone dyes are normally expected to undergo photoreduction at the 9- and 10-positions to give the corresponding hydroquinone. On the other hand, it has been also reported that 1,4-dihydroxy- and 1,4,5,8-tetrahydroxyanthraquinones appear to undergo photoreduction at the 2- and 3-positions, giving two tautomers, as illustrated in Scheme 6.11. However, further evidence for this mechanism is required.

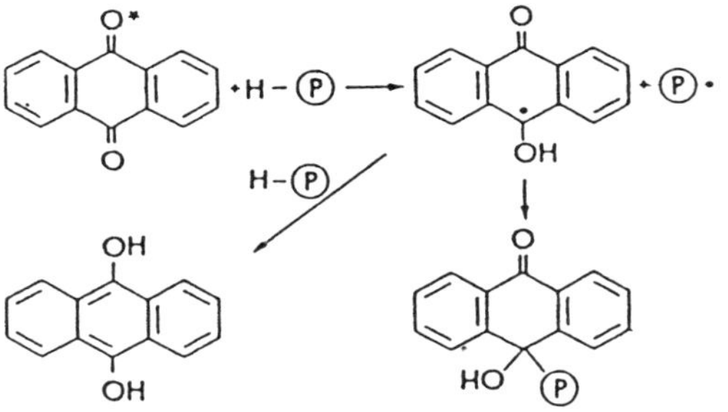

(P)–H represents the C–H linkage at the carbon α to the amide nitrogen on the nylon polymer backbone[159]

Scheme 6.10

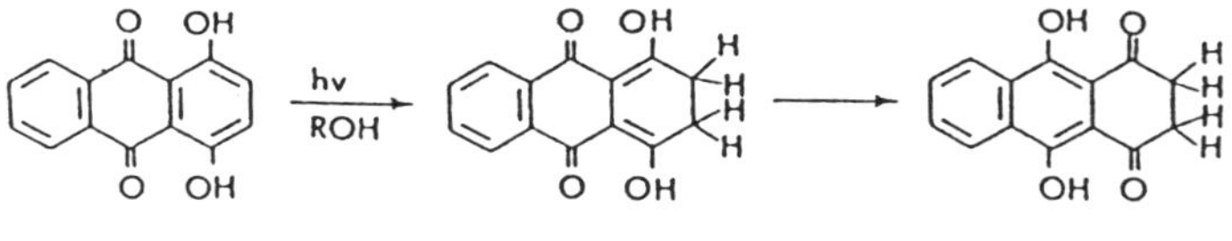

Scheme 6.11

Kuramoto and Kitao[122] demonstrated the contribution of singlet oxygen to the photofading of certain aminoanthraquinone dyes in dichloromethane/methanol (4:1, v/v) or dimethyl sulphoxide (DMSO). The rates of photofading were found to be accelerated in the presence of sensitizers of singlet oxygen, e.g. Rose Bengal or Methylene Blue, and retarded by addition of an effective singlet oxygen quencher, such as nickel dimethyldithiocarbamate (NMC) or DABCO. Certain anthraquinone dyes are oxidized by singlet oxygen formed from hydrogen peroxide and sodium hypochlorite.[49] From additional experiments examining the effect of a deuterated solvent and the competitive photooxidation of the dyes and NMC, the involvement of singlet oxygen was recognized. It was shown that 1-amino-4-hydroxy-, 1-amino-2-methoxy-4-hydroxy-, 1-amino-4-methylamino-, 1,4-bis(methylamino)-, 1-methylamino-4-hydroxyethylamino-, 1,4-bis-toluidino- and 1,4,5,8-tetraaminoanthraquinones underwent self-sensitized or dye-sensitized photooxidation in solution. Colorless products formed were identified as phthalic acid and its anhydride by comparison with authentic samples using GLC or TLC analyses. However, compounds having only one substituent group on the anthraquinone ring were stable to irradiation, and were not oxidized even in the presence of a chemical source of singlet oxygen.

More recently, Freeman et al.[136] have investigated the photodegradation of C.I. Acid Green 25 (**35**), in dimethylformamide and on nylon 66 fiber. They indicate that the process proceeds by hydrogen atom abstraction from the solvent and polymer to produce the leuco structure (**36**). In this case, the leuco compound (**36**) is the only isolated product, characterized by mass (FAB) spectrometric-, TLC- and HPLC analyses. The faded dye undergoes only partial reversion to the starting dye upon prolonged exposure to air.

In other work involving anthraquinone dyes in polymers, Paine et al.[194] found that 1,5-dihydroxy-2,6-diisobutyl-4-thiophenylanthraquinones (**37**) faded up to 100 times faster in styrene–butadiene copolymer than in

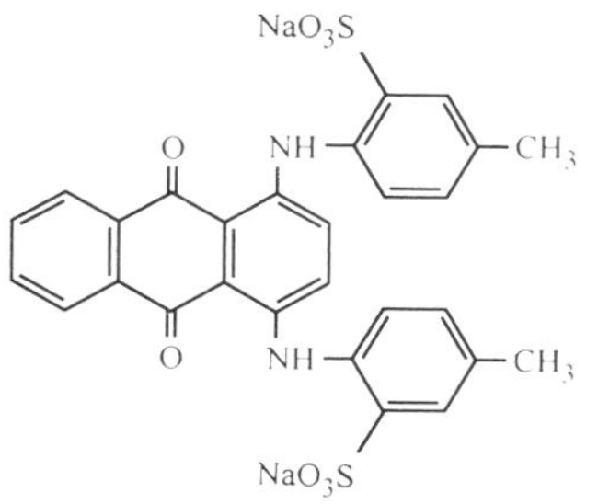

(**35**) C.I. Acid Green 25 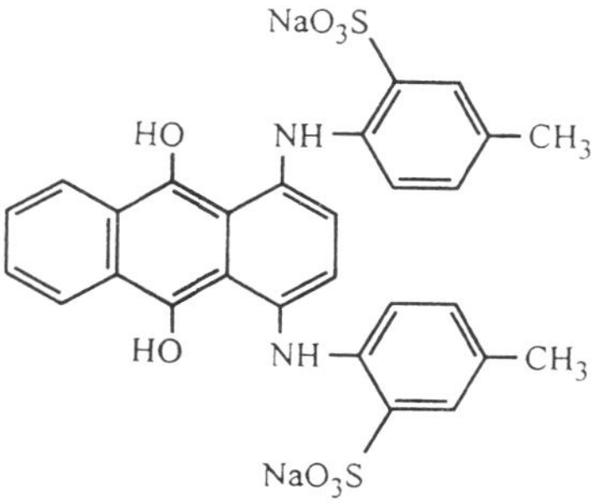   (**36**)

polystrene, styrene/n-butylmethacrylate copolymer, and polyester resins. A yellow product (**38**) was isolated and characterized as the sulfoxide compound. Mechanistic investigation of the accelerated photofading of the thioether **37a** (λ_{max} 518 nm) concluded that the dye sensitizes formation of singlet oxygen, which attacks the double bonds in the styrene–butadiene copolymer, resulting in extensive cleavage and peroxide formation. The major process was oxidation of the polymer, involving both Type I (radical process) and Type II ('ene' reaction by singlet oxygen), as illustrated in Scheme 6.12. The reaction of singlet oxygen with sulfides is complex, generally producing two sulfoxide molecules (Scheme 6.13). In this case, dye fading in a styrene–butadiene resin, whereby the dye is attacked by polymer peroxides, is a minor process because the rate of the chain scission in the polymer is greater than that of fading. However, more detailed study is required since the photofading behavior of thioether substituted anthraquinone depends on the nature of the polymer substrates.

The photochemical reactions of 6-substituted benzanthrones and violanthrone have also been studied. The former undergo a primary photoreduction mechanism which involves hydrogen atom abstraction from solvent saturated with nitrogen.[195,196] On the other hand, 16,17-dimethoxyviolanthrone (C.I. Vat Green 1) and pyranthrone (C.I. Vat Orange 9) undergo photoreversible photooxidation to the corresponding endoperoxide by singlet oxygen, which was generated via the processes of self-sensitized or Rose Bengal sensitized photoreactions.[197]

An undesirable property of many anthraquinone dyes is their ability to sensitize the photochemical degradation of polymer substrates and results from many investigations have been reported in this area.[3–5,10,11,16] Zweig and Henderson[100,192] and the other workers[43,101–103,198,199] have shown that some aminoanthraquinone dyes are able to produce singlet oxygen in

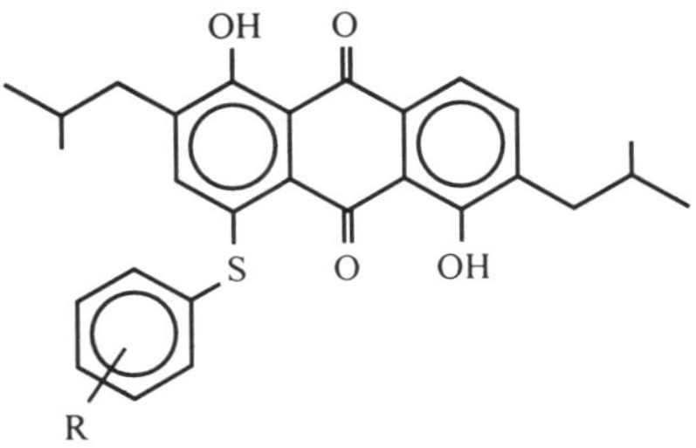

(37)

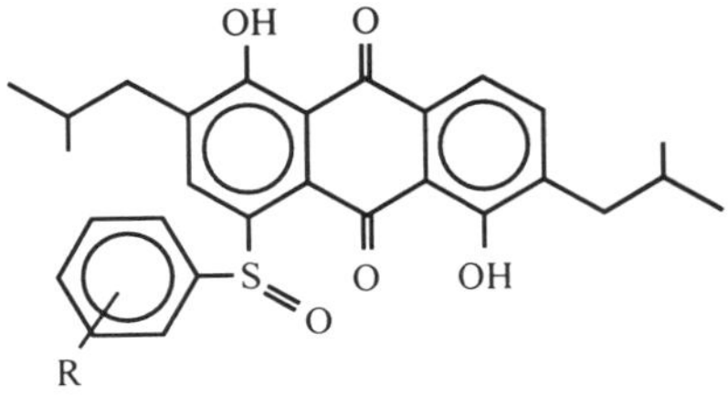

(38)

(a) R=H
(b) R=p—OCH
(c) R=p—C(CH)$_3$

(d) R=m—NH$_2$
(e) R=p—Cl
(f) R=p—Br

(g) R=m—Cl
(h) R=p—NO$_2$

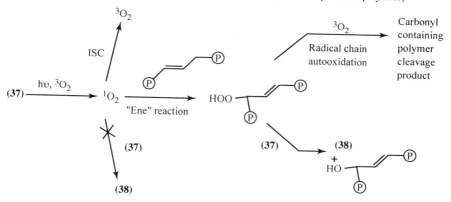

Scheme 6.12

a polymer matrix. It is well known also that certain anthraquinone dyes catalyze the photofading of other dyes in mixed systems and this catalytic fading phenomenon is discussed in a later section.

6.4.3 Triphenylmethane dyes

Triphenylmethane dyes are a major sub-class of basic dyes, and many of them are still widely used because of their high tinctoral strength. Unfortunately, however, their lightfastness leaves much to be desired, although some dyes have an acceptable stability in acrylic polymers.

Henriquer found that an aqueous solution of Crystal Violet turned red after exposure for a few days to a mercury lamp.[200] The fading product

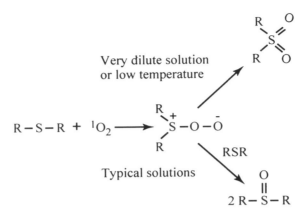

Scheme 6.13

was Fuchsine, formed by elimination of alkyl groups from the dimethylamino substituents on the phenyl rings; a similar observation was made for Malachite Green.[201]

These dyes can be further photodecomposed to yield colorless products. Iwamoto[202] isolated Michler's ketone [4,4'-bis(dimethylamino)benzophenone) from the photofading products of Crystal Violet in the presence of air. Porter and Spears[203] exposed Malachite Green to a carbon arc source in aqueous solution and on carboxylated cellulose film, and isolated 4-dimethylaminobenzophenone, 4-dimethylaminophenol and an additional product, the carbinol form of Malachite Green.

Kuramoto and Kitao[124,125] have investigated the oxidative mechanism, and shown that the photodegradation of Crystal Violet (**39**) and of Malachite Green (**40**) takes place mainly via a 'dioxcetane' intermediate by a singlet oxygen mechanism (self-sensitized), as illustrated in Scheme 6.14. The rates of the photofading in solution were accelerated in the presence of a singlet oxygen sensitizer and as Methylene Blue, and were retarded by addition of an effective singlet oxygen quencher such as β-carotene or nickel dimethyldithiocarbamate. Identification of the products using TLC showed that Crystal Violet was mainly photooxized to Michler's ketone (**43**) and 4-dimethylaminophenol (**45**) via the reaction with singlet oxygen, which can be generated by the dye itself or by an added sensitizer. A minor dealkylation product, Parafuchsine, was also obtained in smaller amounts. Malachite Green underwent degradation in a similar manner.

Oster et al.[204] had previously reported that triarylmethane dyes may undergo a photoreduction process which involves radical formation. Using flash photolysis, Allen et al.[205,206] also showed that the presence of a triplet donor (benzophenone) and a hydrogen atom donor (benzhydrol)

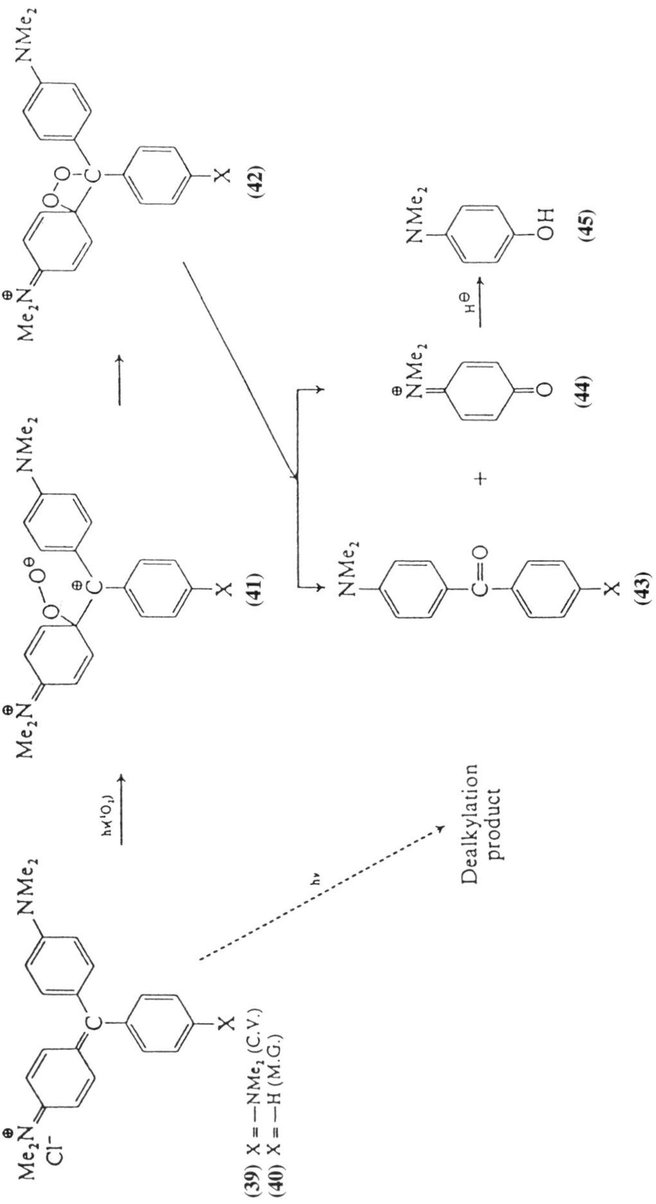

PHOTODEGRADATION OF SYNTHETIC COLORANTS

Scheme 6.14

accelerated the fading of triphenylmethane dyes in non-reductive environments such as poly(acrylonitrile), in which these dyes normally have high light stability. They proposed that the dyes fade through a photoreductive mechanism involving electron abstraction by the triplet state of the dye molecule. An intermediate, which they proposed to be the triarylmethyl radical (D·), was also detected by its absorption during flash photolysis of a 2-propanol solution in the dye. However, the intensities of the ESR signals of irradiated poly(methylmethacrylate) films were very weak. Additionally, a one-electron reduction of Crystal Violet (CV^+) by diphenylketyl (KH·) followed by hydrogen transfer to the intermediate (CV·) with formation of the leuco dye (CVH), has also been proposed.[205,207,208] In flash photolysis studies of Crystal Violet chloride (CV^+Cl^-), the contributions of the triarylmethyl radical in 2-propanol,[205,206] and of the diphenylketyl radical in acetonitrile containing benzophenone (Ph_2CO) as a sensitizer,[209,210] have been investigated. In the presence of oxygen, triphenylmethane dyes undergo photooxidation which is initiated by the ejection of an electron from the photoexcited dye cation to form a radical di-cation and a hydrated electron.[201,203,204,211] The radical, radical di-cation and electron can recombine to restore the original dye, or to generate various irreversible reactions.[203,204,211,212]

There are also several reports indicating that fluorescent brightening agents sensitize the photodecomposition of triphenylmethane dyes by a triplet–triplet resonance energy transfer process.[213–215]

6.4.4 Indigoid dyes

Hibbert[216] and other workers[217] have shown that when dyeings of indigo are exposed to light, the dye is oxidized to isatin. Scholefield *et al.*[218] similarly isolated dibromoisatin from exposed dyeings of tetrabromoindigo. Other work[219] has reported that the photolysis of indigo in tetrahydrofuran containing 5% water using ultraviolet light gives 2-hydroxy-2'-hydro-3,3'-dioxo-diindolinyl-(2,2') by a hydration process, in which hydroxide ions from the solvent attacked the excited indigo molecule. These investigations did not, however, give a detailed mechanism for the photooxidative reaction of indigoid dyes.

Kuramoto *et al.*[92,138] later found that the photofading of indigo in solution or on silica gel may be due to a self-sensitized photooxidation involving singlet oxygen as follows: attack of singlet oxygen on indigo could lead to the unstable dioxetane intermediate, which can then produce isatin by fission of the carbon–carbon bond. The presence of the dioxetane intermediate in the reaction is assumed by analogy with the known reaction with singlet oxygen[55,57] of simple olefins, which have no allylic hydrogen. The results of competitive photooxidation experiments in solution, and determinations of the relative rate constant for the

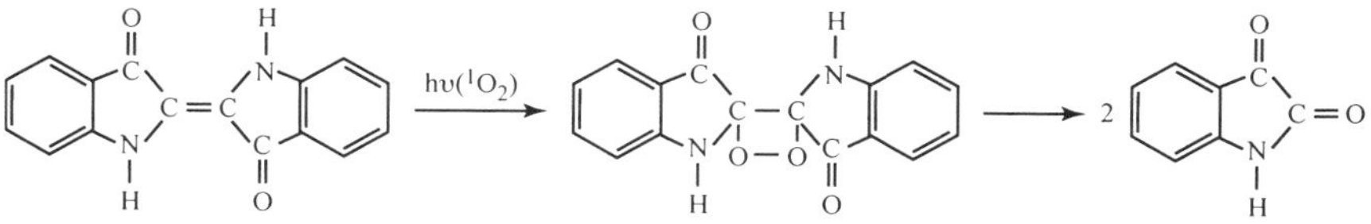

(46) → (47) → (48)

Scheme 6.15

reaction support this mechanism.[92] The high light stability of indigoid dyes on wool and silk has been attributed to a quenching of singlet oxygen by the amino acid residue in these protein polymers.[220]

Several studies of the photoisomerization of indigo and thioindigo have been also reported.[221–227] Thioindigo and N,N'-diacetyl- or N,N'-dibenzoly-indigo dyes show interesting photochemical behavior with respect to their *trans* ⇌ *cis* isomerization.[228–233] Their decolorization involves a reversible pathway of the excited dye, which has been proposed for use in solar energy conversion and storage.[228–232]

6.4.5 Quinophthalone dyes

Dyes and pigments in this series are used as colorants for plastic and paper,[234–238] and in ink-jet recording systems. Kuramoto and Kitao[119,120,138,238] have investigated the photochemistry of quinophthalone dyes and pigments in solutions or polymer substrates.

Quinophthalone [2-(2-quinolyl)indan-1,3-dione] (**49a**) (λ_{max} 418 and 442 nm), exists in equilibrium with the tautomeric form (**49b**).[119,235] A solution of **49** fades in the presence of oxygen upon irradiation with filtered visible radiation or unfiltered radiation from a high-pressure mercury lamp. The reaction is accelerated by the addition of Rose Bengal or Methylene Blue, as is found with arylazonaphthols and indigo. The photooxidation proceeds smoothly with filtered radiation ($\lambda > 520$ nm) which is absorbed by the sensitizer only, and analysis of the products showed that the reaction was identical to the slower self-sensitized process. The products in dichloromethane–methanol (9:1, v/v) were phthalic acid (28%), dimethylphthalate (9%), quinoline-2-carboxaldehyde (trace), quinoline-2-carboxylic acid (5%) and an unidentified material which appeared to be identical to the compound derived from irradiation of quinoline-2-carboxaldehyde alone under these conditions.[119,120] Similar products were obtained by oxidation using a chemical source of singlet oxygen.

The most probable mechanism for the photodecomposition of **49** is shown in Scheme 6.16, and involves attack by photochemically generated

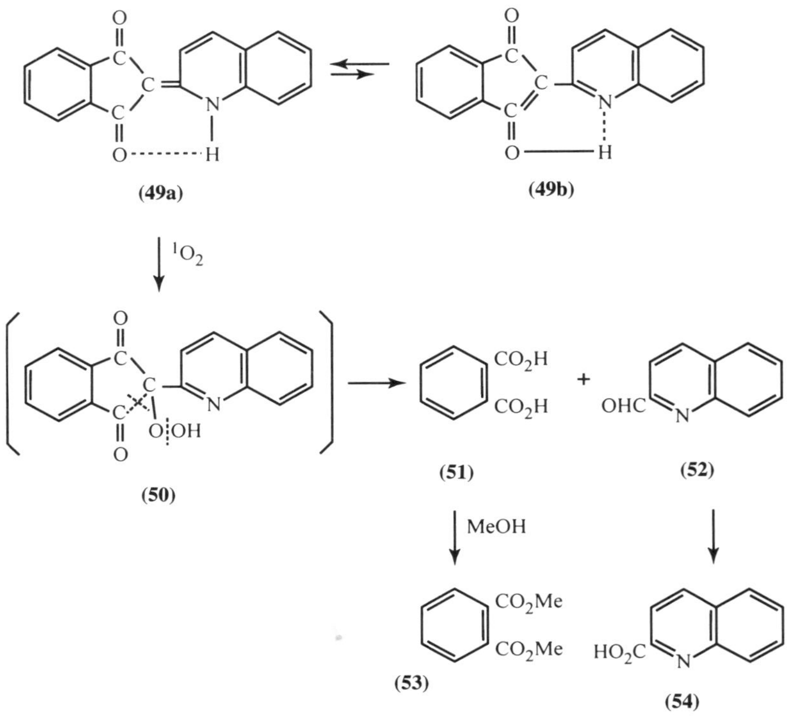

Scheme 6.16

singlet oxygen (presumably $^1\Delta_g O_2$) on the central double bond in the tautomeric form (**49a**) in a concerted 'ene' reaction, or possibly via a 'peroxide' type intermediate. This would lead to the hydroperoxide, which could decompose homolytically or heterolytically to phthalic acid (**51**) and quinoline-2-carboxaldehyde (**52**).[120]

Many quinophthalone dyes show greatly improved photochemical stability when a hydroxy group is introduced into the 3'-position (e.g. 2-(3-hydroxy-2-quinolyl)indan-1,3-dione (**55**)) or a halogeno group into 8'-position (e.g. 2-(8-chloro-2-quinolyl)indan-1,3-dione (**56**).[237] The effect of the hydroxy group is related to its ability to form an intramolecular hydrogen bond with an adjacent carbonyl oxygen atom (**55b**), and that of the halogeno group is related to the formation of an intramolecular hydrogen bond[146,147,237] involving the N–H hydrogen, halogen, and carbonyl oxygen (**56b**), thereby decreasing the contribution of the central double bond in (**55a**) or (**56a**).[120] Compound **55** has better photochemical stability than **49** by a factor of *ca* 3–5 in solution and on polyethylene substrate, as established empirically.[238]

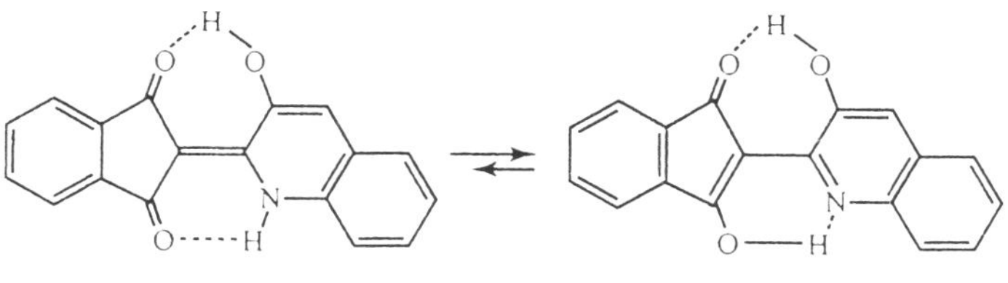

Scheme 6.17

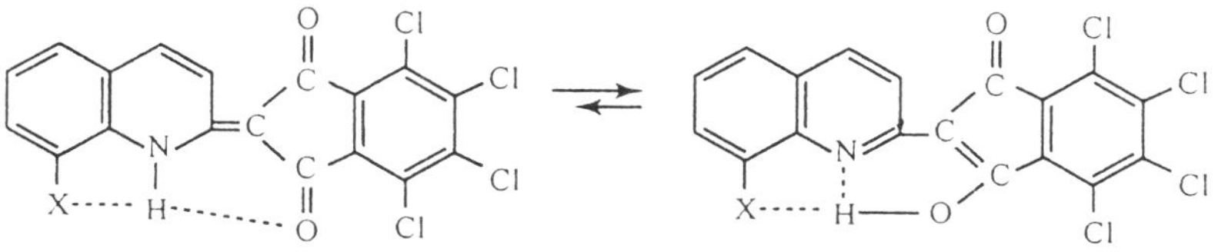

(55) (a) ⇌ (b)

(56) (a) ⇌ (b)

Scheme 6.17

6.4.6 Fluorescent dyes

The photofading behavior of some coumarins and related compounds in solution, and on nylon and polyester film has been studied.[239–242]

The photochemistry of stilbene derivatives is of particular importance in connection with the fading of stilbene fluorescent whitening agents. Suggested mechanisms for the photofading of stilbene fluorescent agents include cis–trans isomerization,[243,246–250] dimerization[251] and photoreductive cleavage[243,251] and an oxidative reaction involving the ethylene bond.[243,250–253]

Oda and Kuramoto[254] have reported the photochemistry of trans-4,4′-bisacetamidostilbene (trans-BAAS) (57) in solution and on silica gel by irradiation with unfiltered light from a mercury lamp in the presence of oxygen. Rapid isomerization from trans-BAAS to cis-BAAS, and then gradual decomposition of both isomers, results in the formation of 4-acetamidobenzaldehyde (60). It was also found that Rose Bengal accelerated the reaction and increased the yield of 60. The latter photooxidation in the presence of Rose Bengal proceeded smoothly with filtered radiation ($\lambda > 510$ nm), which can be absorbed by the sensitizer only. The photodecomposition of trans-BAAS in acetonitrile containing the sensitizer were 4-acetamidobenzaldehyde, 4-acetamidobenzoic acid (61) and unchanged material. In contrast to trans-BAAS, the photooxidation of cis-BAAS proceeded more slowly than that of trans-stilbene; in the cis-isomer, the reaction is inhibited by the ortho hydrogen atoms.

The most probable mechanism of the photodegradation of trans-BAAS is shown in Scheme 6.18; i.e. attack of singlet oxygen on the compound

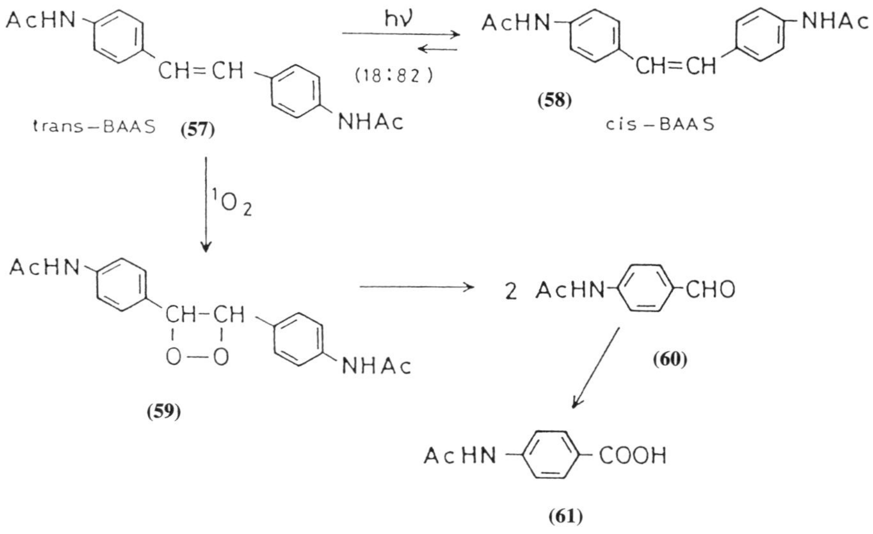

Scheme 6.18

leads to the unstable dioxetan intermediate, which can produce 4-acetamidobenzaldehyde by thermal or photochemical fission of the carbon–carbon bond.[125] Similar observations have been made in the photodegradation of *trans*- and *cis*-bisbenzoylamidostilbenes under the same conditions.[255]

The photofading of the direct dye type fluorescent brightening agent, disodium 4,4'-bis(2-sulfonatostyryl)biphenyl has been studied in aqueous solution, and it is suggested that an initial reversible *trans* ⇌ *cis* isomerization occurs, with a subsequent oxidation stage.[247,256] Further, Yamada et al.[257-259] found that fluorescent brightening agents of the stilbene and pyrazoline type enhance the photofading rate of some mixtures of direct azo dyes in aqueous solution or in cellulose acetate films. They suggested that the photofading mechanism may involve both the transfer of triplet energy and oxidation by singlet oxygen. The photostabilization and photosensitizing properties of triazinylaminostilbenes on cotton fiber have also been reported.[260,261]

6.4.7 Miscellaneous dyes

The photochemistry of rhodamine dyes continues to attract much interest in view of the laser applications of these dyes.[262-265] The reactions shown in Scheme 6.19 have been proposed for the decay of the excited triplet state of Rhodamine 6G (**62**) on the basis of laser flash photolysis in water.[264,265] In Scheme 6.19, Q is oxygen or some impurity; oxygen

$$D \to {}^1D^*$$
$$^1D \to {}^3D$$
$$^3D \to D$$
$$^3D + Q \to D + Q$$
$$^3D + D \to D^{\cdot+} + D^{\cdot-}$$
$$^3D + D \to 2D$$
$$^3D + {}^3D \to D^{\cdot+} + D^{\cdot-}$$
$$^3D + {}^3D \to {}^1D^* + D$$
$$^3D + D^{\cdot+} \to D^{\cdot+} + D$$
$$^3D + D^{\cdot-} \to D^{\cdot-} + D$$
$$D^{\cdot+} + D^{\cdot-} \to 2D$$
$$D^{\cdot-} + D^{\cdot-} \to D^{--} + D$$
$$D^{\cdot+} + D \to \text{Products}$$

Scheme 6.19

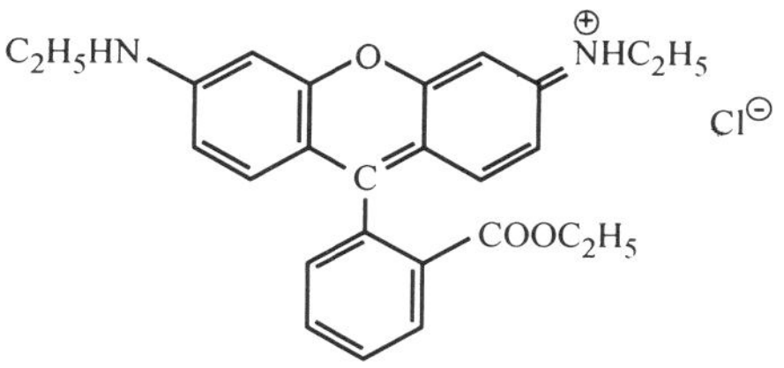

(62)

quenches the dye triplet state. The radical anion (D^-) and the radical cation ($D^{\cdot+}$) species react to regenerate the original dye. Reaction between the radical cation and the dye (D) results in irreversible decomposition of the dye. The effects of the environment on the photofading of Rhodamine 6G have also been reported.[266-268]

The photodecomposition of azomethine dyes has been studied in isopropanol in the presence of sensitizers such as benzophenone and anthracene.[269] In isopropanol, leuco dyes are formed when excluding oxygen. When oxygen is present, the chromophore is decomposed via reductive processes involving free radicals.[270] On the other hand, a photooxidative mechanism involving singlet oxygen has been reported.[271,272] Miwa et al.[273-275] have demonstrated the relevance of oxidative and/or reductive pathways for azomethine dyes designed for photographic applications.

6.5 Catalytic fading

Several authors have observed that the lightfastness of certain dyes applied singly to fibers is much better than when the same dyes are applied in combinations.[276,277] This phenomenon has been termed catalytic fading. In most of the reported cases, the lightfastness of blue, violet or red dyes deteriorates when a yellow dye is added.[276] Asquith et al.[276] reported on this type of fading for acid dyes on wool, nylon and in solution. Scholefield and Goodyear[278] observed decreased lightfastness for blue dyes on cotton after addition of a yellow dye. In view of the technical importance of this phenomenon, several investigations on the mechanism of the catalytic fading have been carried out.[3,101,102,121,125,138,279–283]

Fiebig and Metwally[277] reported the catalytic fading of a blue, red or violet dye in admixture with a yellow or orange dye. One reaction mechanism suggested from this work is that energy-transfer processes from the yellow to the blue, red or violet dye are involved, thus leading to a higher yield of the photoactive state of the latter dyes than generated by its direct excitation. Because of this, the dye mixture may fade faster. Another mechanism proposed is that singlet oxygen is generated in the mixture and reacts with the blue, red or violet dye.

In contrast to the above cases, an increased fading of some yellow azo dyes in admixture with blue, red or violet anthraquinone dyes has been reported by Rembold and Kramer.[101,102,281] Kuramoto and Kitao[121] also reported on the catalytic fading in mixed systems of a yellow quinophthalone dye with certain anthraquinonoid dyes. Energy transfer processes from an excited singlet state of the anthraquinonoid dyes to a singlet state of the azo-dyes can be excluded for energy reasons. Energy transfer between both triplet states might be energetically possible, if the energy of the triplet states of the azo dyes is lower than that of the anthraquinonoid dyes.[101] This condition, however, is usually not fulfilled because the excited singlet states of azo dyes lie much higher than those of anthraquinonoid dyes. With regard to the mechanism of the reaction, Rembold et al.[101,102,281] concluded that this type of fading is most likely to occur via the singlet oxygen mechanism (Type II photooxygenation),[100] with the triplet states of the anthraquinonoid dyes (C.I. Acid Red 143, C.I. Acid Green 25, C.I. Acid Violet 47 or C.I. Disperse Blue 56) producing singlet oxygen and the azo dyes (C.I. Acid Yellow 127, C.I. Acid Yellow 29, or C.I. Disperse Yellow 50) being oxidized by the formed singlet oxygen.

The ability to initiate catalytic fading via a singlet oxygen mechanism is closely related to the quantum yield of the triplet formation in anthraquinone dyes. Rembold et al. have determined the triplet quantum yields of a series of diaminoanthraquinones by laser flash photolysis, and discussed their ability to induce catalytic fading.[102] Whereas, 1,4- and 1,2-diaminoanthraquinones have very small quantum yields (Φ isc, < 0.05), 1,5- and 1,8-diaminoanthraquinones are excited to the triplet state with a fairly high efficiency (0.67, 0.77). These authors concluded that catalytic fading

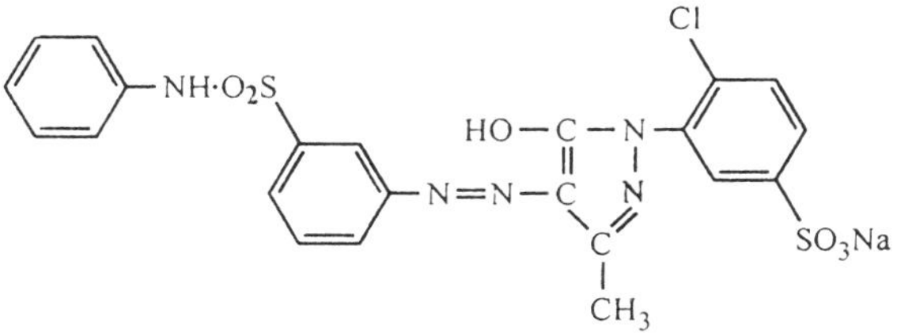

(**63**) C.I. Acid Yellow 29

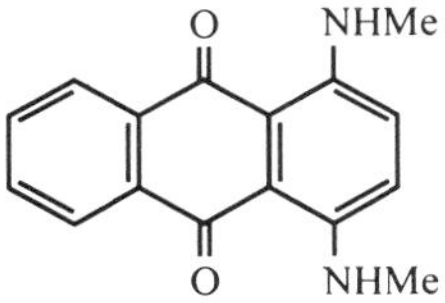

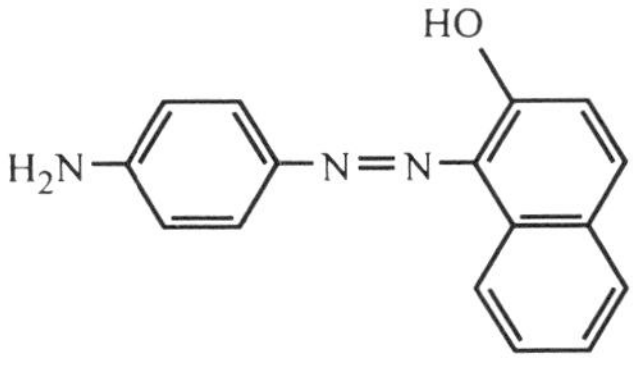

(**64**) C.I. Disperse Blue 14

(**65**) 1-(*p*-aminophenylazo)-2-naphthol

occuring via a singlet oxygen mechanism may be expected in dye mixtures containing 1,5- and 1,8-diaminoanthraquinone and an azo dye which is easily oxidized by singlet oxygen,[101,102,281] a condition which is satisfied by azo dyes of the arylazopyrazolone type (**63**).

Kuramoto et al.[121] have reported that the relative rate constants of individual dyes derived from the Wilson method[284] may be useful in predicting the possibility of catalytic fading in dye combinations. From competitive photooxidation experiments involving dyes with added singlet oxygen quenchers (Q) such as nickel dimethyldithiocarbamate,[45,100,285-289] it is possible to obtain values of the relative rate constant, thus providing evidence for involvement of singlet oxygen in the self-sensitized process. The results showed that the smaller the value of k_Q/k_{Dye} a dye has, the more reactive it is towards singlet oxygen. The fading of quinophthalone (**49**; λ max 418 and 442 nm; $k_Q/k_{Dye} = 14.6$) in acetone increases when 1-amino-4-hydroxyanthraquinone (C.I. Disperse Red 15, λ max 525 and 560 nm; $k_Q/k_{Dye} = 16.4$) is added. However, when quinophthalone mixed with 1,4-bis(methylamino)anthraquinone (**64**; C.I. Disperse Blue 14, λ max 596 and 643 nm; $k_Q/k_{Dye} = 8.7$) was irradiated under the same condition ($\lambda > 500$ nm), no catalytic fading occurred. This suggests that decomposition of both dyes is decreased relative to the individual dyes, and it is reasonable to assume that quinophthalone acts as a filter for C.I.

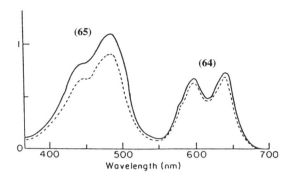

Figure 6.5 Absorption spectra of the mixture of C.I. Disperse Blue 14 (**64**) and 1-(4-aminophenylazo)-2-naphthol (**65**) in acetone; concentration 4.0×10^{-5} mol dm^{-3} for each dye. ——, Before exposure; – – – –, after 8 h of exposure with filtered radiation ($\lambda > 520$ nm).

Disperse Blue 14, whereas the latter acts as an inhibitor of the singlet oxygen photooxidation of the former.

Similar experiments using C.I. Disperse Red 15, C.I. Disperse Blue 14 (**64**) and 1-(4-aminophenylazo)-2-naphthol (**65**; λmax 435 and 478 nm; $k_Q/k_{Dye} = 1.5$) show the catalytic fading of the azo dye resulting from addition of aminoanthraquinone dyes (Figure 6.5). These observations, together with earlier findings suggest that dye combinations undergo catalytic fading when the dye with the greater k_Q/k_{Dye} value is a sensitizer for singlet oxygen, and the dye with the smaller value of k_Q/k_{Dye} is sensitive towards oxidation. Thus, it can be concluded that the singlet oxygen generated predominantly reacts with the dye for which the k_Q/k_{Dye} value is smaller.[121,125]

In other studies of catalytic fading, Shosenji and Yamada[259,290] have reported the accelerated fading of azo dyes[214] and of triphenylmethane dyes[258] in the presence of fluorescent whitening agents of a pyrazoline or stilbene type; they suggest a mechanism involving either triplet energy transfer or singlet oxygen chemistry.

6.6 Photodecomposition and stabilization of dye–polymer matrices

6.6.1 Natural fibers

Undyed cellulosic fibers are relatively stable to ultraviolet and visible light. Intense irradiation with near-ultraviolet radiation, however, results in the initiation of chemical reactions which lead to a deterioration in the physical properties of the fiber, and to changes in the nature and degree of polymerization of the polymer molecule. Exposure to light of various wavelengths also results in changes in the degree of whiteness of fibers.[16] Undyed wool fibers are much less stable to light in comparison with cellulosic fibers. Yellowing occurs on exposure to light of wavelengths shorter than about 380 nm.[9] In the presence of fluorescent brighteners, the fiber is rendered sensitive to yellowing with light of up to about 430 nm. On the other hand, the use of ultraviolet absorbers and various pretreatments can enhance the photostability of the substrate.[17,291]

Vat dye-sensitized photodegradation of cotton and other cellulosic fibers is an important technological problem.[2,292,293] This problem has been reviewed by Bentley,[16] and a number of photochemical studies, and mechanisms thereof, have been reported.[170,171,294–296] The mechanism is complex and not fully understood.[13,17,294] It is thought, however, that the initial attack on the cellulose molecule by the excited dye is either by hydrogen- or electron-abstraction or is indirect, via the formation of a reactive singlet oxygen intermediate.

Two general mechanisms have been outlined to account for the dye-sensitized photodegradation and/or oxidation of the polymer. The first originated from the work of Egerton,[3,5–8] who proposed a mechanism in which the photoactive excited state of the dye (normally the triplet) is quenched by ground state molecular oxygen, thereby producing active singlet oxygen. The singlet oxygen then reacts with the substrate or with water to form hydroperoxide and/or hydrogen peroxide, both of which may then induce oxidative breakdown of the substrate. The basic features of the mechanism are shown in Scheme 6.20, in which D = dye and P—H = polymer

$$D \to {}^1D^* \xrightarrow{iso} {}^3D^* \quad \text{(i)}$$
$$D^* + {}^3O_2 \to D + {}^1O_2 \quad \text{(ii)} \quad \text{dry conditions}$$
$$ {}^1O_2 + P\text{—}H \to \text{Oxidation products} \quad \text{(iii)}$$
$$ {}^1O_2 + 2H_2O \to 2H_2O_2 \quad \text{(iv)}$$
$$ H_2O_2 + P\text{—}H \to \text{Oxidation products} \quad \text{(v)} \quad \text{moist conditions.}$$

Scheme 6.20

An important feature of phototendering is that, in general, the presence of moisture increases the rate of degradation. If only singlet oxygen molecules ($^1\Sigma_g$ or $^1\Delta_g$) are produced by the photoactive form of the dye, it is difficult to see how water vapor increases phototendering, particularly when the more energetic $^1\Sigma_g$ state is so efficiently quenched by water. To explain this behavior, it is considered that interaction of triplet dye and ground state oxygen forms a complex comprising an equilibrium between the non-ionized and ionized forms.[7] In the absence

$$[D^*\text{---}O_2] \rightleftarrows [\cdot D^*\text{---}O_2^-] \quad \text{(vi)}$$

of water, only singlet oxygen molecules are formed via the non-ionized complex, and these attack the polymer by reaction (iii). In the presence of moisture, the ionized complex reacts with hydroxide ions:

$$[\cdot D^+ \text{---} O_2^-] + OH^- \rightarrow \cdot D^+ \text{---} OH + O_2^- \qquad \text{(vii)}$$

The hydrogen peroxide required for reaction (v) is formed by further reactions, i.e.

$$O_2^- + H^+ \rightarrow HO_2 \cdot \qquad \text{(viii)}$$

$$2H_2O \cdot \rightarrow H_2O_2 + O_2 \qquad \text{(ix)}$$

If singlet oxygen is a phototendering intermediate, then three factors become apparent: (a) the energy of the dye molecules must be such that the production of singlet oxygen is favorable, (b) the amount of singlet oxygen that a dye can generate must be related to its ability to degrade the fiber, and (c) the removal of singlet oxygen from the dye–polymer system must reduce the photodegradation of the substrate.

By using tetraphenylcyclopentadienone (TPC) as a probe, Griffiths and Hawkins[170] established that irradiation of some yellow/orange vat dyes (C.I. Vat Orange 9, C.I. Vat Yellow 26 and C.I. Vat Yellow 2) in pyridine-methanol solution, and on cotton fabric, results in the formation of singlet oxygen ($^1\Delta_g$). The singlet oxygen sensitizing efficiencies of the dyes on cotton parralleled their relative photottendering activities towards cellulose. This result is favorable evidence in support of the singlet oxygen mechanism of phototendering. Garston[294] investigated the phototendering mechanism of cotton fabric, and of model cellulose films, dyed with Cibanone Yellow R, a photosensitizing vat dye capable of producing singlet oxygen, using FT/IR spectroscopy. On the basis of a good correlation between the energy of the triplet state of the vat dye and the ability of the dye to photosensitize the degradation of cotton, he proposed that singlet oxygen in its high energy $^1\Sigma^+_g$ state, produced by energy transfer from the triplet state of the dye to ground state molecular oxygen, is most likely to be formed in the vat dye photosensitization

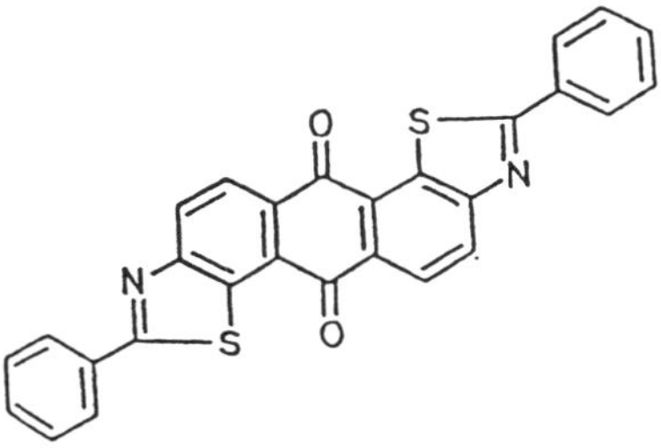

(66) C.I. Vat Yellow 2

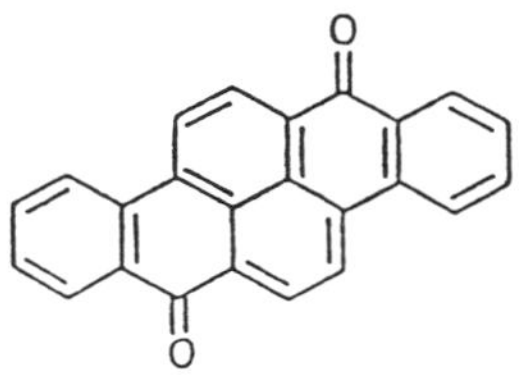

(67) C.I. Vat Orange 9

process, although this would quickly decay to the longer-lived $^1\Delta_g$ state; the pertinent reaction mechanism is shown in Scheme 6.21. Thus, cellobiose (**68**), a dimer of two molecules of β-D-glucose, reacts with singlet oxygen (the higher energy $^1\Sigma^+_g$ form) to give gluconic acid-5-lactone (**69**). It was consequently concluded that the initial step is singlet oxygen, abstraction of a hydrogen atom adjacent to the oxygen bridge. Scission of the carbon–oxygen bridging bond then leads to the formation of a carbonyl group (*cf*. **69**).

Another mechanism for phototendering was proposed by Bamford and Dewar,[297] in which initial interaction between the photoactive dye and

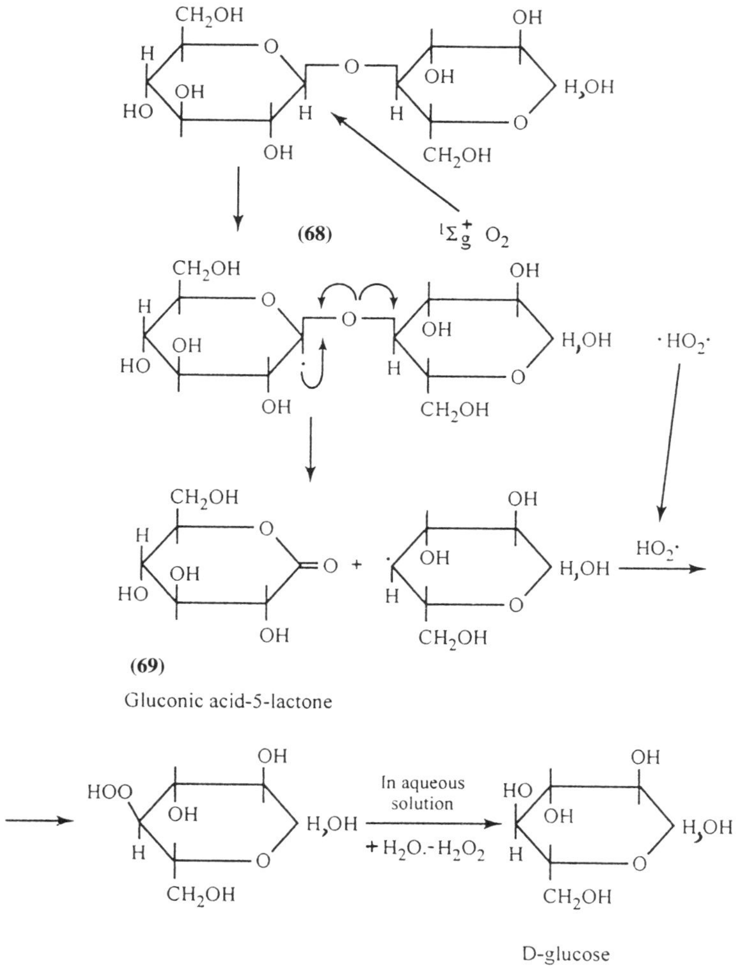

Scheme 6.21

the substrate by a process of hydrogen-atom abstraction is postulated. This generates a free radical center in the substrate, which is then attacked by oxygen in the usual way. The presence of alkali was found to accelerate the process. Under these conditions, the photoactive dye was concluded to abstract an electron from the hydroxyl anion to give active hydroxyl radicals.

$$D + h\nu \rightarrow D^*$$

$$D^* + \text{Cell-H} \rightarrow \text{DH}\cdot + \text{Cell}\cdot$$

$$\text{Cell}\cdot + O_2 \rightarrow \text{Cell-O}_2\cdot$$

$$\text{Cell-O}_2\cdot + \text{Cell-H} \rightarrow \text{Cell-O}_2\text{H} + \text{Cell}\cdot$$

$$\text{Cell-O}_2\cdot + \text{DH} \rightarrow \text{Cell-O}_2\text{H} + D$$

$$D^* + OH^- \rightarrow D^{\overline{\cdot}} + OH\cdot$$

$$OH\cdot + \text{Cell-H} \rightarrow H_2O + \text{Cell}\cdot$$

Scheme 6.22

Instead of the electron transfer step, equations (x) and (xi), which are essentially in accord with the hydrogen atom abstraction theory, have also been reported to account for the accelerating effect of water, and for the regeneration of the dye.[298–299]

$$D^* + H_2O \rightarrow DH\cdot + \cdot OH \quad (x)$$

$$DH\cdot + O_2 \rightarrow D + HO_2\cdot \quad (xi)$$

Several attempts have been made to improve the photostability of some direct dyes on cotton and in cellulose films.[300–304] For example, Kuramoto et al.[300–303] have shown that the addition of a radical scavenger affords little protection against photodegradation, whilst singlet oxygen quenchers such as nickel complexes or salts of dithiobenzil, dialkyldithiocarbamate, mesitylenesulphonate conferred improved resistance of the dyes towards photodegradation. The influence of various additives on the photostability and photoyellowing of jute have been also studied.[305]

6.6.2 Polyamide fibers

The photodegradation of nylon 66 fiber has been extensively reviewed.[14,306,307] Irradiation at wavelengths lower than 300 nm causes direct photolysis of the amide linkages, especially in the absence of oxygen. Photodecomposition at wavelengths greater than 300 nm has been attributed to the presence of metal ions, hydroperoxides, and α, β-unsaturated carbonyl groups introduced during processing.[307,308]

From studies with model amides, e.g. poly(1,3-phenylene isophthalamide),[16,288,307,308] the photodegradation has been considered to involve a photo-Fries rearrangement, as illustrated in Scheme 6.23.

The photostability of polyamide fibers is very dependent upon processing conditions and the presences of additives.[17] Compounds containing carbonyl groups lead to enhanced tendering on exposure to light,[17,307-310] and incorporation of titanium dioxide accelerates phototendering, but the simultaneous incorporation of manganese salts lowers it.[311]

Many compounds have been reported as photostabilizers for nylon 66 fiber.[312] However, UV absorbers are not particularly effective, probably because hydrogen-bonding interactions between the absorber and the amide linkages in the polymer compete with the intramolecular hydrogen bonding system of the absorber.[17]

Many dyes sensitize the photodegradation of nylon 66 fiber, as well as that of other polymers. It has been concluded that both the direct and the dye-sensitized photodegradation of nylon polymer fiber probably involves excited singlet-state oxygen.[7,9,16,188,268] On the other hand, an acid anthraquinone dye (C.I. Acid Blue 40) has been found to form a ground-state complex with nylon 66 fiber upon dyeing.[313] This complex acts as an energy trap which sensitizes the photochemical oxidation of the fiber.

The photodegradation of some monoazo disperse dyes,[89] and of azo and anthraquinone acid dyes[159] in nylon 66 has been examined. In contrast to the oxidative mechanism postulated by Giles et al.[314,315] a

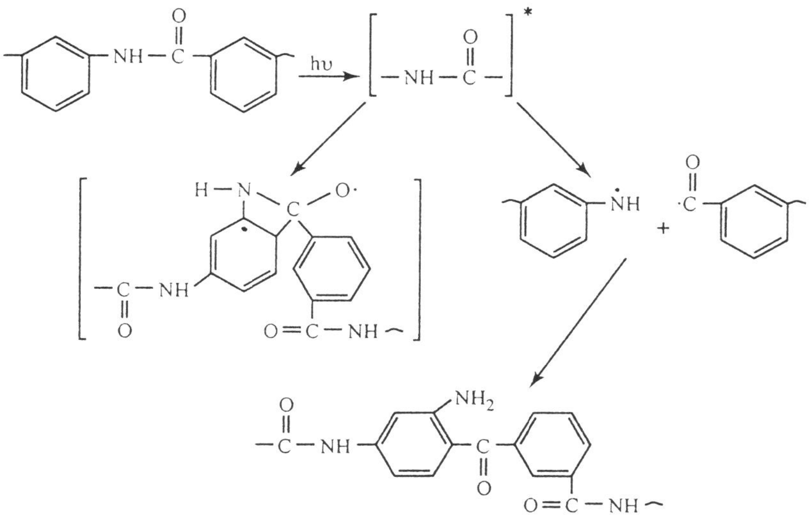

Scheme 6.23

reductive pathway for the azo disperse dyes occurs, and this is confirmed by the formation of amines arising from the reduction of the azo linkage.[89]

Irick and Pacifici,[113] Pietra et al.,[114] Allen et al.[115,116,316] and Chang and Miller[117] have already obtained, in mainly solution-phase studies, evidence for the formation of hydroquinones on UV photolysis of several substituted anthraquinones. Lishan et al.[159] have shown that the photodegradation of azo dyes such as C.I. Acid Red 361 and C.I. Acid Orange 156 involves formation of the hydrazo forms by a reduction process involving hydrogen transfer from the nylon 6 polymer backbone to the excited dye. In this case, the photodegradation process yields two products, as outlined in Scheme 6.24. Amides act as excellent hydrogen donors in the photoreduction, with donation occurring from the carbon atom to the amide nitrogen. The initial step in the reductive process involves hydrogen abstraction by the excited dye from the nylon 6 polymer backbone. Once a radical pair has been formed as a result of hydrogen abstraction, a second hydrogen can then be abstracted to give the hydrazo form of the dye (product A), or the radical pair can collapse to yield a polymer-bound N-alkylhydrazobenzene derivative (product B). The alkyl group is derived from the polymer backbone; product A is oxidized by atmospheric oxygen back to the starting dye, while product B is stable to thermal reactions in the presence of oxygen.[159]

The photodegradation of C.I. Acid Orange 60 (**3**) and C.I. Acid Green 25 (**35**) on nylon 66 fiber has been investigated by Freeman.[136] The process proceeds by hydrogen abstraction from the polymer and solvent. In the case of C.I. Acid Green 25, only the leuco compound is produced and undergoes partial reversion to the starting dye on prolonged exposure to air.

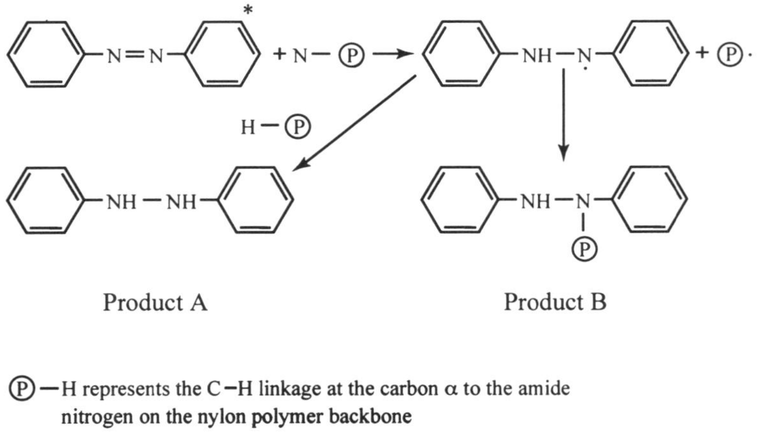

Ⓟ—H represents the C—H linkage at the carbon α to the amide nitrogen on the nylon polymer backbone

Scheme 6.24

6.6.3 Polyester fibers

Photodegradation of poly(ethylene terephthalate) (PET) mainly results by absorption of 290–320 nm light by the ester carbonyl groups.[317] Dull polyester, delustred by incorporation of titanium dioxide, absorbs light of longer wavelengths (320–360 nm) and is much less photostable than bright polyester.[318]

$$H[-OCH_2CH_2O_2C-(C_6H_5)-CO]_n OCH_2CH_2OH \quad (PET)$$

Exposure of PET to light results in cleavage of ester groups and the production of carboxylic, phenolic and ethylene groups, carbon monoxide and carbon dioxide.[317,318] Some terephthalate residues are converted to the corresponding 2-hydroxy- and 2,5-dihydroxyterephthalic acids.[318,319] Simultaneous processes involving a Norrish Type II cleavage reaction, H-atom capture by an excited oligomer chromophore, peroxy radicals and hydroperoxides have been proposed to account for the products and chain fission is believed to occur predominantly in the following way:[320]

$$-\text{C}_6\text{H}_4-\text{COOCH2CH2}\sim\sim\sim \longrightarrow -\text{C}_6\text{H}_4-\text{COOH} + \text{CH2}=\text{CH}\sim\sim\sim$$

A kinetic study of the *cis-trans* isomerization and the photofading behavior of some dialkylaminoazobenzenes in a polyester matrix has been examined by several workers.[152,153,321–324] The dye photofading is found to be strongly dependent on the crystallization conditions and morphology of the PET employed.[323] Photodecomposition mechanisms for aminoanthraquinone disperse dyes on polyester substrates have been reported by many workers, as noted previously.[109,118,179,190] The application of disperse dyes to delustred polyester, which generally improves fiber photostability[318] has been reported and selected disperse dyes used for automotive upholstery fabrics can be applied with a disperse-type UV absorber to achieve adequate lightfastness.[17,325–327]

6.6.4 Polypropylene fibers

Exposure of isotactic polyproplene leads to the production of hydroperoxide and carbonyl groups, and to scission of the polymer chains, resulting in decreased tensile strength and elongation at break. Scheme 6.25 is a simplification of many proposals which have been advanced to account for the oxidative and cleavage reactions.[328,329]

The sensitized photodegradation of polypropylene has also been studied.[16,328] The main initiating process in the benzophenone-sensitized photodegradation is hydrogen-atom abstraction from the polymer matrix by triplet state benzophenone. Other studies have been concerned with the influence of temperature and light intensity on the photodegradation.[330]

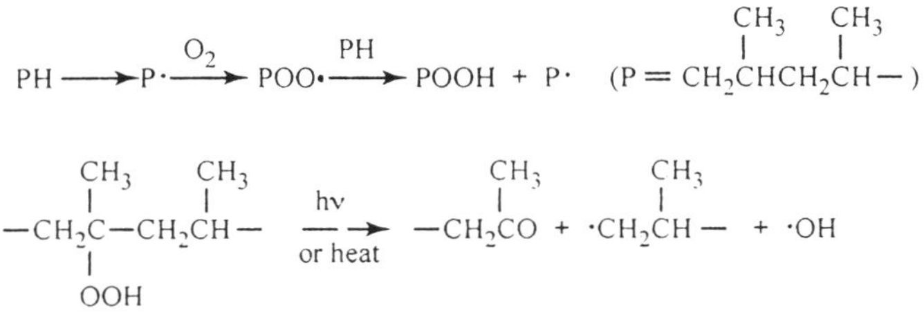

$$PH \xrightarrow{O_2} P\cdot \xrightarrow{PH} POO\cdot \longrightarrow POOH + P\cdot \quad (P = CH_2\overset{\overset{\displaystyle CH_3}{|}}{C}HCH_2\overset{\overset{\displaystyle CH_3}{|}}{C}H-)$$

$$-CH_2\overset{\overset{\displaystyle CH_3}{|}}{\underset{\underset{\displaystyle OOH}{|}}{C}}-CH_2\overset{\overset{\displaystyle CH_3}{|}}{C}H- \xrightarrow[\text{or heat}]{h\nu} -CH_2\overset{\overset{\displaystyle CH_3}{|}}{C}O + \cdot CH_2\overset{\overset{\displaystyle CH_3}{|}}{C}H- + \cdot OH$$

Scheme 6.25

6.6.5 Sensitized photodegradation of dye–polymer systems

Allen et al. have investigated mechanisms of the photosensitized oxidation of polymers by TiO_2 and other white pigments.[13,14,128,331] Polymer oxidation is related to the formation of an oxygen radical anion $O_2^{\cdot -}$ by electron transfer from photoexcited TiO_2 to molecular oxygen.[332] Reactive hydroxyl radicals also form by electron transfer from H_2O to photoexcited TiO_2.[333] These mechanisms are very complex and still require clarification; Allen[13,14] has reviewed more detailed considerations of them.

However, the role of singlet oxygen in the photodegradation of a number of polymers has been established by many workers,[100,192,193] as has its influence on the photofading (tendering) in dye–textile systems. For example, polymers such as polypropylenes, polybutadienes and polyisoprene are somewhat reactive to singlet oxygen produced through dye-photosensitization, generating hydroxy and carbonyl derivatives.[100,192,193] Singlet oxygen produced by a microwave discharge in a flow system reacts with polymers such as polyethylene and nylon 66, resulting in photodegradation.[100,309]

In order to improve the lightfastness of dyes and pigments in polymer substrates, stabilizers such as benzotriazoles (**70**),[334] nickel complexes (**71**),[100,192] hindered phenols (**73**) and hindered amines (**74**, HALS)[334–338] have been introduced into the dye–polymer system. In particular, the photostabilizing efficiencies of nickel complexes of bisditho-α-diketone toward organic dyes and polymer substrates have been widely examined.[59,124,284,302,303]

Kuramoto et al. synthesized a series of bis(ethylenedithiolato) nickel complexes of general formula {$(RCSCSR')_2Ni$, (in which R and R' = phenyl, substituted phenyl, condensed or heterocyclic rings} (**71**)[302,339,340] and reported the photostabilizing efficiencies resultant from the addition of these nickel complexes to various dyes and pigments on fibers,[300] leather,[300] cellulose acetate films,[238,302,340,341] coating films[342–344] and polymers.[303,345] It was observed that substrates containing these complexes showed excellent lightfastness relative to systems without

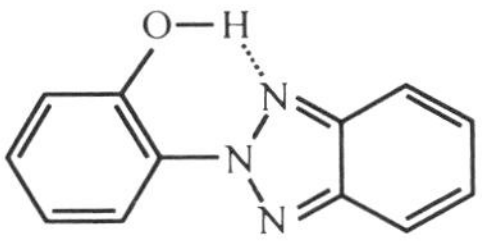

(70) 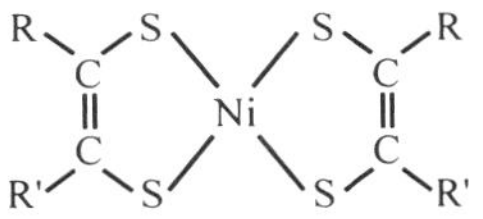 (71)

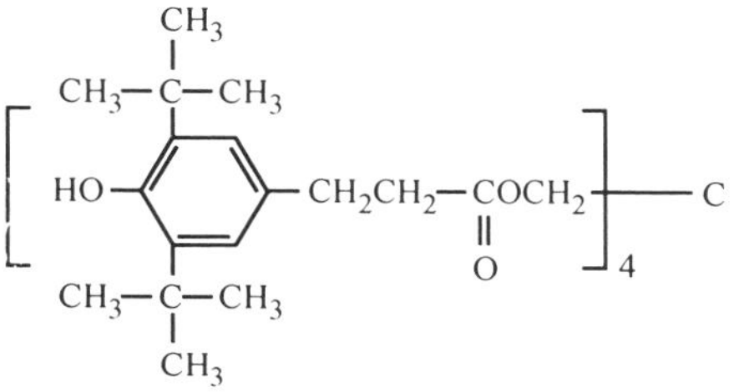

(73) 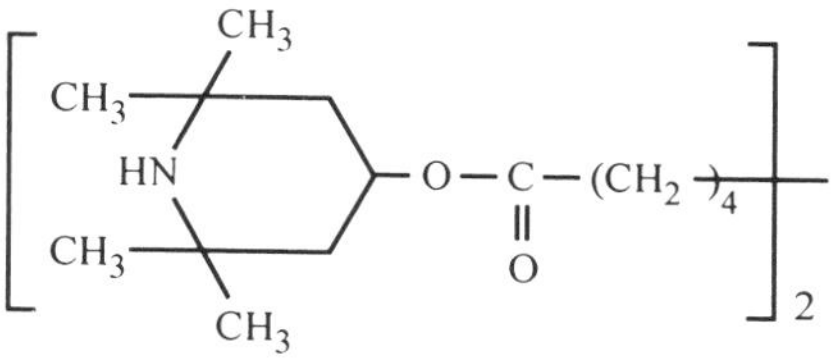 (74)

them. This stabilizing function, based on singlet oxygen quenching, has also been discussed in connection with crystal structure,[340] reduction potential[346] and other properties.[346-349]

The photosensitized oxidation of dyes and/or polymers by zinc oxide or titanium dioxide has received much attention, and the effect seems to be greater in the presence of water.[13,16] Dyeings on some polymers treated with titanium dioxide have an abnormally poor light fastness,[3,16,17] particularly when the dyeings are exposed to light under humid conditions. The active form of titanium dioxide in these experiments appears to have been the anatase form.

6.7 Photochemistry and photostabilization of functional dyes

There are numerous reports outlining the properties,[130,350-352] structural features[353-367] and applications[368-391] of 'functional' dyes. The reader is referred to these sources for detailed background information, as only an outline of specific applications of these dyes is given here.

Some infra-red absorbing dyes such as pentamethinecyanine,[368,369] naphthalocyanines,[370] squarylium[368,371-374] and naphthoquinones dyes[375-377] have been reported as media for optical recording systems. Perylene, phthalocyanine, indigo and azo chromophores have been applied as organic photoconductors in electrophotographic processes. Crystal Violet lactone and fluoranes can be used as heat- and pressure-sensitive dyes for use in carbonless copying paper and in facsimile processes.[378] Quinones, quinophthalones,[379] styryls and indonaphthols are of use in thermal-printer recording systems, giving full color copies from TV and video sets.[365,380,381] Spiropyrans, spirooxazines,[382,383] flugides[384,385] and diarylethenes[386,387] have been proposed as recording media for erasable recording systems. Azo, anthraquinone, naphthoquinone, indigoid and cyanine dichroic dyes can be used in guest–host liquid crystal display systems such as portable TVs and wordprocessors.[376,388–390] Phthalocyanine, trisazo, azulenium and squarylium dyes can be used in the charge carrier generation layer (CGL) in electrophotography and laser printers.[365,376] Some merocyanine, phthalocyanine and squarylium dyes have been proposed as energy-transfer media for solar materials, and the cis-trans photoisomerism of azo dyes, N,N'-diacetylindigo and stilbenes can be utilized in solar energy transfer and in energy storage.[228–230,232] Dithiol metal complexes absorbing at 700–1200 nm have been proposed as infrared absorbing dyes.[391] Some dyes can also be used as probes in medical technology.[350,351] Typical 'functional' dyes are shown in Figure 6.6 and although these dyes need to be sensitive to the action of light, heat or an electric current, they also require a suitable stability against the environment pertinent to their applications.

The lightfastness property of recorded optical disks is a very important parameter. Dyes used in such a system must have a good response towards the laser radiation used in the recording process, while on the recorded layer the disc needs to have good lightfastness properties. Cyanine dyes containing benzothiazole, benzoselenazole, indole and quinoline moieties as terminal heterocyclic rings have been reported in patent specifications for this application. Whilst these dyes have absorption maxima at approximately 800 nm, they generally have poor lightfastness. Although cyanine dyes in solution photodecompose by reaction with singlet oxygen generated via a self-photosensitized process,[392,393] detailed mechanisms of photofading on the optical disk are still unknown. The lightfastness of cyanine dyes is slightly improved by introducing a cyclic unsaturated group, carbonyl group or hetero atom in the center of the linear conjugated system.[381]

In order to study the photodegradation mechanism of squarylium dyes useful for optical data storage systems, Kuramoto[394] carried out photofading experiments on squarylium dyes such as 2,4-bis(4-diethylamino-2-hydroxyphenyl)cyclobutadienediylium-1,3-diolates (**79**), 2,4-bis[(3-ethyl-2-benzothiazolylidene)methyl]cyclobutadienylium-1,3-diolates (**80**) and related dyes, both in solution and in cellulose acetate films. By using both Stern–Volmer analysis and competitive photooxidation experiments on the squarylium dyes with nickel bis(dithiobenzil (**81**), an efficient singlet oxygen quencher, it was found that the photodegradation reactions of the squarylium dyes involved oxidation by excited singlet molecular oxygen in a self-sensitized process, and that singlet oxygen quenchers such as nickel dithiolato complexes had an inhibiting effect on the photoreaction of the dyes. The rate constants (k) for the reaction of squarylium dyes with singlet oxygen were $k = 1.3 \sim 1.8 \times 10^{10}$ l/(mol s), and nickel bis(dithiobenzil) was shown to react with singlet oxygen about 8–13 times faster than the dyes.[341,394] It has also been shown that the use of singlet oxygen quenchers is effective in improving the lightfastness of spin-coated thin layers of squarylium dyes on polycarbonate substrates.[395,396] A similar inhibiting effect by adding nickel dithiolate complexes has been shown for thin layers of cyanine dyes.[397]

Namba[369,398] found that salt formation between a quencher anion and a dye cation could increase the lightfastness of cyanine dyes in a recording layer. The salt formed between the cyanine dye cation and a benzenedithiol Ni complex anion is remarkably effective. This salt (**84**) allows the power with which reproduction could be carried out to be more than 10^6 times increased, i.e. to a practically acceptable level.

Photochromic spiropyranes and spirooxazines have received considerable attention as functional materials for various applications, due to their durability as materials which can reversibly change between colored and non-colored states. The Spiropyrane (**85**) is colorless, but absorbs at 700 nm after UV irradiation to give the merocyanine (**86**), which then reverts to (**85**) on exposure to visible light. Many investigations have been carried out in order to obtain information on the more detailed mechanism of these color changes,[382,383] and on their practical application in erasable optical recording media.[399] The reversible cycle of these dyes, however, takes place together with gradual photo- and/or thermal decomposition, which make the dyes too unstable for use in optical recording media. Flugide (**87**) shows a similar photochromism, to give the colored compound (**88**).

It is thought that the most probable mechanism of the photodecomposition is an 'ene type' oxidation of a central conjugated double bond in open form, by singlet oxygen generated via a self-sensitized process. In fact, the photostability of spiropyranes and spirooxazines in solution or polymer films increases on addition of singlet oxygen quenchers, such as nickel dithiolato complexes and DABCO.

Crystal Violet lactone (**89**) and 3-diethylamino-6-methyl-7-anilinofluorane (**91**) are extensively used as indicator dyes in various business machines. The colored materials (**90** and **92**) derived from these color formers (chromogenic compounds; colorless) show a bluish–violet and black tone. The colored materials are not fast to light, and an improvement of this property is necessary.

An acid catalyst opens the lactone ring in **89**, thereby allowing the violet triarylmethane dye **90** to form. The structure of **90** resembles that of Crystal Violet (C.I. Basic Violet 3). Therefore, it can be assumed that the processes of oxidation by singlet oxygen,[124] and of dealkylation[201–204]

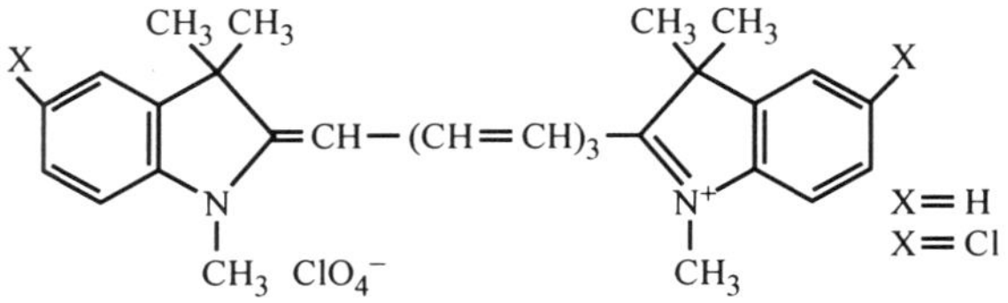

(75)

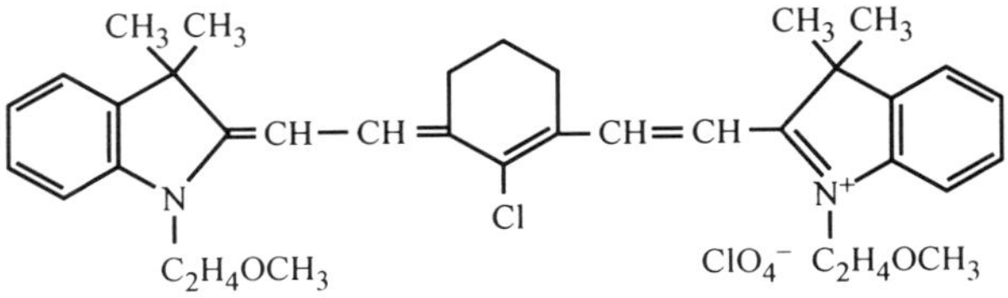

(76)

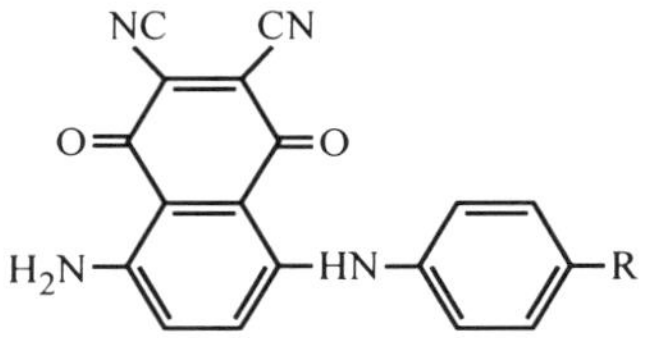

(77)

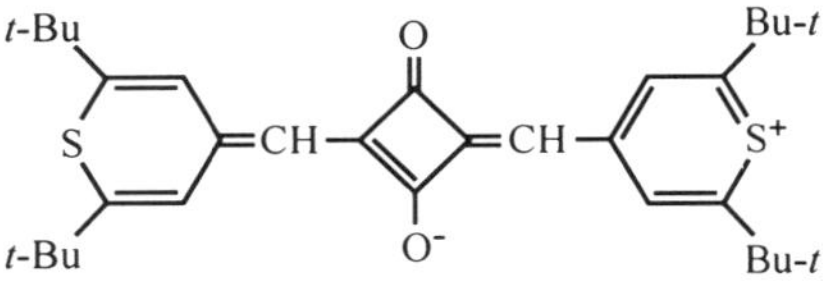

(78)

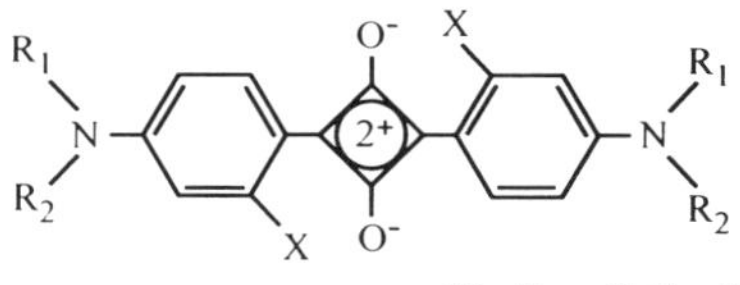

(79) a) $R_1, R_2 = C_2H_5$, $X = OH$
b) $R_1, R_2 = CH_3$, $X = CH_3$

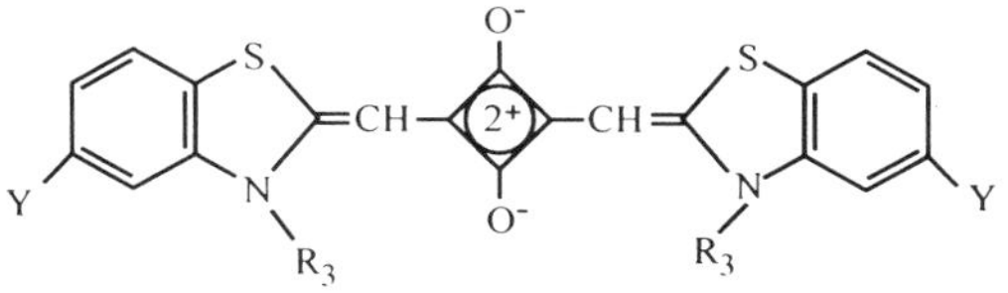

PHOTODEGRADATION OF SYNTHETIC COLORANTS 241

(80) a) $R_3 = C_2H_5$, Y = H
      b) $R_3 = C_2H_5$, Y = CH_3

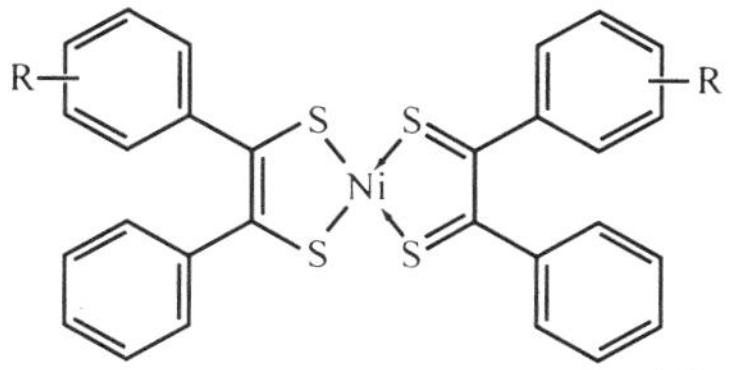

(81) a) R = H
      b) R = 4-OCH_3

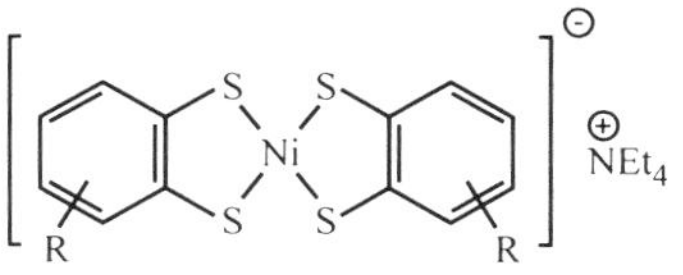

(82)

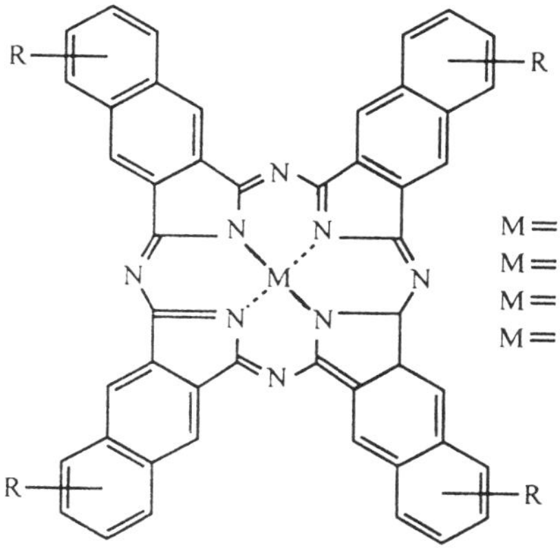

(83) M = Zn, R = H
      M = H, R = F_4
      M = VO, R = t-C_5H_{11}
      M = Si[OSi(CH_3)$_2$OR]$_2$

Figure 6.6 Structures of some functional dyes and related compounds used in the electronics industry.

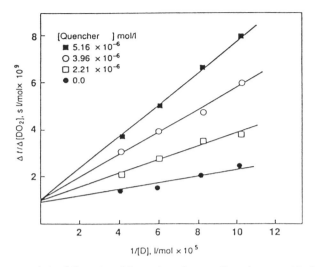

Figure 6.7 Inverse plots of the rate of formation of squarylium dye-peroxide ($\Delta t/\Delta[DO_2]$) as a function of $1/[D]$ for different quencher concentrations; duration of irradiation ($\lambda > 400$ nm) using a 500 W Xenon lamp was 300 s under atmospheric conditions in chloroform (dye: **79a**, quencher: **81a**).

(**84**)

may be involved in the photodecomposition of **90**, as described in a previous section. In fact, when **90** is exposed for 10 h to filtered radiation ($\lambda > 300$ nm) on a silica gel plate in air, it shows 48% conversion, and in the presence of equivalent molar amounts of nickel bis(dithiobenzil) or nickel dibutyldithiocarbamate, both of which are good singlet oxygen quenchers, shows only 13% and 15% conversion, respectively.[124] Although the photodecomposition mechanism of **92** has not been fully defined, similar effects of the additives are observed on a silica gel plate.

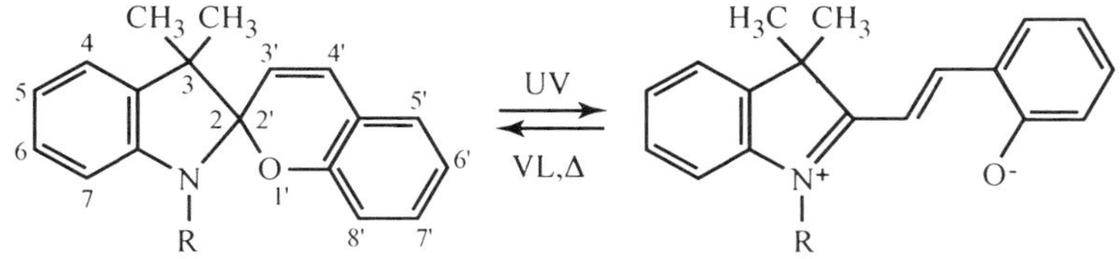

PHOTODEGRADATION OF SYNTHETIC COLORANTS 243

(85) ⇌ (UV / VL,Δ) **(86)**

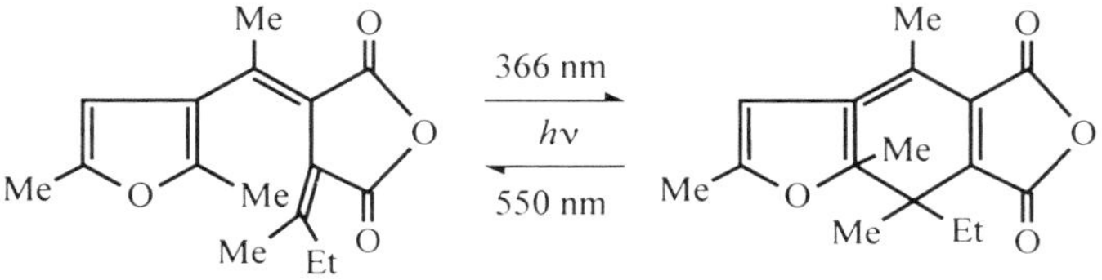

(87) ⇌ (366 nm hν / 550 nm) **(88)**

Scheme 6.26 Photochromic compounds.

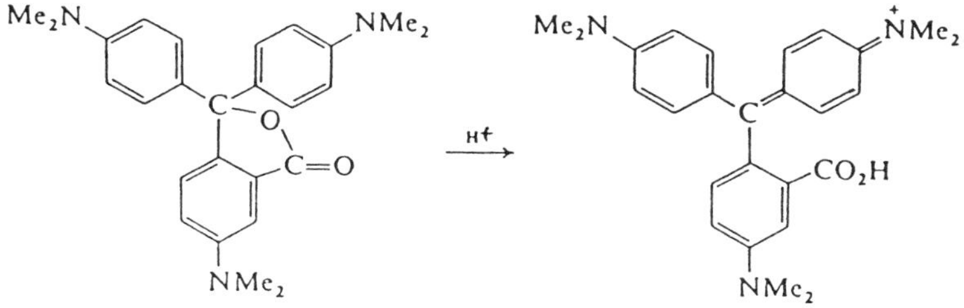

(89) (colourless) →(H+) **(90)** (violet)

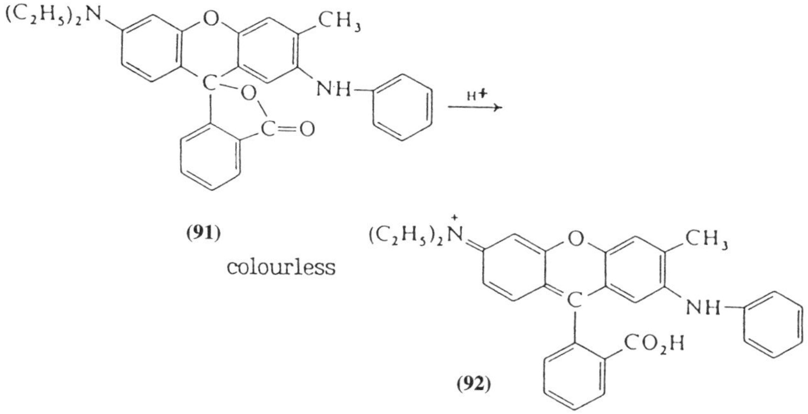

(91) colourless →(H+) **(92)** black

Scheme 6.27

On the other hand, it has also been found that the dye counter ion, e.g. amphoteric counter ion, plays an important role in the suppression of the decomposition of color formers.[400] In this case, Ni and Zn and complexes of 1,2-benzenedicarboxylic acid and 2-hydroxybenzoic acid afford an excellent suppressing effect on the photo- and/or thermal decomposition.[401]

Many other properties and applications of functional dyes, and developments in new electro-optical systems, have been summarized.[352,365–367,402]

6.8 Conclusions

The photodecomposition of synthetic dyes has been widely studied and yet remains a complex and in some cases, little understood subject. This is true, in part, because several different mechanisms of dye decomposition may govern the fading of the same dye on different substrates and in the presence of different light sources. While the extensive studies which have been carried out in this area outline mechanisms thought to be important controlling influences on the light stability of dyed substrates, they still require fuller clarification.

Although functional dyes for use in the electronics industry require special properties, such as near-infrared absorption and photocromism, stability to the light is still a very important feature. A key to the success of the next generation of lightfast dyes will no doubt be the development of colorants that take into account the results outlined in reviews such as this present chapter.

Acknowledgment

The author is pleased to dedicate this review to the late Professor T. Kitao of the University of Osaka Prefecture, who died on 4 January 1992. The author would like to acknowledge the invaluable guidance and encouragement given by Professor Kitao in the conception of the review.

References

1. E. Bancroft, *Experimental Researches Concerning the Philosophy of Permanent Colours*, Cadell and Davies, Eds., 2nd ed., London (1813).
2. C. H. Giles and R. B. McKay, *Text. Res. J.*, **33**, 527 (1963).
3. G. S. Egerton and A. G. Morgan, *J. Soc. Dyers Colour.*, **86**, 242 (1970).
4. N. A. Evans and I. W. Stapleton, *The Chemistry of Synthetic Dyes*, Vol. **8**, K. Venkataraman ed., p. 221, Academic Press, (1978).

5. G. S. Egerton and A. G. Morgan, *J. Soc. Dyers Colour.*, **86** 79 (1970).
6. G. S. Egerton and A. G. Morgan, *J. Soc. Dyers Colour.*, **87** 223 (1971).
7. G. S. Egerton and A. G. Morgan, *J. Soc. Dyers Colour.*, **87** 268 (1971).
8. C. S. Egerton, *Br. Polym. J.*, **3**, 63 (1971).
9. H. Meir, in *The Chemistry of Synthetic Dyes*, Vol. **IV**, K. Venkataraman, ed., Chapter 7, Academic Press, New York (1971).
10. J. Griffiths, *Chem. Soc. Rev.*, **1**, 481 (1972).
11. C. H. Giles, D. G. Duff and R. S. Sinclair, *Rev. Prog. Coloration*, **12**, 58 (1982).
12. R. S. Sinclair, *Photochem. Photobiol.*, **31**, 627 (1980).
13. N. S. Allen, *Rev. Prog. Coloration*, **17**, 61 (1987).
14. N. S. Allen and J. F. McKellar, eds., *Photochemistry of Dyed and Pigmented Polymers*, Elsevier Applied Science, London (1980).
15. C. H. Giles and S. D. Forrester, *Photochemistry of Dyed and Pigmented Polymers*, N. S. Allen and J. F. McKellar, eds., Chapter 2, p. 51, Elsevier Applied Science, London (1980).
16. P. Bentley, J. F. McKellar and G. O. Phillips, *Rev. Prog. Coloration*, **5**, 33 (1974).
17. B. Miligan, *Rev. Prog. Coloration*, **16**, 1 (1986).
18. T. Matsuura, *Organic Photochemistry*, p. 23, Kagakudojin, Kyoto (1970).
19. H. H. Jaffé and M. Orchin, *Theory and Applications of Ultraviolet Spectroscopy*, Wiley, New York (1962).
20. C. A. Parker, in *Advances in Photochemistry*, Vol. **II**, W. A. Noyes Jr., G. S. Hammond and J. N. Pitts Jr., eds., Interscience, New York (1964).
21. M. Koizumi, S. Kato, N. Mataga, T. Matsuura and Y. Usui, *Photosensitized Reactions*, p. 7, Kagakudojin, Kyoto (1978).
22. H. Nakata, *Organic Photochemistry*, p. 11, Maruzen, Tokyo (1968).
23. G. S. Egerton and A. G. Morgan, *J. Soc. Dyers Colour.*, **86**, 79 (1970).
24. H. C. A. van Beek and P. M. Heetjes, *J. Soc. Dyers Colour.*, **79**, 661 (1963).
25. H. C. A. van Beek and P. M. Heetjes, *J. Phys. Chem.*, **70**, 1704 (1966).
26. J. G. Neevel, H. C. A. van Beek and B. van de Graaf, *J. Soc. Dyers Colour.*, **107**, 110 (1991).
27a J. G. Neevel, H. C. A. van Beek and B. van de Graaf, *J. Soc. Dyers Colour.*, **108**, 150 (1992).
27b J. G. Neevel, M. Peereboom and B. van de Graaf, *J. Soc. Dyers Colour.*, **109**, 116 (1993).
28. Y. Okada, M. Hirose, T. Kato, H. Motomura and Z. Morita, *Dyes Pigm.*, **14**, 265 (1990).
29. Y. Okada, E. Satoh, H. Motomura and Z. Morita, *Dyes Pigm.*, **19**, 1 (1992).
30. Y. Okada, T. Kato, H. Motomura and Z. Morita, *Sen'i Gakkaishi*, **46**, 346 (1990).
31. Y. M. A. Naguib, S. G. Cohen and C. Steel, *J. Am. Chem. Soc.*, **108**, 128 (1986).
32. H. Kautsky, H. de Brujn, H. Neuwirth and W. Baumeister, *Chem. Ber.*, **66**, 1588 (1933).
33. H. Kautsky, *Trans. Faraday Soc.*, **35**, 216 (1939).
34. Y. Usui and M. Koizumi, *Bull. Chem. Soc. Jpn.*, **40**, 440 (1967).
35. Y. Usui, K. Itoh and M. Koizumi, *Bull. Chem. Soc. Jpn.*, **38**, 1015 (1965).
36. G. S. Egerton, *J. Soc. Dyers Colour.*, **65**, 764 (1949).
37. G. S. Egerton, E. Attle, F. Guirguis and M. A. Rathor, *J. Soc. Dyers Colour.*, **79**, 49 (1963).
38. N. Kuramoto and T. Kitao, *Nippon Kagaku Kaishi*, **258** (1977).
39. R. F. Weiner and H. H. Seliger, *Photochem. Photobiol.*, **4**, 1207 (1965).
40. V. Kasche and L. Lindqvist, *Photochem. Photobiol.*, **4**, 923 (1965).
41. C. Balny and P. Darzon, *Biochem. Biophys. Res. Commun.*, **56**, 386 (1974).
42. H. Kautsky, *Biochem. Z.*, **291**, 271 (1937).
43. H. Kautsky and A. Hirsch, *Chem. Ber.*, **65B**, 401 (1932).
44. C. S. Foot and S. Wexler, *J. Am. Chem. Soc.*, **86**, 389 (1964).
45. C. S. Foot and S. Wexler, *J. Am. Chem. Soc.*, **86**, 3880 (1964).
46. C. S. Foot, *Acc. Chem. Res.*, **1**, 104 (1968).
47. E. J. Gorey and W. C. Talor, *J. Am. Chem. Soc.*, **86**, 3881 (1964).
48. C. S. Foot, S. Wexler and W. Ando, *Tetrahedron Lett.*, 4111 (1965).

49. C. S. Foot, S. Wexler, W. Ando and R. Higgins, *J. Am. Chem. Soc.*, **90**, 975 (1968).
50. E. J. Corey and W. C. Taylor, *J. Am. Chem. Soc.*, **86**, 3881 (1964).
51. A. U. Khan and M. Kasha, *J. Chem. Phys.*, **39**, 2105 (1963).
52. G. O. Schenck and E. Koch, *Z. Elektrochem.*, **64**, 170 (1960).
53. H. H. Wasserman and J. R. Scheffer, *J. Am. Chem. Soc.*, **89**, 3073 (1967).
54. P. B. Merkel, R. Nilsson and D. R. Kearns, *J. Am. Chem. Soc.*, **94**, 1030 (1972).
55. R. W. Denny and A. Nickon, *Organic Reactions*, Vol. **20**, p. 133, John Wiley, New York (1973), and references therein.
56. K. Kawaoka, A. U. Khan and D. R. Kearns, *J. Chem. Phys.*, **46** 1842 (1967).
57. D. R. Kearns, *Singlet Oxygen*, H. H. Wasserman, ed., p. 115, Academic Press, New York (1979), and references therein.
58. E. Wasserman, R. M. Murray, M. L. Kaplan and W. A. Yager, *J. Am. Chem. Soc.*, **90**, 4160 (1968).
59. M. Kasha and D. E. Brabham, *Singlet Oxygen*, H. H. Wasserman, ed., p. 24, Academic Press, New York (1979), and references therein.
60. G. Herzberg, *Spectra of Diatomic Molecules*, 2nd ed., p. 560, D. van Nostrand, Princeton, New York (1950).
61. A. M. Winer and K. D. Bayes, *J. Phys. Chem.*, **70**, 302 (1966).
62. K. Kear and E. W. Abrahamson, *J. Photochem.*, **3**, 409 (1974).
63. R. P. Wayne, *Adv. Photochem.*, **7**, 311 (1969).
64. K. Gollnick and G. O. Schenck, *Pure Appl. Chem.*, **9**, 507 (1964).
65. C. S. Foote, Y. C. Chang and R. W. Denny, *J. Am. Chem. Soc.*, **92**, 5216, 5218 (1970).
66. P. B. Merkel and D. R. Kearns, *J. Am. Chem. Soc.*, **94**, 1029 (1972).
67. D. R. Adams and F. Wilkinson, *J. Chem. Soc., Faraday Trans.*, **I**, 586 (1972).
68. C. A. Long and D. R. Kearns, *J. Am. Chem. Soc.*, **97**, 2018 (1975).
69. T. Kajiwara and D. R. Kearns, *J. Am. Chem. Soc.*, **95**, 5886 (1973).
70. K. McLaren, *Can. Text. J.*, **76**, 41 (1959).
71. K. McLaren, *J. Soc. Dyers Colour.*, **78**, 34 (1962).
72. T. Hirashima, O. Manabe and H. Hiyama, *Kogyo Kagaku Zasshi*, **70**, 1533 (1967).
73. British Standard 1006 (1978), British Standards Institution (London).
74. L. M. Lock and G. C. Frank, *Text. Res. J.*, **43**, 502 (1973).
75. M. J. Wall and G. C. Frank, *Text. Res. J.*, **41**, 32 (1971).
76. J. Park, *J. Soc. Dyers Colour.*, **95**, 304 (1979).
77. L. A. Holt and P. J. Waters, *Proc. Int. Wool Text. Res. Conf.*, Tokyo, **IV**, 1 (1985).
78. Japanes Standard JIS-L0843 (1971).
79. S. V. Vaeck, *J. Soc. Dyers Colour.*, **94**, 301 (1978).
80. K. McLaren, *J. Soc. Dyers Colour.*, **72**, 527 (1956).
81. F. Gasser and H. Zukriegel, *Melliand Textilber*, **33**, 44 (1952).
82. C. H. Giles and R. B. Mckay, *Text. Res. J.*, **33**, 527 (1963).
83. A. H. Little and J. W. Clayton, *J. Soc. Dyers Colour.*, **79**, 671 (1963).
84. T. Toda and M. Hida, *Sen'i Gakkaishi*, **46**, 155 (1990).
85. K. Nakatsuka, H. Hattori and S. Ikeo, *Senryo to Yakuhin*, **33**, 255 (1988).
86. C. H. Giles, C. D. Shah, W. E. Watts and R. S. Sinclair, *J. Soc. Dyers Colour.*, **88**, 433 (1972).
87. M. Kamel, R. M. Issa, M. F. Abdel-Wahab, S. Shakra and A. M. Osman, *Textilveredlung*, **6**, 226 (1971).
88. A. Arcoria and G. Parisi, *J. Soc. Dyers Colour.*, **100**, 13 (1984).
89. A. Arcoria, M. L. Long and G. Parisi, *J. Soc. Dyers Colour.*, **100**, 339 (1984).
90. J. Griffiths and C. Hawkins, *J. Chem. Soc. Chem. Commun.*, (1972) 463.
91. J. Griffiths and C. Hawkins, *J. Chem. Soc. Perkin II*, (1977) 747.
92. N. Kuramoto and T. Kitao, *J. Soc. Dyers Colour.*, **95**, 257 (1979).
93. N. Kuramoto and T. Kitao, *Senryo to Yakuhin*, **24**, 91 (1979).
94. N. Kuramoto and T. Kitao, *J. Soc. Dyers Colour.*, **96**, 529 (1980).
95. H. Gruen, H. Steffen and Schulte-Frohlinda, *J. Soc. Dyers Colour.*, **97**, 430 (1981).
96. P. B. Merkel and W. F. Smith Jr., *J. Phys. Chem.*, **83** 2834 (1979).
97. R. Haessner, H. Mustroph and J. Epperlein, *J. Prakt. Chem.*, **325**, 943 (1983).
98. I. M. Byteva, O. L. Golomb, G. P. Gurinovich and V. V. Karpov, *Zh. Prikl. Spektrosk.*, **36**, 770 (1982).

99. J. Griffiths and C. Hawkins, *J. Soc. Dyers Colour.*, **89**, 173 (1973).
100. A. Zweig and W. A. Henderson, *J. Polym. Sci. Polym. Chem. Ed.*, **13**, 717 (1977).
101. M. W. Rembold and H. E. A. Kramer, *J. Soc. Dyers Colour.*, **94**, 12 (1978).
102. M. W. Rembold and H. E. A. Kramer, *J. Soc. Dyers Colour.*, **96**, 122 (1980).
103. G. Schwen and G. Schmidt, *J. Soc. Dyers Colour.*, **75**, 101 (1959).
104. C. H. Giles and R. H. Haslam, *Text. Res. J.*, **47**, 348 (1977).
105. F. M. Tera, L. A. Aldon, M. N. Michael and A. Hebeish, *Polym. Photochem.*, **6**, 373 (1985).
106. P. G. Wang and I. J. Wang, *Text. Res. J.*, **61**, 162 (1991).
107. Y. Honma, N. Choji and M. Karasawa, *Sen'i Gakkaishi*, **42**, T514 (1986).
108. T. H. Morton, *J. Soc. Dyers Colour.*, **65**, 597 (1949).
109. C. H. Giles, *J. Soc. Dyers Colour.*, **73**, 127 (1957).
110. G. Baxter, C. H. Giles, M. K. McKee and N. Macaulay, *J. Soc. Dyers Colour.*, **71**, 218 (1955).
111. M. Hayashi, *J. Chem. Soc. Jpn, Ind. Eng. Sect.*, **63**, 118 (1960).
112. C. H. Giles, A. Yabe and C. D. Shah, *Text. Res. J.*, **38**, 467 (1968).
113. G. Irick and J. G. Pacific, *Tetrahedron Lett.*, 2207 (1969).
114. A. Albini, E. Fasani and S. Pietra, *J. Chem. Soc. Perkin II*, 1021 (1983).
115. N. S. Allen, B. Harwood and J. F. McKellar, *J. Photochem.*, **10**, 187 (1979).
116. N. S. Allen, J. M. Mckellar and B. M. Moghaddam, *J. Photochem.*, **10**, 193 (1979).
117. I. Y. Chang and I. K. Miller, *J. Soc. Dyers Colour.*, **102**, 46 (1986).
118. C. H. Giles and R. S. Sinclair, *J. Soc. Dyers Colour.*, **88**, 109 (1972).
119. N. Kuramoto and T. Kitao, *J. Chem. Soc. Chem. Commun.*, 379 (1979).
120. N. Kuramoto and T. Kitao, *J. Chem. Soc. Perkin II*, 1569 (1980).
121. N. Kuramoto and T. Kitao, *J. Chem. Tech. Biotechnol.*, **30**, 129 (1980).
122. N. Kuramoto and T. Kitao, *Dyes Pigm.*, **2**, 133 (1981).
123. N. Kuramoto and T. Kitao, *J. Soc. Dyers Colour.*, **98**, 159 (1982).
124. N. Kuramoto and T. Kitao, *Dyes Pigm.*, **3**, 49 (1982).
125. N. Kuramoto and T. Kitao, *J. Soc. Dyers Colour.*, **98**, 334 (1982).
126. H. Mustroph, J. Potocnak and N. Grossmann, *J. Prakt. Chem.*, **326**, 979 (1984).
127. H. Mustroph and C. Weiss, *J. Prakt. Chem.*, **328**, 937 (1986).
128. N. S. Allen, *Rev. Plast. Mod.*, **55**, 221 (1988).
129. T. Kitao, *Kagaku Kogyo*, **31**, 1035 (1980).
130. S. Ohokawara, Y. Kuroki and T. Kitao, *Chemistry of Kinousei-Shikiso*, CMC Publ. Co. Ltd., p. 65, Osaka (1981).
131. T. Kitao, *Sensyoku Kogyo*, **24**, 501 (1976).
132. C. D. Shah and R. Srivastava, *J. Appl. Chem. Biotechnol.*, **27**, 429 (1977).
133. N. G. Makhviladze, V. I. Pergushov, V. S. Gurman and G. E. Krichevskii, *Zh. Prikl. Khim*, **5**, 1331 (1984).
134. J. N. Pitts Jr., E. A. Schuck and J. K. S. Wan, *J. Am. Chem. Soc.*, **86**, 296 (1964).
135. C. A. Parker, *J. Phys. Chem.*, **63**, 26 (1959).
136. H. S. Freeman and J. Sokolowska-Gajda, *Text. Res. J.*, **60**, 221 (1990).
137. J. Griffiths and C. Hawkins, *J. Appl. Chem. Biotechnol.*, **27**, 558 (1977).
138. N. Kuramoto and T. Kitao, *Rep. Osaka Pref. Ind. Res. Inst.*, **80**, 9 (1982).
139. N. Kuramoto and M. Wakae, *Yuki Gosei Kagaku Kyoukaishi*, **32**, 118 (1974).
140. N. V. Koval'chuk, A. V. Anisimov, O. A. Tambieva and V. V. Titov, *Zh. Org. Khim.*, **22**, 1944 (1986).
141. A. Albini, E. Fasani, S. Pietra and A. Sulpizio, *J. Chem. Soc., Perkin Trans II*, 1689 (1984).
142. P. Ball and C. H. Nicholls, *Dyes Pigm.*, **5**, 437 (1984).
143. H. G. O. Becker and J. Franze, *J. Prakt. Chem.*, **323**, 171 (1981).
144. C. Muller, *Am. Dyest. Rep.*, **59**, 37 (1970).
145. T. Kitao, Y. Watada, M. Matsuoka and K. Konishi, *Nippon Kagaku Kaishi*, 757 (1974).
146. T. Urbanski, *Hydrogen Bonding*, D. Hadzi, ed., Pergamon Press, London p. 145 (1959).
147. A. Whitaker, *J. Soc. Dyers Colour.*, **94**, 431 (1978).
148. G. M. Wyman, *Chem. Rev.*, **55**, 625 (1955).
149. P. Jacques, *Dyes Pigm.*, **5**, 351 (1984).

150. H. Yamamoto and A. Nishida, *Nippon Kagaku Kaishi*, 2338 (1985).
151. F. K. Sutcliffe, H. M. Shahidi and D. Patterson, *J. Soc. Dyers Colour.*, **94**, 306 (1978).
152. E. Dubini-Paglia, P. L. Beltrame, B. Marcandalli and P. Carniti, *J. Appl. Polym. Sci.*, **31**, 1251 (1986).
153. P. L. Beltrame, E. Dubini-Paglia and B. Marcandalli, *J. Appl. Polym. Sci.*, **33**, 2965 (1987).
154. P. L. Beltrame, E. Dubini-Paglia and B. Marcandalli, *J. Appl. Polym. Sci.*, **38**, 755 (1989).
155. E. Dubini-Paglia, P. L. Beltrame, A. Seves and G. Prati, *J. Soc. Dyers Colour.*, **105**, 107 (1989).
156. H. Inoue, T. D. Tuong, H. Hida and T. Murata, *J. Chem. Soc., Chem. Commun.*, 1347 (1971).
157. C. V. Stead, *Rev. Prog. Coloration*, **6**, 1 (1975).
158. C. H. Giles, B. T. Hojiwala and C. D. Shah, *J. Soc. Dyers Colour.*, **88**, 403 (1972).
159. D. G. Lishan, P. W. Harris, K. Brahim and R. L. Jackson, *J. Soc. Dyers Colour.*, **104**, 33 (1988).
160. G. Irick and J. G. Pacifici, *Tetrahedron Lett.*, 1303 (1969).
161. A. Albini and E. Fasani and S. Pietra, *J. Chem. Soc., Perkin II*, 1393 (1982).
162. N. S. Allen, B. Harword and J. F. McKellar, *J. Photochem.*, **9**, 559 (1978).
163. G. Irick and E. G. Boyd, *Text. Res. J.*, **43**, 238 (1973).
164. A. K. Davies, G. A. Gee, J. F. McKellar and G. O. Phillips, *J. Chem. Soc., Perkin II*, 1742 (1973).
165. N. S. Allen, J. P. Binkley, B. J. Parsons and G. O. Phillips, *Dyes Pigm.*, **4**, 1 (1983).
166. G. A. Horsfall, *Text. Res. J.*, **52**, 197 (1982).
167. B. V. Rao, V. Choudhary and I. K. Varma, *J. Soc. Dyers Colour.*, **106**, 388 (1990).
168. A. K. Davies, R. Ford, G. A. Gee, J. F. McKellar and G. O. Phillips, *J. Chem. Soc., Chem. Commun.*, (1972) 873.
169. A. K. Davis, G. A. Gee, J. F. McKellar and G. O. Phillips, *Chem. Ind.*, 431 (1973).
170. J. Griffiths and C. Hawkins, *J. Soc. Dyers Colour.*, **89**, 173 (1973).
171. B. Garston, *J. Soc. Dyers Colour.*, **96**, 535 (1980).
172. A. T. Peters and N. O. Soboyejo, *J. Soc. Dyers Colour.*, **105**, 79 (1989).
173. A. T. Peters and N. O. Soboyejo, *J. Soc. Dyers Colour.*, **105**, 315 (1989).
174. A. T. Peters and Y. C. Chao, *J. Soc. Dyers Colour.*, **104**, 435 (1988).
175. G. E. Egerton and N. E. N. Assaad, *J. Soc. Dyers Colour.*, **83**, 85 (1967).
176. G. E. Egerton, J. M. Gleadle and A. G. Roach, *J. Soc. Dyers Colour.*, **82**, 369 (1966).
177. K. Yamada, H. Shosenji, S. Fukunaga and K. Hirahara, *Bull. Chem. Soc. Jpn.*, **49**, 3701 (1976).
178. N. K. Bridge, *J. Soc. Dyers Colour.*, **76**, 484 (1960).
179. E. McAlpine and R. S. Sinclair, *Text. Res. J.*, **47**, 283 (1977).
180. D. Wegerle, *J. Soc. Dyers Colour.*, **89**, 54 (1973).
181. N. S. Allen, K. O. Fatinikum, A. K. Davies, B. J. Parsons and G. O. Phillips, *Dyes Pigm.*, **2**, 219 (1981).
182. C. Couper, *Text. Res. J.*, **21**, 720 (1951).
183. V. S. Salvin and P. Walker, *Text. Res. J.*, **30**, 381 (1960).
184. C. H. Giles and R. S. Sinclair, *J. Soc. Dyers Colour.*, **88**, 109 (1972).
185. N. S. Allen and B. Harwood, *Polym. Deg. Stab.*, **4**, 319 (1982).
186. N. S. Allen, P. Bentley and J. F. McKellar, *J. Photochem.*, **5**, 225 (1976).
187. N. S. Allen, J. F. McKellar and B. M. Moghaddam, *J. Chem. Technol. Biotechnol.*, **29**, 119 (1979).
188. N. S. Allen, J. F. McKellar, B. M. Moghaddam and G. O. Phillips, *J. Photochem.*, **11**, 101 (1979).
189. N. S. Allen, J. F. McKellar and S. Protopappas, *J. Appl. Chem. Biotechnol.*, **27**, 269 (1977).
190. N. S. Allen, J. F. McKellar and D. Wilson, *Makromol. Chem.*, **179**, 269 (1978).
191. N. S. Allen, J. F. McKellar, B. M. Moghaddam and G. O. Phillips, *Chem. Ind.*, (London) (1979) 593.
192. A. Zweig, W. A. Henderson, *J. Polym. Sci. Polym. Chem. Ed.*, **13**, 993 (1975).
193. J. P. Gullory and C. F. Cook, *J. Polym. Sci. Polym. Chem. Ed.*, **11**, 1927 (1973).
194. A. J. Paint, F. M. Winnik and V. Bocking, *J. Appl. Polym. Sci.*, **36**, 1205 (1988).

195. P. Bentley, J. F. McKellar and G. O. Phillips, *J. Chem. Soc., Perkin II*, 523 (1974).
196. P. Bentley, J. F. McKellar and G. O. Phillips, *J. Chem. Soc., Perkin II*, 1259 (1975).
197. N. Kuramoto, *Rep. Osaka Pref. Ind. Res. Inst.*, **91**, 22 (1987).
198. G. Oster, *J. Polym. Sci.*, **B2**, 891 (1964).
199. McVie, R. S. Sinclair, Truscott, *Photochem. Photobiol.*, **29**, 395 (1979).
200. P. C. Henriquer, *Rec. Trav. Chim.*, **52**, 998 (1933).
201. C. M. Desai and B. K. Vaidya, *J. Indian Chem. Soc.*, **31**, 261 (1970).
202. K. Iwamoto, *Bull. Chem. Soc. Jpn.*, **10**, 420 (1935).
203. J. J. Porter and S. B. Spears, *Text. Chem. Color.*, **2**, 191 (1970).
204. G. Oster and J. B. Bellin, *J. Am. Chem. Soc.*, **79**, 294 (1957).
205. N. S. Allen, J. F. McKellar, B. Mohajerani, *Dyes Pigm.*, **1**, 49 (1980).
206. N. S. Allen, B. Mohajerani and J. T. Richards, *Dyes Pigm.*, **2**, 31 (1981).
207. R. Nakamura and H. Hida, *Sen'i Kagakushi*, **39**, T125, T360 (1983).
208. R. Bangerb, W. Aichelele, E. Schollmeyer, B. Weiman and H. Herlinger, *Melliand Textilber*, **58**, 399 (1977).
209. Y. M. A. Naguib, S. G. Cohen and C. Steel, *J. Am. Chem. Soc.*, **108**, 128 (1986).
210. S. G. Cohen, A. W. Rose, P. G. Stone and A. Ehret, *J. Am. Chem. Soc.*, **101**, 1827 (1979).
211. C. M. Desai and B. K. Vaidya, *J. Ind. Chem. Soc.*, **31**, 265 (1954).
212. S. G. Cohen and A. D. Litt, *Tetrahedron Lett.*, 837 (1970).
213. H. Shosenji, K. Gotoh, C. Watanabe and K. Yamada, *J. Soc. Dyers Colour.*, **99**, 98 (1983).
214. K. Yamada, H. Shosenji and K. Gotoh, *J. Soc. Dyers Colour.*, **94**, 219 (1977).
215. S. Tazuke and R. Takasaki, *J. Polym. Sci. Polym. Chem. Ed.*, **21**, 1529 (1983).
216. E. Hibbert, *J. Soc. Dyers Colour.*, **43**, 204 (1929).
217. R. Haller, O. Hackl and M. Frankfurt, *Melliand Textilber.*, **9**, 415 (1928).
218. F. Scholefield, E. Hibbert and C. F. Patel, *J. Soc. Dyers Colour.*, **44**, 236 (1928).
219. Yamada, Konakahara and Iida, *J. Fac. Eng. Chiba Univ.*, **21**, 157 (1970).
220. I. B. C. Matheson and J. Lee, *Photochem. Photobiol.*, **29**, 879 (1979).
221. A. D. Kirch and G. M. Wyman, *J. Phys. Chem.*, **81**, 413 (1977).
222. G. M. Wyman and B. M. Zarnegar, *J. Phys. Chem.*, **77**, 831 (1973).
223. C. R. Giuliano, L. D. Hess and J. M. Margerun, *J. Am. Chem. Soc.*, **90**, 587 (1968).
224. G. Haucker and R. Paetzold, *J. Prakt. Chem.*, **321**, 978 (1979).
225. C. P. Klages, K. Kobs and R. Memming, *Chem. Phys. Lett.*, **90**, 46 (1982).
226. Y. Maeda, T. Okada, N. Mataga and M. Irie, *J. Phys. Chem.*, **88**, 1117 (1984).
227. G. Engi, *Z. Angew. Chem.*, **27**, 146 (1914).
228. J.-i. Setsune, H. Wakemoto, K. Matsukawa, S. Ishihara, R. Yamamoto and T. Kitao, *J. Chem. Soc., Chem. Commun.*, 1022 (1982).
229. J.-i. Setsune, H. Wakemoto, T. Matsueda, T. Matsuura, H. Tajima, S. Ishihara, R. Yamamoto and T. Kitao, *J. Chem. Soc. Perkin Trans.* I, 2305 (1984).
230. J.-i. Setsune, T. Fujiwara, K. Murakami, Y. Mizuta and T. Kitao, *Chem. Lett.*, 1393 (1986).
231. H. Görner and D. S. Frohlinde, *J. Mol. Struct.*, **84**, 227 (1982).
232. H. Görner, J. Pouliquen and J. Kossanyi, *Can. J. Chem.*, **65**, 708 (1987).
233. K. Fukunishi and M. Nomura, *Senryo to Yakuhin*, **32**, 283 (1987).
234. B. K. Manukian and A. Mangini, *Chimia*, **24**, 328 (1970).
235. H. Oda, M. Matsuoka and T. Kitao, *Nippon Kagaku Kaishi*, 1513 (1976).
236. T. Kitao, Y. Watada, M. Matsuoka and K. Konishi, *Nippon Kagaku Kaishi*, 939 (1974).
237. H. Oda, K. Shimada, M. Matsuoka and T. Kitao, *Nippon Kagaku Kaishi*, 1186 (1977); 1191 (1977).
238. N. Kuramoto, *J. Soc. Dyers Colour.*, **103**, 318 (1987).
239. K. Priyadarsini, J. T. Kunjiappu and P. N. Moorthy, *Indian J. Chem.*, **26A**, 899 (1987).
240. T. Moriya, *Bull. Chem. Soc. Jpn.*, **60**, 4462, 3855 (1987).
241. O. V. Zin'kovskaya, N. A. Kuznetsova and O. L. Kaliya, *Zh. Prikl. Spectrosk.*, **41**, 626 (1984).
242. G. Jones and W. R. Bergmark, *J. Photochem.*, **26**, 179 (1984).
243. B. Milligan and L. A. Holt, *Aust. J. Chem.*, **27**, 195 (1974).

244. H. Theidel, *Melliand Textilber.*, **45**, 514 (1964).
245. M. Komaki and A. Yabe, *Nippon Kagaku Kaishi*, 859 (1982).
246. G. Montaudo, *Gazz. Chim. Ital.*, **94**, 127 (1964).
247. H. Ikuno, M. Honda, M. Komaki and A. Yabe, *Nippon Kagaku Kaishi*, 1603 (1985).
248. R. Anliker, G. Müler, *Fluorescent Whitening Agents*, George Thieme Publisher, Stuttgurt (1975).
249. K. Yagami, K. Yamada and H. Shosenji, *Sen'i Gakkaishi*, **38**, T254 (1982).
250. Yashida and Yamashita, *Yuki Gosei Kagaku*, **36**, 406 (1978).
251. A. Mustafa, *Chem. Rev.*, **51**, 1 (1952).
252. M. Matsuo and T. Sakaguch, *Nippon Kagaku Kaishi*, 1994 (1972).
253. G. Rio and J. Berthelot, *Bull. Soc. Chem. Fr.*, 3609 (1969).
254. H. Oda, N. Kuramoto and T. Kitao, *J. Soc. Dyers Colour.*, **97**, 462 (1981).
255. H. Oda and T. Kitao, *Sen'i Gakkaishi*, **42**, T102 (1986).
256. K. Suganuma, A. Yabe and M. Hida, *Sen'i Gakkaishi*, **35**, T388 (1979).
257. K. Yamada and H. Shosenji, *J. Soc. Dyers Colour.*, **93**, 219 (1977).
258. K. Yamada, H. Shosenji, Y. Nakano, M. Uemura, S. Uto and M. Fukushima, *Dyes Pigm.*, **2**, 21 (1981).
259. H. Shosenji, K. Gotoh, C. Watanabe and K. Yamada, *J. Soc. Dyers Colour.*, **99**, 98 (1983).
260. R. P. Hurd and B. M. Reagan, *J. Soc. Dyers Colour.*, **106**, 49 (1990).
261a R. S. Davidson, G. M. Ismail and D. M. Lewis, *J. Soc. Dyers Colour.*, **103**, 261 (1987).
261b R. S. Davidson, D. King, D. M. Lewis and S. K. R. Jones, *J. Soc. Dyers Colour*, **101**, 291 (1985).
262. E. Varga, L. Kozma, E. Farkas and M. Molnar, *Acta, Phys. Chem.*, **24**, 371 (1978).
263. V. V. Ryl'kov and Sh. Z. Krymiski, *Khim. Vip. Energ.*, **12**, 237 (1978).
264. V. E. Korobov and A. K. Chibisov, *J. Photochem.*, **9**, 411 (1978).
265. V. E. Korobov, V. V. Shulin and A. K. Chibisov, *Chem. Phys. Lett.*, **45**, 498 (1977).
266. N. A. Evans and P. J. Waters, *J. Soc. Dyers Colour.*, **94**, 252 (1978).
267. N. A. Evans, D. E. Rivett and P. J. Waters, *Text. Res. J.*, **46**, 214 (1976).
268. N. A. Evans, in *Photochemistry of Dyed and Pigmented Polymers*, Applied Science Publisher, London, (1980).
269. N. Grossmann, V. Wehner, A. Weise, B. Winnig and E. Fanghaenel, *J. Prakt. Chem.*, **330**, 204 (1988).
270. A. S. Tatikolov, E. A. Shipina, V. I. Sklyarenko, M. A. Al'perovich and V. A. Kuz'min, *Izv. Akad. Nauk SSSR, Ser. Khim.*, **11**, 2637 (1984).
271. P. Douglas, S. M. Townsend and R. Ratcliffe, *J. Imaging Sci.*, **35**, 211 (1991).
272. N. Grossmann, B. Winnig, A. Weise, E. Fanghanel, *J. Prakt. Chem.*, **329**, 767 (1987).
273. T. Miwa, A. Mouri and Y. Asahi, *Nippon Shashin Gakkaishi*, **50**, 128 (1987).
274. T. Miwa, T. Oshimi and Y. Asahi, *Nippon Shashin Gakkaishi*, **52**, 532 (1989).
275. T. Miwa, T. Oshimi, T. Nakabayashi and Y. Asahi, *Nippon Shashin Gakkaishi*, **52**, 304 (1989).
276. R. S. Asquith and P. S. Ingham, *J. Soc. Dyers Color.*, **89**, 81 (1973).
277. D. Fiebig and S. A. Metwally, *Melliand Textilber.*, **56**, 637 (1975).
278. F. Scholefield and E. H. Goodyear, *Melliand Textilber.*, **11**, 867 (1927).
279. H. E. A. Kramer and A. Maute, *Photochem. Photobiol.*, **17**, 413 (1973).
280. G. S. Egerton, *Br. Polym. J.*, **3**, 63 (1971).
281. M. W. Rembold and H. E. A. Kramer, *Org. Coat. Plast. Chem.*, **42**, 703 (1980).
282. H. Oda and T. Kitao, *J. Soc. Dyers Color.*, **102**, 305 (1986).
283. H. Oda and T. Kitao, *Chem. Express*, **1**, 407 (1986).
284. T. Wilson, *J. Am. Chem. Soc.*, **88**, 2898 (1966).
285. A. Zweig and W. A. Henderson, *J. Polym. Sci. Polym. Chem. Ed.*, **13**, 993 (1975).
286. J. P. Guillory and C. F. Cook, *J. Polym. Sci. Polym. Chem. Ed.*, **11**, 1927 (1973).
287. J. P. Guillory and C. F. Cook, *J. Am. Chem. Soc.*, **95**, 4885 (1973).
288. D. J. Carlsson, G. D. Mendenhall, T. Suprunchuk and D. M. Wiles, *J. Am. Chem. Soc.*, **94**, 8960 (1972).
289. A. T. Trozzolo and F. H. Winslow, *Macromolecules*, **1**, 98 (1968).
290. H. Shosenji, H. Igarashi and K. Yamada, *Nippon Kagaku Kaishi*, 271 (1978).
291. C. C. Cook, *Rev. Prog. Coloration*, **12**, 73 (1982).

292. N. K. Bridge, *J. Soc. Dyers Color.*, **76**, 484 (1961).
293. P. G. Baugh, G. O. Phillips and N. W. Worthington, *J. Soc. Dyers Color.*, **85**, 19 (1970).
294. B. Garston, *J. Soc. Dyers Color.*, **100**, 376 (1984).
295. J. Griffiths and C. Hawkins, *Polymer*, **17**, 113 (1976).
296. D. R. Carlsson, T. Suprunchuk and M. Wiles, *J. Polym. Sci., Polym. Lett.*, **14**, 493 (1976).
297. C. H. Bamford and J. S. Dewar, *J. Soc. Dyers Colour.*, **106**, 214 (1949).
298. J. J. Moran and H. I. Stonehill, *J. Chem. Soc.*, 779 (1957).
299. J. J. Moran and H. I. Stonehill, *J. Chem. Soc.*, 788 (1957).
300. N. Kuramoto, I. Satake, K. Natsukawa, K. Yamamoto, H. Hirota, K. Shiozaki, Y. Murakami, T. Kitao and R. Yamamoto, *Sen'i*, **37**, 49 (1985); *Sen'i*, **37**, 148 (1985).
301. N. Kuramoto, *J. Soc. Dyers Colour.*, **103**, 318 (1987).
302. N. Kuramoto and K. Asao, *Chem. Express*, **2**, 437 (1987).
303. Japanese Patent, No. 1506466 (1989).
304. N. Nakamura, I. Okagawa, T. Katayama, M. Noda, F. Sakane and S. Nakanishi, *Sen'i*, **37**, 155 (1985).
305. A. B. M. Abdullan and L. W. C. Miles, *Text. Res. J.*, **84**, 415 (1984).
306. B. S. Stowe, R. E. Fornes and R. D. Gilbert, *Polym. Plast. Technol. Eng.*, **3**, 159 (1974).
307. N. S. Allen and J. F. McKellar, *J. Polym. Sci., Macromol. Rev.*, **13**, 241 (1978).
308. A. Anton, *J. Appl. Polym. Sci.*, **9**, 1631 (1965).
309. T. Yamagata, T. Ishii, Y. Takanaka and T. Handa, *Nippon Kagaku Kaishi*, 1079 (1979).
310. Z. Osawa, E. L. Chen and K. Nagashima, *J. Polym. Sci.*, **15**, 445 (1977).
311. H. A. Taylor, W. C. Tincher and W. F. Hamner, *J. Appl. Polym. Sci.*, **14**, 141 (1970).
312. W. B. Hardy, in *Developments in Polymer Photochemistry-3*, N. S. Allen, ed., Applied Science Publ., p. 287, London (1982).
313. H. S. Koenig and C. W. Roberts, *J. Appl. Polym. Sci.*, **19**, 1847 (1975).
314. C. H. Giles, C. D. Shah, W. E. Waltts and R. S. Sinclair, *J. Soc. Dyers Colour.*, **88**, 433 (1972).
315. C. H. Giles, B. T. Hojiwala, C. D. Shah and R. S. Sinclair, *J. Soc. Dyers Colour.*, **90**, 45 (1974).
316. N. S. Allen, B. Harwood and J. H. McKellar, *J. Photochem.*, **10**, 193 (1979).
317. B. Ranby and J. F. Rabek, in *Photodegradation, Photooxidation and Photostabilisation of Polymers*, John Wiley, London (1975).
318. G. A. Horsfall, *Text. Res. J.*, **52**, 197 (1982).
319. J. P. Pacifici and J. M. Staley, *J. Polym. Sci., Polym. Lett.*, **7**, 7 (1969).
320. M. Day and D. M. Wiles, *J. Appl. Polym. Sci.*, **16**, 203 (1972).
321. C. S. Paik and H. Morawetz, *Macromolecules*, **5**, 171 (1972).
322. F. P. Chernyakovskij, K. A. Chernyakovskaya and L. A. Blyumenfel'd, *Russ. J. Phys. Chem.*, **47**, 3 (1973).
323. P. L. Beltrame, E. Dubini-Paglia and B. Marcandalli, *J. Appl. Polym. Sci.*, **38**, 763 (1989).
324. K. S. Schanze, T. F. Mattox and D. G. Whitten, *J. Org. Chem.*, **48**, 2808 (1983).
325. R. Eichler, P. Richter and P. Vonhone, *Textilveredlung*, **20**, 126 (1985).
326. J. Park, *Rev. Prog. Coloration*, **11**, 19 (1981).
327. J. R. Bond, *Rev. Prog. Coloration*, **12**, 17 (1982).
328. A. Garton, D. J. Carlsson and D. M. Wiles, in *Developments in Polymer Photochemistry-1*, N. S. Allen, ed., Appl. Sci. Publ., London, pp. 93–124 (1981).
329. J. Lemaire and R. Arnaud, *Polym. Photochem.*, **5**, 243 (1984).
330. J. Czerny, *J. Appl. Polym. Sci.*, **16**, 2623 (1972).
331. N. S. Allen, D. J. Bullen and J. F. McKellar, *J. Mater. Sci.*, **12**, 1130 (1977).
332. G. E. Egerton and K. M. Shah, *Text. Res. J.*, **38**, 130 (1968).
333. H. G. Voltz, G. Kampf and H. G. Fitsky, *Prog. Org. Coat.*, **3**, 223 (1974).
334. F. Gugmus, *Developments in Polymer Stabilization-1*, G. Scoott, ed., Appl. Science Publ., p. 261 (1979).
335. K. Murakami, *Yuki Goseikagaku*, **31**, 198 (1973).
336. N. S. Allen, A. Chirinos-Padron and T. J. Henman, *Polym. Deg. Stab.*, **13**, 31 (1985).

337. J. Sedlar, J. Marchal and J. Pertruj, *Photochemistry*, **2**, 175 (1982).
338. T. Kurumada, *Shikizai Kyokaishi*, **62**, 215 (1989).
339. N. Kuramoto and K. Asao, *Rep. Osaka Pref. Ind. Inst.*, **89**, 1 (1986).
340. N. Kuramoto and K. Asao, *Dyes Pigm.*, **12**, 65 (1990).
341. N. Kuramoto, *Senryo to Yakuhin*, **37**, 8 (1992).
342. N. Kuramoto, K. Natsukawa and M. Hirota, *Dyes Pigm.*, **9**, 319 (1988).
343. N. Kuramoto, *Sen'i*, **37**, 197 (1985).
344. N. Kuramoto, *J. Soc. Dyers Colour.*, **106**, 135 (1990).
345. Japanese Patent, No. 1710190 (1992).
346. H. Shiozaki, H. Nakazumi and T. Kitao, *J. Soc. Dyers Colour.*, **104**, 173 (1988).
347. H. Nakazumi, T. Ueyama, T. Kitaguch and T. Kitao, *Phosphorus and Sulfur*, **16**, 59 (1983).
348. H. Shiozaki, H. Nakazumi and T. Kitao, *Shikizai Kyokaishi*, **60**, 415 (1987).
349. H. Shiozaki, H. Nakazumi, Y. Nakado, T. Kitao and M. Ohizumi, *Chem. Express*, **3**, 61 (1988).
350. T. Kitao, *Kagaku to Kogyo*, **41**, 1029 (1988).
351. T. Kitao, *Sensyoku-kogyo*, **33**, 218 (1985).
352. M. Matsuoka, *J. Soc. Dyers Colour.*, **105**, 167 (1989).
353. J. Griffiths, *Colour and Constitution of Organic Molecules*, Academic Press (1976).
354. J. Fabin and H. Hartmann, *Light Absorption of Organic Colorants*, Springer-Verlag (1980).
355. R. Pariser and R. G. Parr, *J. Chem. Phys.*, **21**, 466, 767 (1953).
356. J. A. Pople, *Trans. Faraday Soc.*, **49**, 1375 (1953).
357. K. Y. Chu and J. Griffiths, *J. Chem. Soc., Perkin Trans. I*, 696 (1979).
358. M. Matsuoka, M. Kishimoto and T. Kitao, *J. Soc. Dyers Colour.*, **94**, 435 (1978).
359. Y. Kogo, H. Kikuchi, M. Matsuoka and T. Kitao, *J. Soc. Dyers Colour.*, **96**, 526 (1980).
360. J. Griffiths and K. J. Pender, *Dyes Pigm.*, **2**, 213 (1982).
361. M. Matsuoka, *Sensyoku Kogyo*, **35**, 22 (1987).
362. G. Hallas, *J. Soc. Dyers Colour.*, **95**, 285 (1979).
363. S. Yasui, M. Matsuoka and T. Kitao, *Dyes Pigm.*, **10**, 13 (1988).
364. H. Zollinger, *Color Chemistry*, VCH, Weinheim (1987), revised version (1991).
365. M. Ohkawara, M. Matsuoka, T. Hirashima and T. Kitao, *Kinousei-shikiso*, Kodansya-Elsevier, Tokyo (1992).
366. Z. Yoshida and T. Kitao, eds., *Chemistry of Functional Dyes*, Vol. **1**, Mita Press, Tokyo (1989), and references therein.
367. Z. Yoshida and Y. Shirota, eds., *Chemistry of Functional Dyes*, Vol. **2**, Mita Press, Tokyo (1993), and references therein.
368. V. B. Jipson and C. R. Jones, *J. Vac. Sci. Technol.*, **18**, 105 (1981).
369. Tokyo Denki Kagaku (TDK), Japanese BP 60-73892 (1985).
370. Japanese P, 61-25886.
371. K. Y. Law and F. C. Bailey, *Can. J. Chem.*, **64**, 2267 (1986).
372. K. Y. Law, S. Kaplan and R. K. Crandall, *Dyes Pigm.*, **9**, 187 (1988).
373. N. Kuramoto, K. Natsukawa and K. Asao, *Dyes Pigm.*, **11**, 21 (1989).
374. D. Keil, H. Hartmann and T. Moschny, *Dyes Pigm.*, **17**, 19 (1991).
375. S. H. Kim, M. Matsuoka and T. Kitao, *Chem. Lett.*, 1351 (1985).
376. M. Ohkawara, T. Kitao, T. Hirashima and M. Matsuoka, *Organic Colorants—A Handbook of Data of Selected Dyes for Electro-Optical Application*, Kodansya-Elsevier, Tokyo (1988).
377. M. Matsuoka, *Senryo to Yakuhin*, **30**, 308 (1985).
378. K. Yamamoto, *Senryo to Yakuhin*, **19**, 230 (1974).
379. K. Yoshida, *Shikizai Kyokaishi*, **61**, 338 (1987).
380. M. Umehara, T. Abe and H. Oba, *Yuki Gosei Kagaku Kyokaishi*, **43**, 334 (1985).
381. M. Itoh, S. Esho, K. Nakagawa and M. Matsuoka, *Optical Storage Media*, SPIE-420, A. E. Bell and A. A. Jamberdino, eds., p. 332 (1983).
382. R. Heilingman-Rim, Y. Hirschberg and E. Fischer, *J. Phys. Chem.*, **62**, 2471 (1962).
383. A. Miyashita, *Shikizai Kyokaishi*, **64**, 76 (1991).
384. H. J. Heller, P. J. Darcy, S. Patharakorn, R. D. Piggott and J. Whittall, *J. Chem. Soc. Perkin I*, 315 (1986).

385. H. J. Heller, Proceeding in Reference No. 366, p. 267 (1989).
386. M. Irie and M. Mohri, *J. Org. Chem.*, **53**, 803 (1988).
387. K. Uchida and M. Irie, *Bull. Chem. Soc. Jpn.*, **65**, 430 (1992).
388. N. Matsumura, *Sensyoku Kogyo*, **32**, 215 (1984).
389. S. Yasui, M. Matsuoka, M. Takao and T. Kitao, *J. Soc. Dyers Colour.*, **104**, 284 (1988).
390. S. Yasui, *Shikizai Kyokaishi*, **60**, 212 (1987).
391. S. H. Kim, M. Matsuoka, M. Yomoto, Y. Tsuchiya and T. Kitao, *Dyes Pigm.*, **8**, 381 (1987).
392. E. S. Voropai and M. P. Samtsov, *Opt. Spektrosk.*, **62**, 64 (1987).
393. N. Kuramoto, Unpublished data.
394. N. Kuramoto, *J. Soc. Dyers Colour.*, **106**, 181 (1990).
395. N. Kuramoto, in Reference No. 366, p. 178 (1989).
396. N. Kuramoto, K. Natsukawa, Y. Sakurai, M. Fujishima, A. Aoki and M. Hirota, *Rep. Osaka Pref. Ind. Tech. Res. Inst., (Gijyutushiryo)*, Vol. **1**, 45 (1991).
397. H. Nakazumi, R. Takamura, T. Kitao, K. Kashiwagi, H. Harada and H. Shiozaki, *J. Soc. Dyers Colour.*, **106**, 363 (1990).
398. K. Namba, in Reference No. 366, p. 349 (1989).
399. E. Ando, J. Miyazaki, K. Morimoto, H. Nakahara and K. Fukuda, *Proc. Int. Symp. Future Electron Devices*, Tokyo, p. 47 (1985).
400. H. Oda and T. Kitao, *Dyes Pigm.*, **12**, 97 (1990).
401. H. Oda and T. Kitao, *J. Soc. Dyers Colour.*, **105**, 257 (1989).
402. M. Matsuoka, *Shikizai Kyokaishi*, **63**, 609 (1990).

7 Genotoxicity of azo dyes: Bases and implications
H. S. FREEMAN, D. HINKS and J. ESANCY

7.1 Introduction

Genotoxicity is a general term employed by genetic toxicologists when referring to adverse interactions between DNA and various substances to produce a hereditable change in the cell or organism. In humans, such changes are associated with birth defects, carcinogenesis, teratogenesis, and other types of diseases. It is generally believed that interactions with DNA which cause mutations to occur constitute the early events leading to hereditable changes. Consequently, much of the experimental work in this area has been devoted to mutagenicity testing. This approach is used as a cost-effective and relatively quick way to predict the potential carcinogenicity of organic substances. Although the genotoxicity of azo dyes has been the subject of numerous publications since the carcinogenicity of benzidine towards humans was first confirmed, this subject continues to be extremely important. It is interesting to note, however, that very little of the reported work on this subject appears to be aimed at developing data useful to dyestuff chemists in the design of non-genotoxic dyes. Rather, it is evident that much of the generated data are intended for use in formulating databases that can be used by regulatory agencies in predicting the potential health risks of proposed (new) and existing commercial dyes. This chapter, while by no means presented as an exhaustive treatment of the subject, contains a summary of recent literature, along with examples of how this information has been used to design non-mutagenic azo dyes and aromatic amines, and will be presented more from the perspective of a dyestuff chemist rather than a genetic toxicologist.

More recently, the ecological impact of synthetic dyes has become a matter of some concern. Therefore, a brief summary of efforts directed towards addressing environmental problems arising from residual color, toxic metals, and salt in industrial wastewater will be presented, with emphasis placed on ecological issues pertaining to metallized dyes.

7.2 Test methods

The most directly useful data is obtained via measurement of genotoxicity in whole living organisms. However, such *in vivo* experiments are time consuming, requiring up to 3 years for comprehensive testing, and are exorbitantly expensive. For instance, it has been suggested[1] that the total cost of testing 33 benzidine-based azo dyes would be at least US $100 million. Hence, there has been considerable energy devoted to developing alternative testing procedures that are effective, rapid, inexpensive and do not require the sacrifice of large numbers of animals. Over the past two decades a number of *in vitro* mutagenicity screening methods have been developed that use microorganisms (bacteria) or isolated tissues as substitutes for whole animal studies. The following are examples of tests that are being used to provide mutagenicity test data on an increasing number of azo dyes.

7.2.1 *The* Salmonella *mutagenicity assay*

Introduced in 1975,[2] and later revised by Maron and Ames[3] in 1983, the *Salmonella/mammalian* microsome assay, often referred to as the Ames test, is now widely used as an initial screening test procedure for new experimental compounds. The test is an *in vitro* method commonly involving two or more of five strains of the microorganism *Salmonella typhimurium*. The bacterial strains, each designed to detect a specific type of mutation (e.g. TA 98 and TA 1538 for frameshift mutations; TA 100 and TA 1535 for base pair substitution mutations), are developed without the amino acid histidine, a component essential for growth. Thus, the bacteria is unable to multiply unless it is incubated with a suitable mutagen. A quantitative measure of mutagenic activity can be obtained by simply counting the number of colonies present after incubating the bacteria with the test compound (and other necessary test additives) for a standard length of time. The change in the bacteria is called a reverse mutation, since the deliberate histidine mutation has been reversed, and the colonies that form are called revertant colonies. In addition to the histidine mutation, the *Salmonella* strains have been developed with two additional features that enhance their sensitivity to mutagenic compounds.[3]

For each bacterial strain employed, the test is repeated using varying doses of the test compound. Also, other components are added to the test mixture to further optimize the mutagenic activity of the test compound. A solvent is required to facilitate adequate mixing, and a metabolizing enzyme is also added. The latter component is especially important because most compounds are not themselves directly mutagenic. Rather they are chemically altered, or metabolized, in the body (by the liver and other tissues) into a different compound that is the actual mutagen. In these cases the parent compound is called a promutagen and the metabolized compound is the mutagen (Figure 7.1).

Maron and Ames[3] found that effective *in vitro* metabolic activation

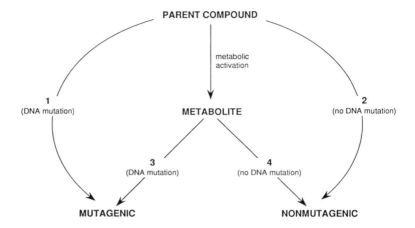

Figure 7.1 Schematic diagram of pathways leading to mutagenic and non-mutagenic compounds. 1, Parent compound: direct mutagen; 2, parent compound: non-mutagen; 3, parent compound: promutagen; metabolite: mutagen; 4, parent compound: non-mutagen; metabolite: non-mutagen.

could be produced by incorporating rat liver microsomal enzymes, S9, into the assay. The requisite enzymes in the liver of rats are induced by administering a mixture of polychlorinated biphenyls prior to sacrificing the animals to isolate the required tissues. *Salmonella* mutagenicity testing on dyes prepared in our laboratories is typically carried out as shown in Figure 7.2.

Depending on the potency of the test compound, dose levels vary from 0 to 5 mg. Each test is carried out in triplicate, both in the presence and absence of S9. The test compound is shown to be a direct mutagen if it is mutagenic in the absence of metabolic activation. Also, since the solvent used may affect the growth of bacteria, a control test is carried out using all the components except the test compound. The number of revertant colonies that form in the control test is called the base count, and is referred to as the number of spontaneous revertants. Generally, if the

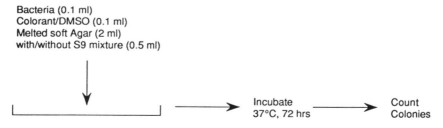

Figure 7.2 Schematic diagram of the method used in a mutagenicity assay.

number of revertant colonies for a test compound is more than twice that of the base count, the compound is considered mutagenic. The compound is characterized as a weak to strong mutagen, depending upon the magnitude of the revertant colonies produced.

Following the initial protocol of Ames *et al.*,[2] a number of variations have been developed,[3-5] each of which is designed for a particular class of test compounds. An important advance in the mutagenicity testing of azo dyes has resulted from the work of Prival and Mitchell,[5] in which modifications to the basic mutagenicity test were made in order to enhance the metabolic breakdown of azo dyes specifically. This method is discussed next.

7.2.2 *The Prival modification*

Introduced in 1982 by Prival and Mitchell,[5] this method emerged from a search for a procedure suitable for testing benzidine-based dyes. The original test method was modified to insure the liberation of the parent diamines and the maximum possible mutagenic activity in each of the four dyes initially studied. There are essentially five modifications to the standard mutagenicity test:

(1) uninduced hamster liver S9 is employed instead of induced rat liver S9,
(2) three times the amount of S9 used in the standard assay is employed,
(3) a reducing agent, flavin mononucleotide (FMN), is added to facilitate the reductive-cleavage of the azo bonds,
(4) exogenous glucose 6-phosphate dehydrogenose (NADH) and four times the normal amount of glucose 6-phosphate are added to further aid in reductive cleavage, and,
(5) a 30 min pre-incubation step is employed before the addition of top agar.

All five changes are deemed necessary for optimal mutagenic activity to be observed. Of particular importance is the inclusion of the reducing agent FMN to the assay, since the reduction of azo compounds can occur in mammals.[6-10] It may be argued that the reduction products could be synthesized and simply tested using the conventional Ames procedure. With many azo dyes, however, this is not practical since their reduction products are unstable (as is the case with *ortho*-hydroxyaniline compounds, for example). Hence the Prival method, which enables the reduction and metabolic activation to be carried out *in situ*, is a valuable modification when testing such compounds. This test is now being actively employed for mutagenicity testing of a range of azo dyes and pigments. For instance, Robertson *et al.*[11] reported the importance of including flavins such as FMN in experiments involving liver enzyme (particularly S9) conversion of benzidine-based dyes (Trypan Blue, Evans Blue, and Congo Red) to mutagenic metabolites. They concluded that FMN effects the reductive-cleavage of azo bonds and that this is followed by S9-mediated oxidative metabolism of the resulting aromatic amines to the ultimate mutagenic species.

In general, the Prival variant is used in conjunction with the standard assay. If a compound displays a negative or weak mutagenic response when rat liver S9 is employed in the standard mutagenicity assay, the Prival method should also be conducted, thereby lowering the risk of a false negative response.

7.2.3 Rat bacterial reduction system

Reid et al.[12] used this type of method to determine the mutagenicity of a group of benzidine-based disazo dyes. In this method, the dyes were first reduced, using either rat cecal flora (bacteria with reducing capabilities) or hamster S9 mix containing FMN. Then, the reduction products were extracted from the crude mixture and subjected to oxidative metabolism using either induced rat liver S9 or uninduced hamster liver S9. Mutagenic activity was measured using *S. typhimurium* strain TA 1538 in the standard mutagenicity protocol.

In an earlier study, the same group had reported[13] that 95+% reduction of Congo Red could be achieved using rat cecal bacteria. Of the 17 azo dyes tested 15 were found to be mutagenic and nine were decolorized by more than 80% using rat cecal flora. The study showed that both cecal bacteria and FMN were efficient at providing reductive metabolism of the azo dyes tested. As might be anticipated, the mutagenicity data arising from the use of this protocol were found to not only depend on the reductive step but also on the type of S9 (rat or hamster) employed when the reduction products were incubated with *Salmonella* bacteria.

7.2.4 Other mutagenicity test methods for azo dyes

A number of other methods have been utilized for the reduction of azo dyes prior to their incubation with *Salmonella* bacteria, including chemical reduction using sodium dithionite[14-15] and the use of FMN-cell free intestinal bacteria.[16] The work of Hartman et al.[15] is particularly interesting since the 'reducing agent' employed was a cell-free extract of *Fusobacterium* sp.2, a human intestinal anaerobe. Although this bacteria has been shown to reduce a number of monoazo dyes, the authors discovered that, in order for efficient reduction of the test dye (Trypan Blue) to take place, FMN was required in the assay.

Dillon et al.[17] followed up on earlier studies which showed D&C Red 9 to be carcinogenic in rats but not mice and non-mutagenic in the standard *Salmonella* mutagenicity assay, by examining the behavior of this dye in a medium containing rat intestinal anaerobes. When D&C Red 9 was incubated in this medium in the absence of oxygen, the major metabolite was 1-amino-2-naphthol. The mutagenicity of this metabolite, when tested separately, paralleled that observed following the incubation of the parent dye with the intestinal anaerobes. The authors concluded that previous negative mutagenicity results reported for 1-amino-2-naphthol may have been a consequence of too low a level of reductive-cleavage, and that the carcinogenicity of D&C Red 9 could well be a consequence of a prior mutagenic process.

Kornbrust and Barfknecht[18] studied a number of azo dyes and their azo reduction products using rat and hamster liver hepatocytes (whole cells). The advantage of using the whole cells rather than the corresponding supernatant fraction (S9) is that the hepatocytes provide the required metabolic activation and serve also as the target cell in which DNA damage is measured. It was found that hamster hepatocytes provide a more efficient system than rat hepatocytes, although both systems generally provided equivocal mutagenic responses. More importantly, perhaps, was the conclusion that this test method might better be suited for testing monoazo dyes since the azo reduction activity is not as high as in other methods. In this study, the authors noted the clear problem that emerges when attempting to assess the potential carcinogenicity of compounds in humans, i.e. that large differences in susceptibility to carcinogens exists among species.[19-20] Thus, a compound that is genotoxic in test methods employing rat or hamster liver enzyme as a metabolic activator or employing whole animals (*in vivo*) may not cause cancer in humans. Despite this inherent imperfection, it was found that protocols for short-term screening of compounds for genotoxicity using *in vitro* bacterial assays correlate reasonably well with known human carcinogens. Moreover, any positive mutagenic activity discovered as a result of *in vitro* testing could always be followed by more intensive, possibly whole animal testing.

Regardless of the test method one chooses, it is imperative to employ purified dyes, if one is to adequately verify the suitability of the method for a specific test compound. Otherwise, an observed genotoxic response may actually be due to one or more impurities in the sample. This has often been the case when commercial dyes were used as received from chemical suppliers.

7.3 Aromatic amines

The research published in this area pertains largely to the behavior of anilines, including phenylenediamines, and benzidine/benzidine-like compounds. The genotoxicity of aniline derivatives has been summarized in some detail in IARC (International Agency for Research on Cancer) reports[21-23] and will not be repeated here. Instead, emphasis will be placed on data from the published literature we believe to be potentially useful in the design of non-genotoxic dyes.

7.3.1 Benzidines and related amines

In a paper presented more than a decade ago, Horning[24] indicated that published data regarding carcinogenicity testing of *ortho*-tolidine and dianisidine based dyes do not provide a basis for excluding such dyes from commerce. He pointed out also that purified *ortho*-tolidine gives only a weak indication of being a liver carcinogen, and that no dyes derived from this diamine have been shown to be carcinogenic in adequately conducted oral testing. Moreover, the commercially used dyes derived from dianisidine are converted to bis-copper complexes of type **1**, preempting the metabolic generation of dianisidine since the non-mutagenic compound 3,3'-dihydroxybenzidine[12] would be formed. We will return to a discussion of the properties of selected metal complexed azo dyes later in this chapter.

Freeman *et al.*[25] summarized work in this area from their laboratory covering a 10-year period that began in 1982 which included the synthesis of acid and direct dyes derived from non-mutagenic analogs of benzidine and 2-naphthylamine. Examples of non-mutagenic dyes prepared include the Direct Black 38 analog **2**, the Congo Red analog **3**, both of which were derived from the non-mutagenic diamine **4**, the Acid Red 33 analog **5**, and a family of reactive (**6**) and disperse (**7**) dyes derived from non-mutagenic amines (**8**) based on the 5,10-dihydrophenophosphazine ring system.

From the work of Bauer and Ritter,[26] it is clear that the incorporation of a sterically bulky alkyl or alkoxy group *ortho* to each amino group in the benzidine structure (*cf*. **9**) is effective in removing the genotoxicity of

1 C.I. Direct Blue 218

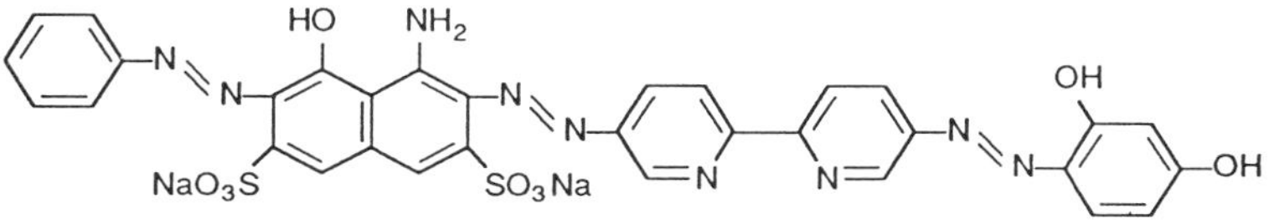

2

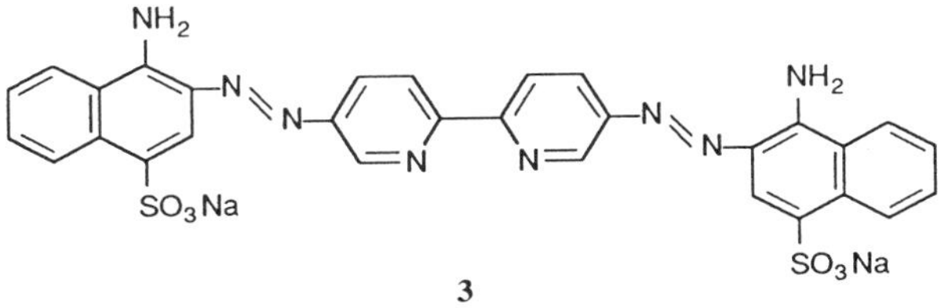

3

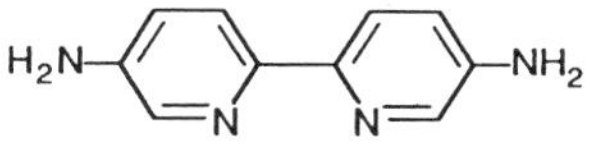

4

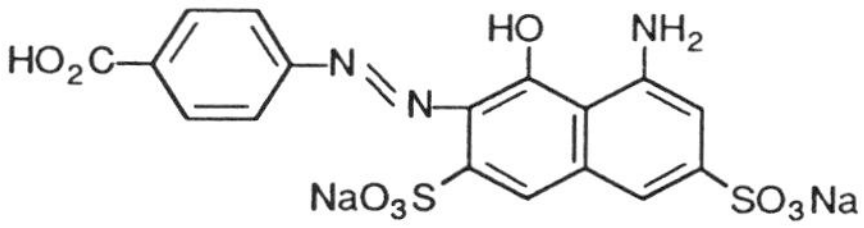

5

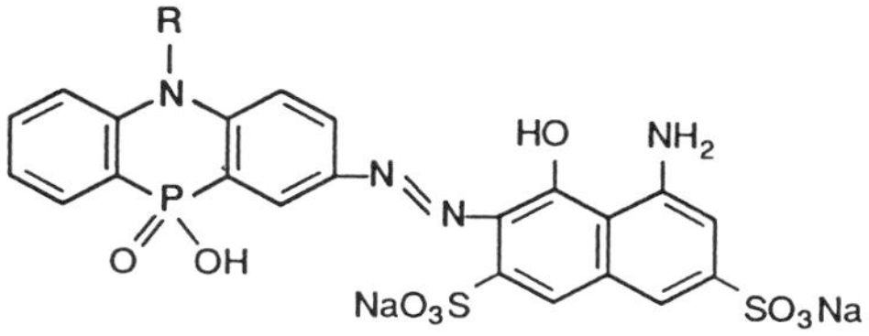

6a R = H, Me

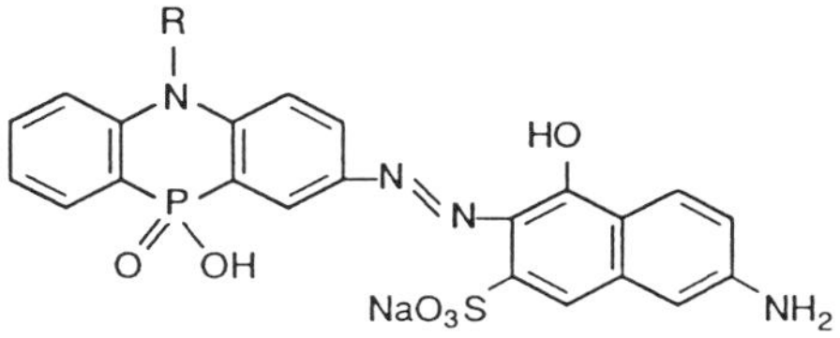

6b R = H, Me

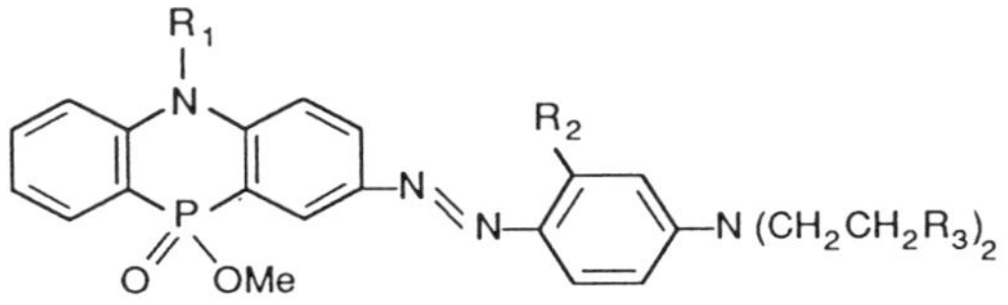

7
$R_1 = H, Me; R_2 = H, Me, NHAc; R_3 = OH, CN, Oac$

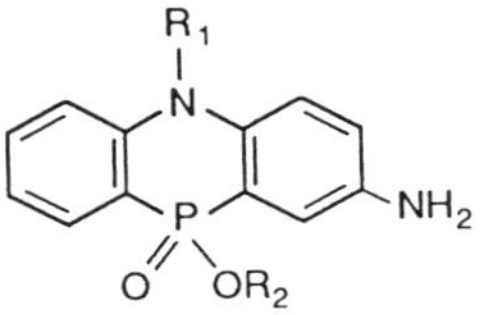

8 $R_1 = H, Me; R_2 = H, Me$

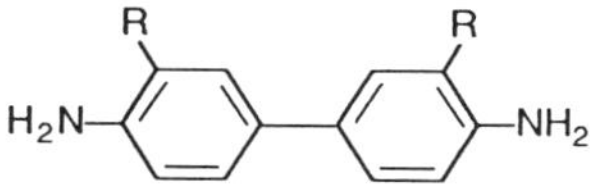

9
$R = OPr, i-OPr, OBu, OCH_2CH_2OH, Bu$

the parent molecule (R=H), and has led to the development of non-genotoxic waterfast tetrakisazo dyes (**10**) for ink-jet printing applications. Other non-genotoxic diamines reported include tetramethylbenzidine **11**[27] and the benzotriazole derivative **12**. Although azo dyes can be prepared from both intermediates, the current synthetic routes to **11** are not sufficiently economical to permit its use in the manufacture of textile dyes. An example of a dye prepared from **12** is the direct blue dye **13**.[28]

Della Porta and Dragani[29] studied 4,4'-diaminobenzanilide (**14**) and 4,4'-diaminoazobenzene (**15**) in mice *in vivo* over a 60-week period and found both compounds to be non-carcinogenic. Neither diamine had an adverse effect on either body growth or survival or caused tumors. With regard to tumor production, it was particularly noteworthy that the compounds were not liver carcinogens, since test animals were used which were susceptible to liver cancer and in view of the propensity of hydrophobic aromatic diamines and aminoazobenzenes to migrate to the liver.

In view of the rather negative connotation now associated with the benzidine name, one wonders whether the dye application community will ever widely accept dyes based on benzidine-type structures even if a 'mountain' of positive and technically sound safety data is provided. This

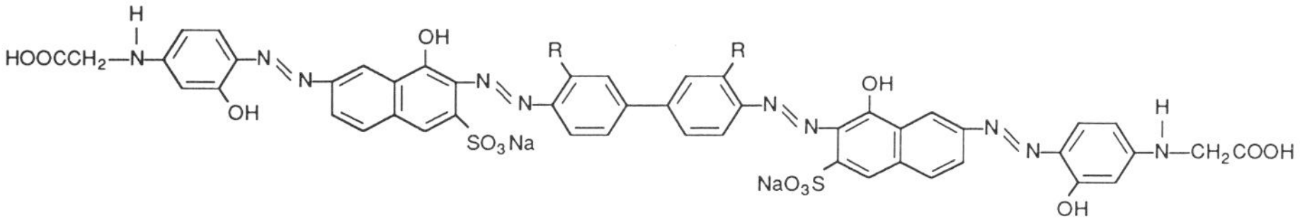

10

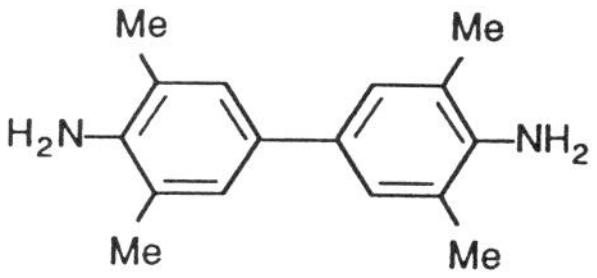

11

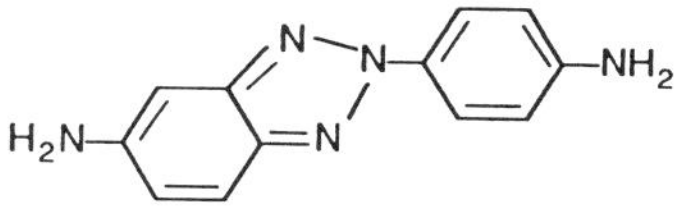

12

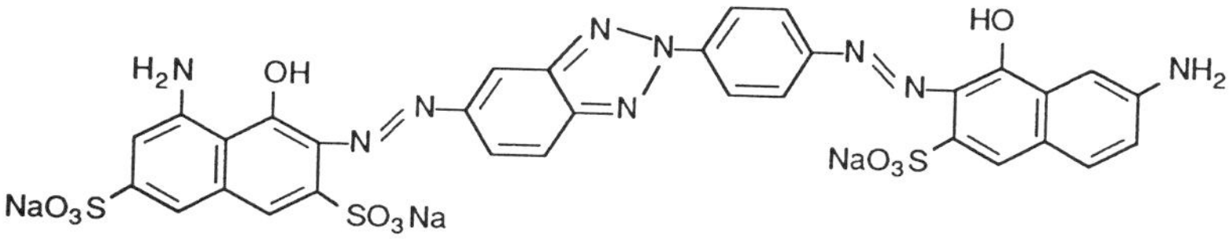

13

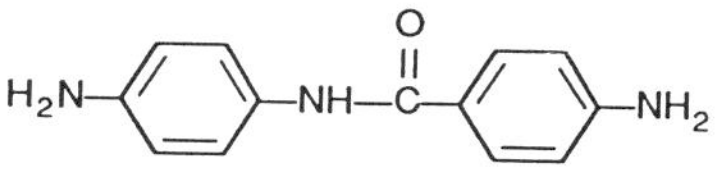

14

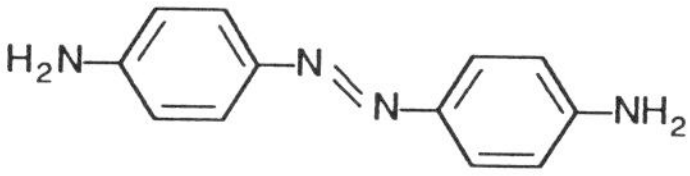

15

point probably makes diamines such as **12** and **14** more logical targets for further evaluation in dye synthesis. However, later in this chapter we cite examples of non-genotoxic dyes derived from hydrophilic derivatives of **15** also.

7.3.2 Anilines and related amines

Freeman et al.[30] have also reported mutagenicity data for a heterogeneous group of aromatic amines which included 2-aminobenzothiazole, 3-aminoquinoline, and 2,6-diaminopyridine, each of which was non-mutagenic in the standard *Salmonella* mutagenicity assay. In the text of this paper, the authors inadvertently referred to 2,6-diaminopyridine and 2-amino-6-hydroxypyridine as mutagenic. A group of letter acids was also evaluated, with H-acid and Chromotropic acid being inactive and the remaining letter acids producing a very weak mutagenic response. In this study, *para*-aminobenzoic acid was shown to be a potential alternative to aniline in the synthesis of non-mutagenic arylazonaphthol dyes. The key point of this work was the development of non-genotoxic azo dyes using non-mutagenic diazo components and coupling components. It should be noted that this approach also produced mutagenic dyes **16** (6-amino- and 7-amino-substituted), despite the use of non-mutagenic intermediates for their synthesis. This observation is one of many which re-emphasize the complex nature of the mechanism of genotoxicity.

Chung et al.[31] investigated the mutagenicity of Methyl Orange (**17**) and attributed its mutagenicity to the formation of N,N-dimethyl-1,4-diaminobenzene (**18**). In their study, the parent dye was incubated *in vitro* with known intestinal bacteria, and the resulting culture filtrate was found to induce frameshift mutations. The investigated filtrate contained sulfanilic acid and N,N-dimethyl-1,4-dimethylaminobenzene, the latter being the only one of the two that is mutagenic. On the other hand,

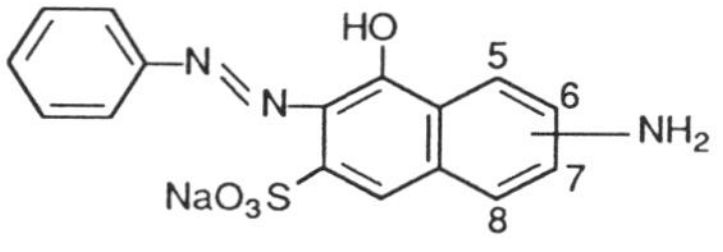

16

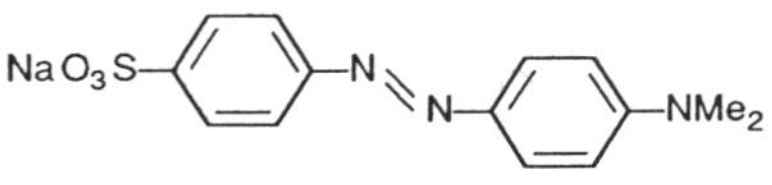

17 Methyl Orange
(C.I. Acid Orange 52)

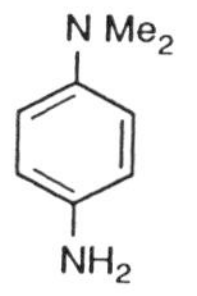

18

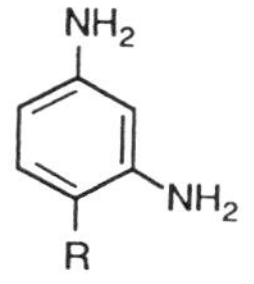

19

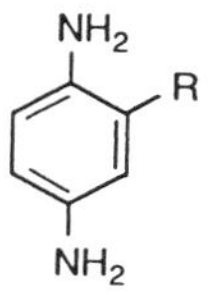

20

R = OPr, Obu, i-OPr, Pr, Bu, OCH$_2$CH$_2$OH

Shahin et al.[32-33] and later Freeman et al.[34-35] have demonstrated that the mutagenicity of phenylenediamines can be removed by incorporating bulky alkyl or alkoxy groups *ortho* to one of the amino groups (cf. **19-20**).

7.4 Mechanisms and metabolism

An overwhelming majority of the experimental work published about the mechanism of azo dye genotoxicity arises from studies conducted on benzidine-based dyes such as **21-25** and hydrophobic monoazo and disazo dyes such as **26-28**. Interestingly, these dyes have continued to be the target of investigations despite the fact that most, if not all, are no longer commercially important. Based on those investigations, it is clear that azo dyes require some type of activation step in order to react with DNA. Activation is known to occur via a number of reduction and

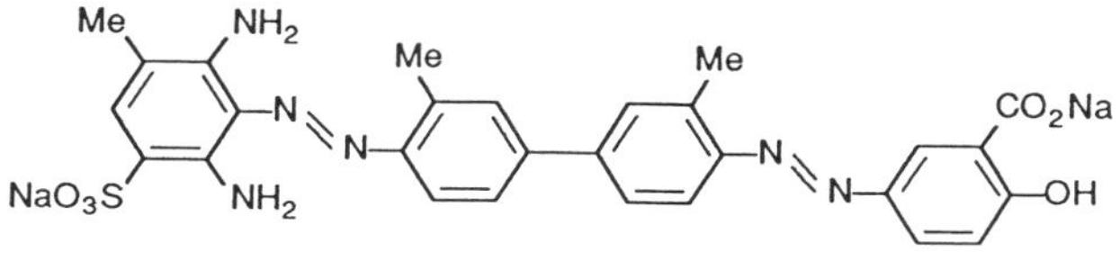

21 C.I. Direct Orange 6

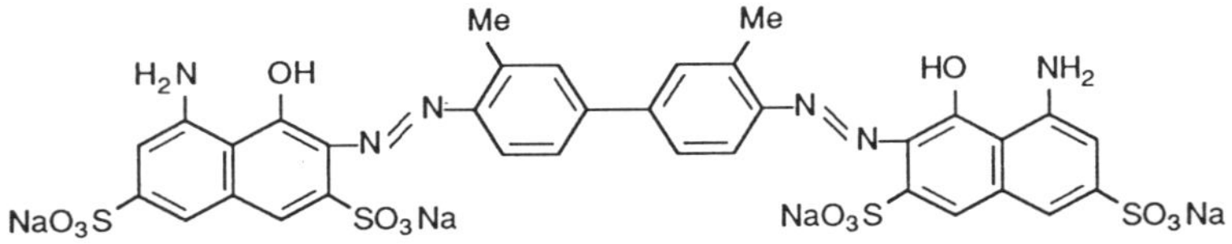

22 Trypan Blue (C.I. Direct Blue 14)

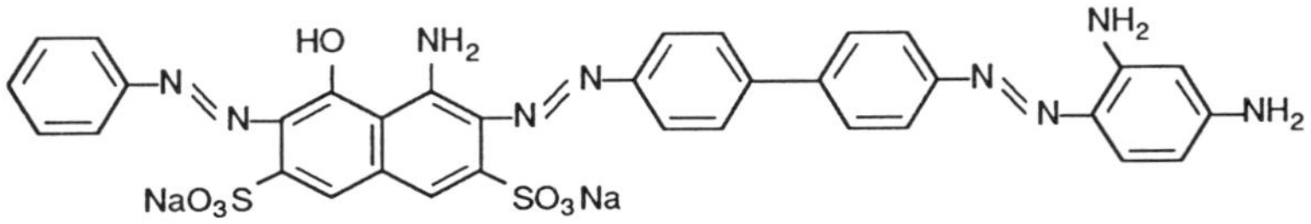

23 C.I. Direct Black 38

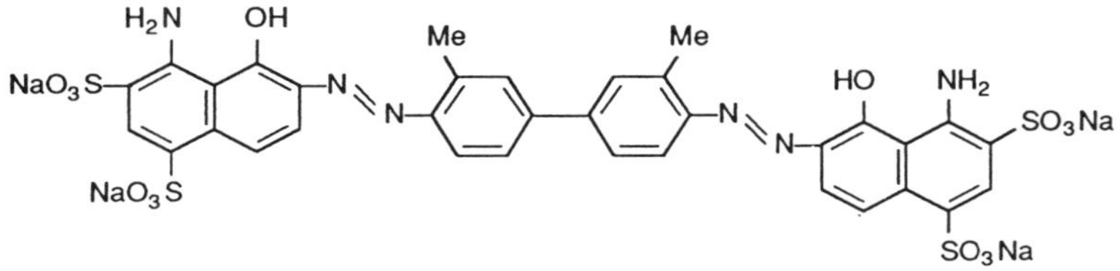

24 C.I. Direct Blue 53

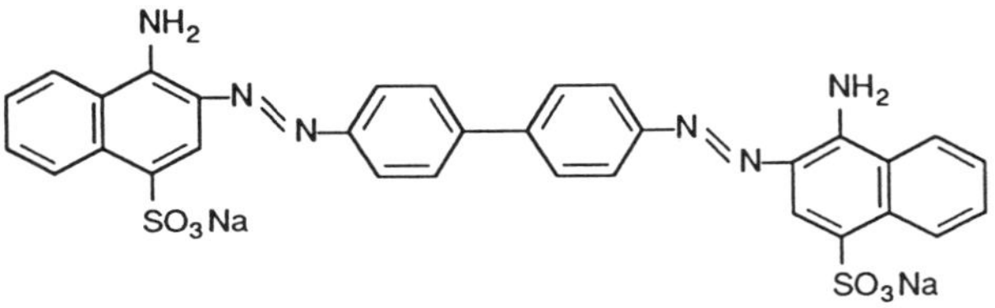

25 Congo Red (C.I. Direct Red 28)

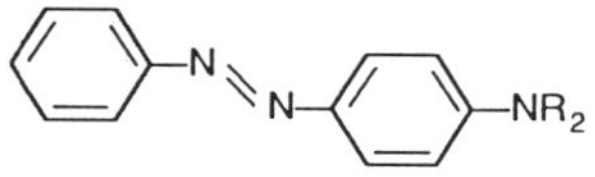

26 R = H; Me (C.I. Solvent Yellow 2)

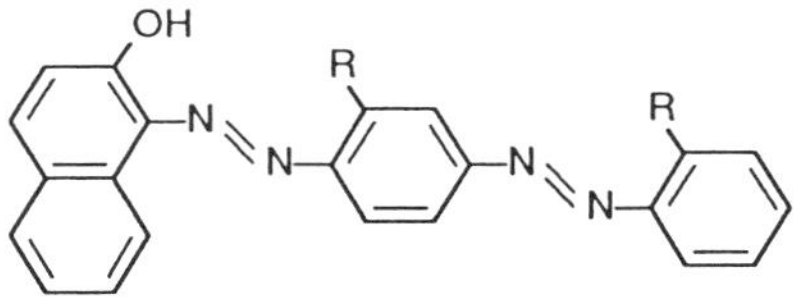

27 R = H (Sudan III)

28 R = Me (Sudan IV)

oxidation processes[36] which include reductive-cleavage of the azo linkage to generate the corresponding aromatic amines, oxidation of any free NH_2 groups present (e.g. N-hydroxylation), and oxidation of the azo linkage or an aromatic ring to produce molecules which are metabolized further to produce DNA-reactive electrophilic species. Not surprisingly, the mechanism by which azo dyes exhibit genotoxicity varies with molecular structure. The behavior of hydrophobic aminoazobenzene type dyes will be presented first.

7.4.1 Monoazo dyes

Much of the research in this area has been directed towards defining the role of enzyme-mediated azo bond reductive-cleavage and N-hydroxylation in azo dye genotoxicity. For instance, Yahagi et al.[4] published one of the early papers which demonstrated that enzymatic reductive-cleavage of Butter Yellow-type dyes (cf. **29**) is not required for mutagenicity. They also suggested that the carcinogenicity of azo dyes may involve an interaction with the base pairs of DNA, and implicated N-acetoxy metabolites as the actual carcinogenic species. Kimura et al.[37] later studied the rate of N-hydroxylation of aminoazo dyes using ESR spectrometry and correlated it with carcinogenic activity. They found that the rate of N-hydroxylation (y) and relative carcinogenic activity (x) could be defined by equation (7.1).

$$y = 0.016x + 0.080 \tag{7.1}$$

The value obtained for non-carcinogenic dyes such as **29** (where R = $4'$-NO_2) was less than 0.080. The authors concluded that the carcinogenicity of C.I. Solvent Yellow 2 (Butter Yellow)-type dyes arises from the

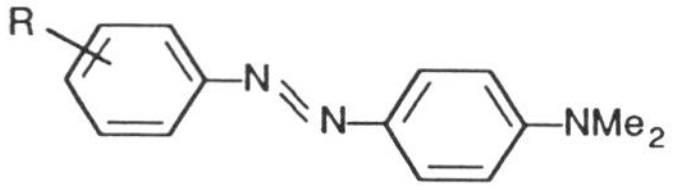

29

action of enzymes responsible for *N*-hydroxylation. Similarly, Kimura *et al.*[38] studied the metabolic activation of the pair of carcinogenic aminoazobenzene dyes **30**, in order to generate evidence for the presumed required *N*-hydroxylation step. In this paper, the authors described the detection and characterization of a nitroxide radical intermediate **31** following *in vitro* rat liver enzyme studies and *in vivo* studies in the whole animal (*cf*. Figure 7.3). The authors indicated that while it is difficult to be certain of what role such radicals play in azo dye carcinogenicity, the fact that the dyes studied are easily converted to nitroxide radicals *in vivo* suggests that their role may be significant. In related work, Brown and Dietrich[39] studied a group of 16 sulfonated monoazo and disazo dyes and developed a mechanism (*cf*. Figure 7.4) to account for the action of riboflavin in the activation of azo dyes under aerobic conditions. The authors proposed the formation of hydrazyl radicals (**33**) which either form DNA adducts directly or undergo cleavage of the nitrogen–nitrogen bond to afford a DNA reactive iminyl radical (**34**). Presumably, in an anaerobic environment, the required radical anion (**32**) is less readily formed or stabilized. Results from other laboratories, however, show that while the reductive-cleavage of azo bonds appears to be an important step in the activation of water soluble dyes, it is less important in the activation of hydrophobic aminoazobenzene dyes. In this regard, Degawa *et al.*[40] evaluated the mutagenicity of 4′-hydroxy-*N*-methyl-4-aminoazobenzenes and a group of its metabolites and found that while the parent dye and its *O*-sulfate and *O*-glucuronide derivatives are mutagenic in TA100, the metabolites are not. They concluded that ring hydroxylation and reductive-cleavage of the azo bond can be ruled out as causes for mutagenicity towards TA100. Similarly, Joachim *et al.*[41] investigated the effects of chemical ($Na_2S_2O_4$) and *in vivo* reduction of azo linkages. In addition to determining that both protocols produce the same genotoxic products, the authors developed an analytical method for isolating dye

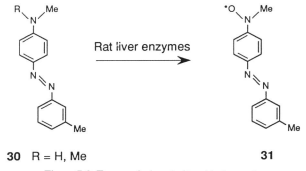

Figure 7.3 Enzyme-induced nitroxide formation.

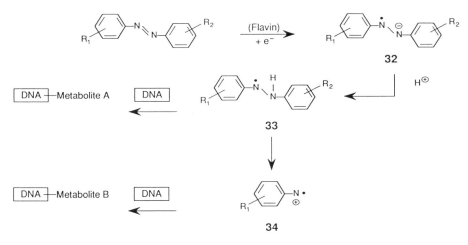

Figure 7.4 Riboflavin-induced activation of an azo dye.

metabolites from rat urine. The lone monoazo and hydrophobic dye in the study, C.I. Solvent Yellow 3 (**35**), was *less* genotoxic following reductive-cleavage. These results led to the conclusion that the observed genotoxicity of this dye is most likely due to a metabolite having an intact azo bond.

Studies conducted in our laboratories[25] with dyes **36** showed the reductive-cleavage products of the mutagenic dyes in this series to be either non-mutagenic or considerably less mutagenic than the dyes themselves when they were evaluated in the standard mutagenicity assay. More recently, we have found dye **37** (mutagenic in the standard *Salmonella* mutagenicity assay) to be less mutagenic in the Prival

35 C.I. Solvent Yellow 3

36
R_1 = OMe, OEt, OPr, OBu, OCH$_2$CH$_2$OH
R_2 = H, NEt$_2$, N(CH$_2$CH$_2$OH)$_2$

modification, an observation consistent with an activation mechanism that does not involve reductive-cleavage of the azo bond.[42]

There are also several interesting papers which outline results of mechanistic studies not involving N-hydroxylation. For example, Winton et al.[43] have presented results from a determination of the chronic effects of oral administration of **29** (R = 3-Me) and related dyes to male rats. In this study, the authors used electron microscopy to study the ultrastructure of liver cells in the endoplasmic reticulum (ER) following a feeding regimen, and observed that liver carcinogens such as **29** (R = 3-Me) cause a significant level of disaggregation of ER liver cells, while the non-carcinogenic derivatives (e.g. **29**, R = 3-CF$_3$) did not cause disaggregation.

A variety of DNA-binding studies have been conducted to further probe the mechanism of azo dye genotoxicity. Among them is the work of Stiborova et al.[44] who treated Sudan I (**38**) with rat liver microsomal enzymes and found that the principal active metabolite was the benzenediazonium ion and that this electrophile binds to deoxyguanosine groups of calf thymus DNA.

Kojima et al.[45] used isomeric dyes **39** and **40** to investigate the properties of DNA adducts of this pair of carcinogenic (**39**) and non-

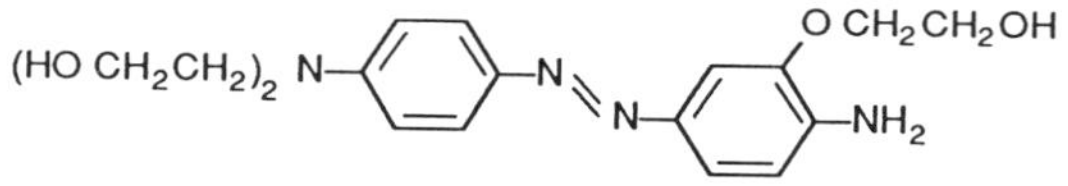

37

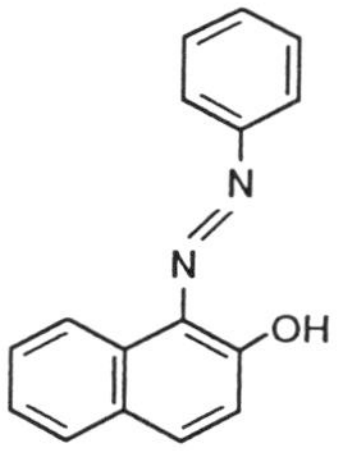

38 Sudan I (C.I. Solvent Yellow 14)

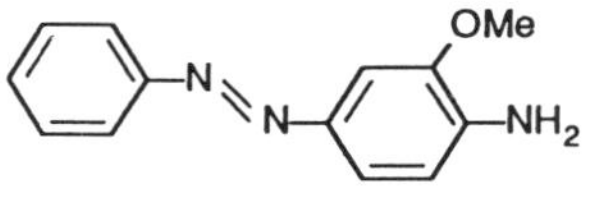

39

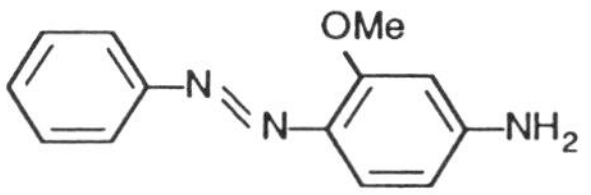

40: PhN=N-C₆H₃(OMe)-NH₂ (structure shown)

carcinogenic (**40**) dyes *in vitro*, and found that while both adducts inhibited DNA synthesis, the former monoazo dye, a rat liver carcinogen, blocked DNA synthesis to a much higher degree. It was suggested that dye **40** is non-carcinogenic *in vivo* because it does not distort the DNA structure to an extent sufficient to block DNA synthesis following adduct formation.

In a study involving heterocyclic azo dyes **41** and **42**, a limitation in the use of short-term *in vitro* DNA assays for predicting the carcinogenicity of azo dyes was observed.[46] The authors were following up on an earlier report which showed that while dye **42** is a reproducibly more potent mutagen than **41**, it is much less active *in vivo* in inducing unscheduled DNA synthesis. To account for the differences in the *in vivo* and *in vitro* behavior, DNA binding studies were conducted using rat liver *in vivo* and calf-thymus DNA *in vitro*. Results from the rat liver protocol showed that **41** was bound more efficiently than **42**, though both dyes reached binding sites equally well. On the other hand, when calf-thymus was used *in vitro*, no differences in binding efficiency were observed.

In studies pertaining to the involvement of specific types of enzymes in azo dye genotoxicity, Mori *et al.*[47] studied dye **29** (R = 3-Me) and its eight metabolites in rats and concluded that rat liver cytochrome *P*-450 enzymes are involved in the metabolic activation of aminoazobenzene dyes. These results were consistent with previous work which showed cytochrome *P*-450 to be involved in the activation of promutagens, and connected these enzymes to the generation of mutagenic and/or tumorogenic compounds **43** and **44**. Similarly, in a 1983 review published by

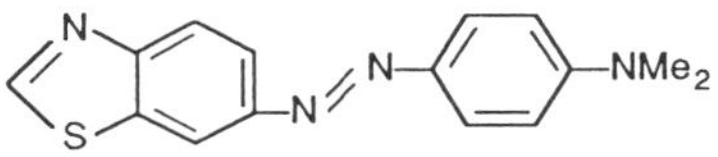

41: benzothiazole-N=N-C₆H₄-NMe₂ (structure shown)

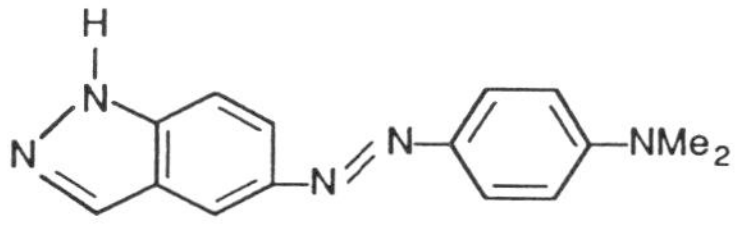

42: 1H-indazole-N=N-C₆H₄-NMe₂ (structure shown)

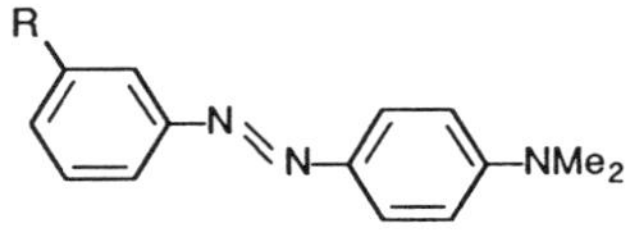

43 R = CH$_2$OH, CHO, CO$_2$H

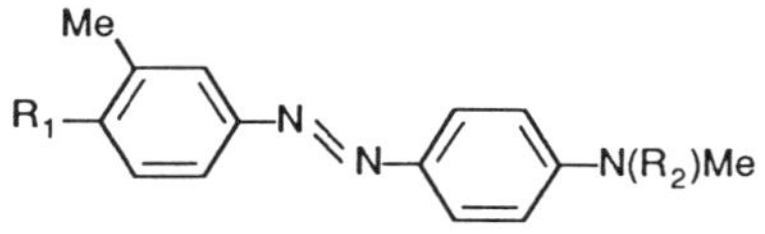

44 R$_1$ = H, OH
R$_2$ = H, Me

Chung,[48] the author described the significance of intestinal microflora-induced reductive-cleavage to the mutagenesis and carcinogenesis of azo dyes. He also pointed out that the aromatic amines produced by this process could then be reabsorbed and interact with body tissue, giving rise to tumor formation. The review covered more than 20 monoazo dyes (e.g. **17** and **45**) and a few disazo dyes (e.g. **22**, **24**, and **28**).

Stiborova et al.[49] conducted a study in which rat liver S-30 (a mixture of enzymes derived from the region of cells known as the mitochondria) and a supernatant containing transfer-RNA (tRNA; the cellular component responsible for the delivery of amino acids during protein synthesis) were used to treat the carcinogenic and non-carcinogenic aminoazobenzenes **46**. Following the required incubations, the tRNA was isolated and characterized. Results of the analyses showed that only the tRNA obtained following incubations with carcinogenic azo dyes displayed enhanced acceptance of methionine. Similarly, lysyl- and alanyl-tRNA adduct formation was stimulated in experiments involving the carcinogenic dyes. All of the dyes used inhibited the formation of leucyl-tRNA,

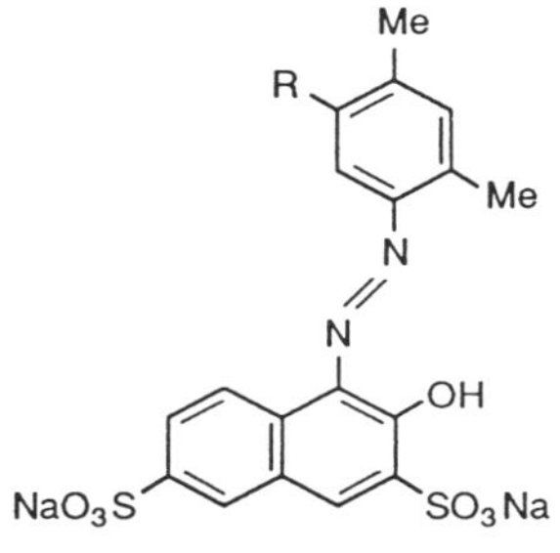

45 R = H, Me

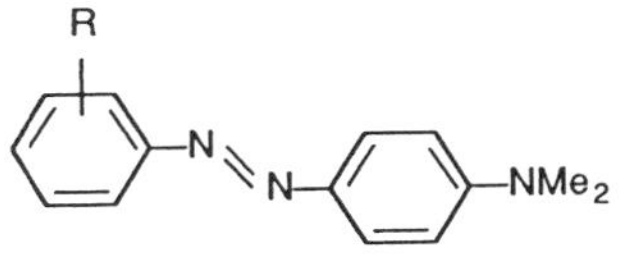

46 R = 2-OH; 3-Me, —OMe, —NO$_2$, —CO$_2$H
—Cl, Br; 4-Me, —OH; H

with the carcinogenic dyes more effective in this regard. The authors found also that none of the dyes prevented the attachment of amino acids to tRNA when they were treated with S-30 prior to introducing the tRNA. On the other hand, the 4'-hydroxy derivative, the suggested metabolite of C.I. Solvent Yellow 2 (**26**), enhanced significantly the attachment of amino acids to tRNA when preincubated with S-30 enzymes or with tRNA itself. The authors concluded that carcinogenic aminoazobenzenes are converted into their active forms by liver enzymes and suggested that the ultimate carcinogens react with tRNA and related species to modify their acceptance of amino acids during protein synthesis. However, it is unclear as to whether this process also leads to abnormal proteins.

7.4.2 Disazo dyes

Most of the published research in this area has involved benzidine- and phenylenediamine-based direct dyes. Unlike their hydrophobic counterparts, these dyes generally require reductive-cleavage of their azo linkages (cf. Figure 7.5) to give a genotoxic response. These studies include the work of Prival et al.[5,50] and later Freeman et al.[51] who showed that Congo Red and its homologs are non-mutagenic when a rat liver S9 enzyme system is employed but mutagenic following incubation with hamster liver S9, and the work of Kennelly et al.[52] who reported the identification of benzidine following oral administration of commercial benzidine-based dyes to male rats.

Cerniglia et al.[53] reported the metabolism of purified C.I. Direct Black 38 (**23**), C.I. Direct Red 2 (**47**), and C.I. Direct Blue 15 (**48**) by intestinal bacteria to generate the corresponding benzidines. This study provided early evidence for the ability of intestinal microorganisms to effect the reductive-cleavage of azo bonds under anaerobic conditions, and in this case form carcinogenic aromatic diamines.

The mutagenicity of some of the proposed metabolites of C.I. Direct Black 38 (**23**) and C.I. Pigment Yellow 12 (**49**) has been investigated.[54] In this study, benzidine, dichlorobenzidine, and their acetylated derivatives were tested using *Salmonella* bacteria with and without S9 activation. It was found that the mutagenicity of mono- and diacetylated benzidine was

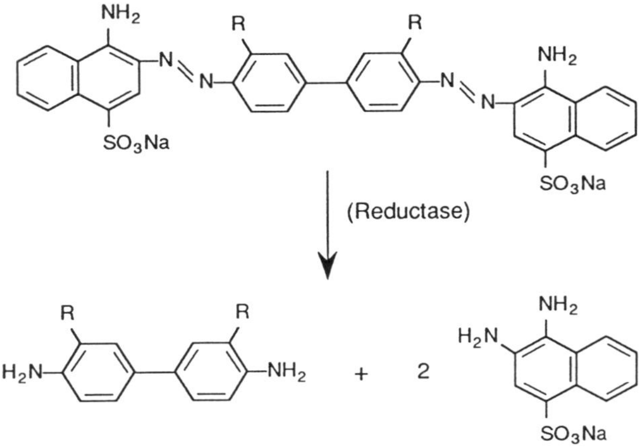

Figure 7.5 Reductive-cleavage of a disazo dye.

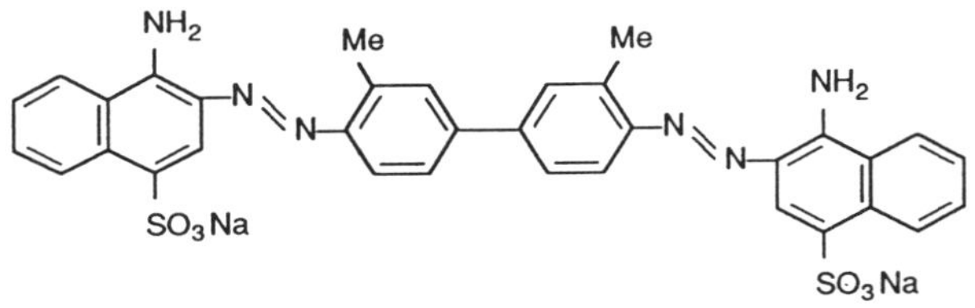

47 C.I. Direct Red 2

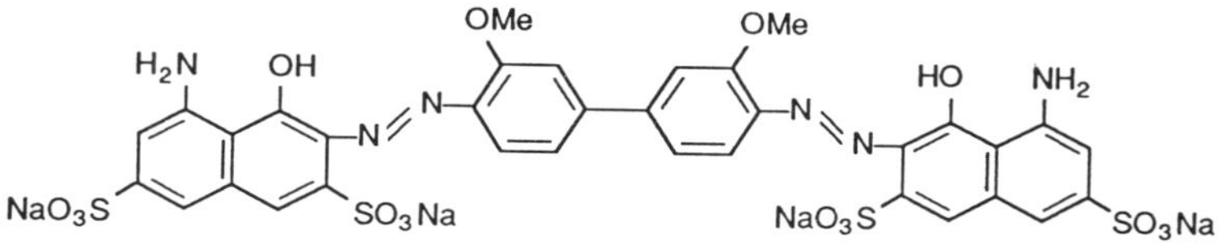

48 C.I. Direct Blue 15

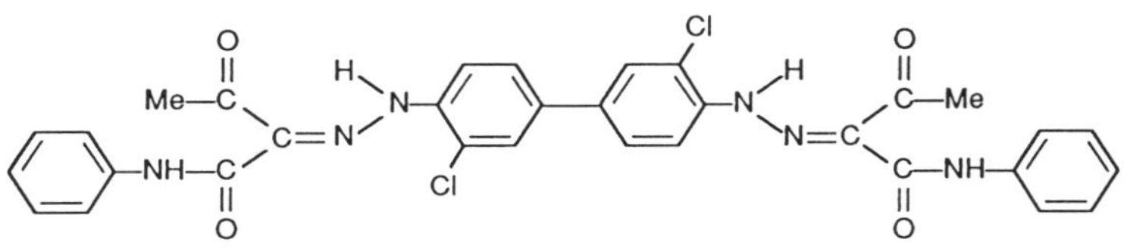

49 C.I. Pigment Yellow 12

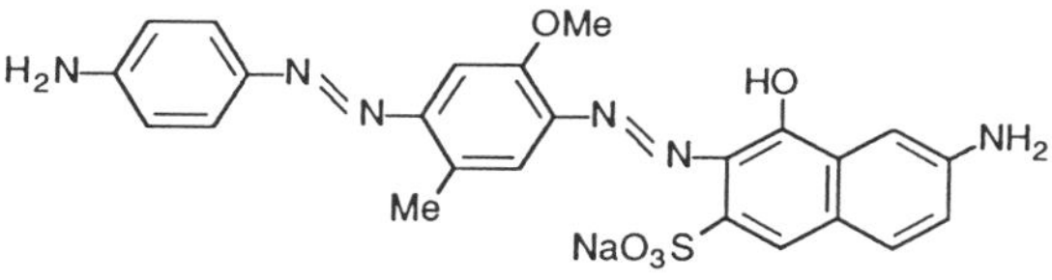

50 C.I. Direct Black 17

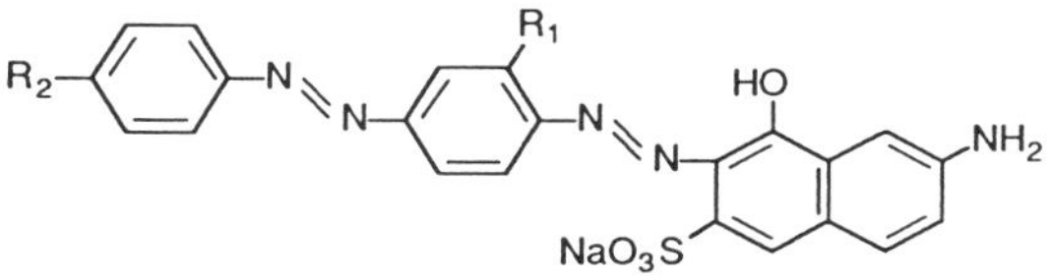

51 C.I. Direct Black 17 Analogs
R_1 = OMe, OEt, OPr, OBu, OCH$_2$CH$_2$OH
R_2 = H, NEt$_2$, N(CH$_2$CH$_2$OH)$_2$

considerably higher than the parent compound following incubation with S9, and that dichlorobenzidine was mutagenic without S9 enzyme activation while the mono- and diacetylated derivatives were non-mutagenic under those conditions. Clearly, these results raise questions about the role of *N*-acetylated derivatives in the genotoxicity of benzidines.

In a study incorporating synthetic work, Freeman *et al.*[55] reported results from efforts aimed at developing a non-mutagenic analog of C.I. Direct Black 17 by building bulky alkoxy substituents into the aminoazobenzene moiety (*cf*. **50** and **51**). Results of the study showed dyes (**51**) having R_2 = H, N(CH$_2$CH$_2$OH)$_2$ were generally more mutagenic than C.I. Direct Black 17; but, when R_2 = NEt$_2$, all five dyes were less mutagenic than the parent dye, with the dye having R_1 = OCH$_2$CH$_2$OH being least mutagenic.

7.5 Structure–activity relationships

7.5.1 *Empirical*

Chung and Cerniglia[56] examined a group of azo dyes in an attempt to establish structure–activity relationships. They observed that the genotoxic dyes examined fell largely into two categories, i.e. *para*-phenylenediamine-based and benzidine-based. These investigators also reported that genotoxicity arising from the phenylenediamine moiety could be decreased by the addition of sulfonic acid or carboxylic acid groups, and that a decrease in mutagenicity among benzidine-based dyes results when such dyes exist as a copper complex. Probably the best known example of the latter observation is seen in the conversion of C.I. Direct Blue 15 (**48**) to C.I. Direct Blue 218 (**1**). Similar observations have been noted by Freeman's group[57] who recently found the 1:2 metal complex C.I. Acid Yellow 151 (**52**) to be less mutagenic than its unmetallized precursor. In the former paper,[56] the authors also mentioned that while the placement of Me, OMe, and Cl groups into the benzidine molecule does not have a beneficial affect on mutagenicity, the introduction of an *N*-beta-hydroxyethyl group does decrease mutagenic activity.

In a closely related investigation, Lin and Solodar[58] reported SAR studies involving analogs of C.I. Direct Black 19 (**53**) and C.I. Direct Black 38 (**23**). The synthesis and evaluation of dyes such as **54–59** demonstrated that the mutagenicity of the prototype dyes could be removed by incorporating a carboxylic acid or sulfonic acid group into the *para*-phenylenediamine residues that would be generated following

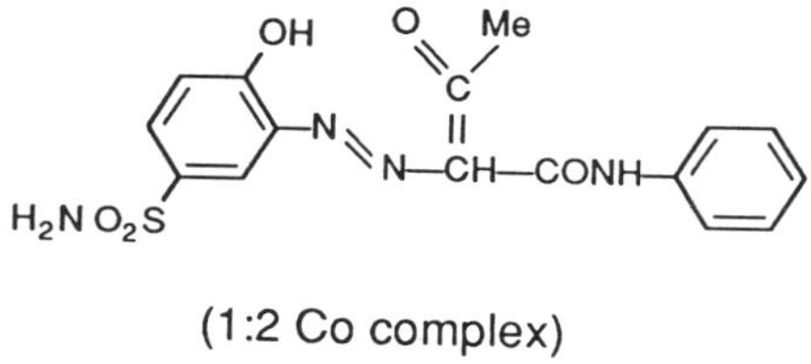

(1:2 Co complex)

52 C.I. Acid Yellow 151

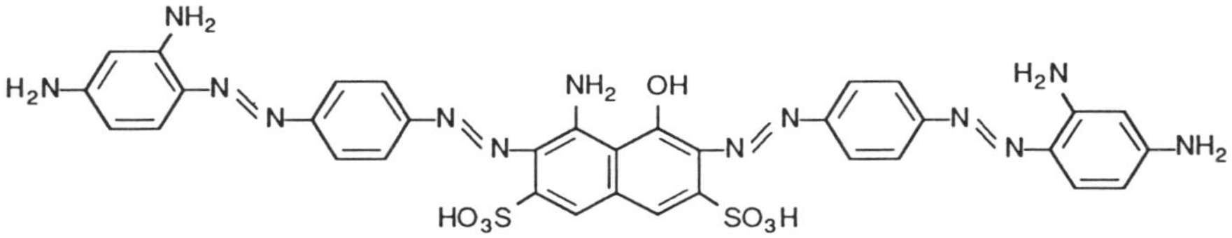

53 C.I. Direct Black 19

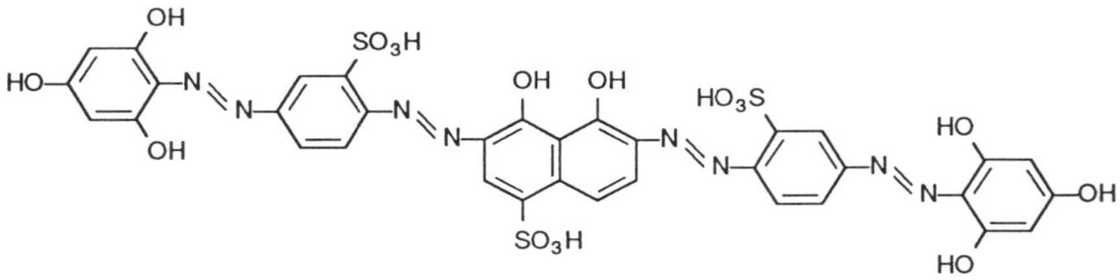

54

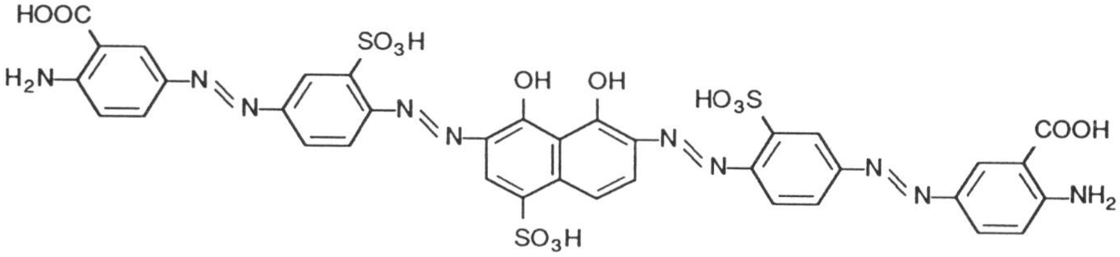
HOOC
H_2N
N
N
SO_3H
N
N
OH
OH
HO_3S
N
N
N
N
COOH
NH_2
SO_3H

55

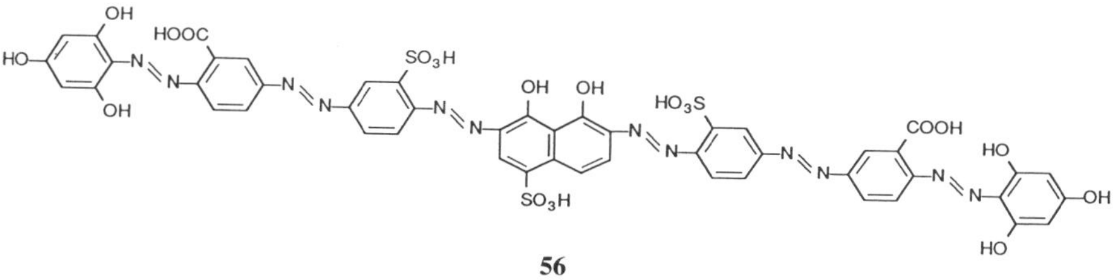
OH
HOOC
HO
OH
SO_3H
OH
OH
HO_3S
COOH
HO
OH
SO_3H
HO

56

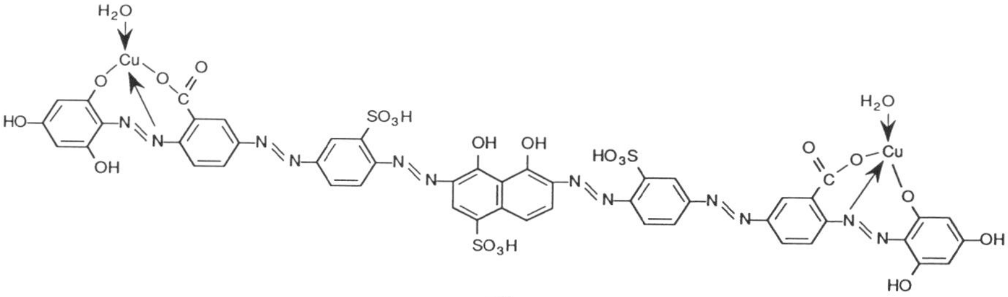
H_2O
Cu
O
O
C
O
HO
OH
SO_3H
OH
OH
HO_3S
H_2O
Cu
O
C
O
O
OH
SO_3H
HO

57

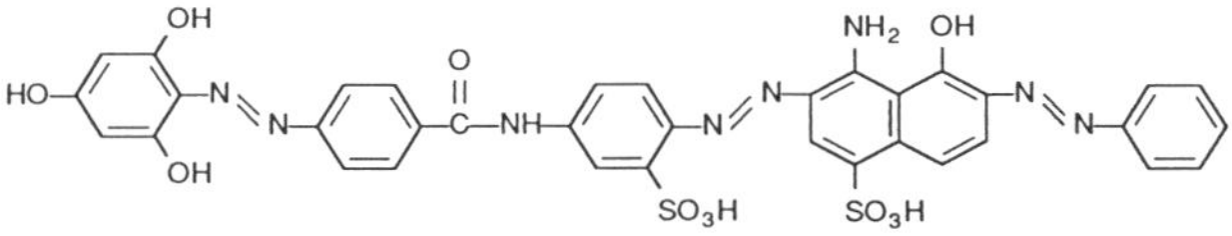
OH
HO
OH
O
C
NH
NH_2
OH
SO_3H
SO_3H

58

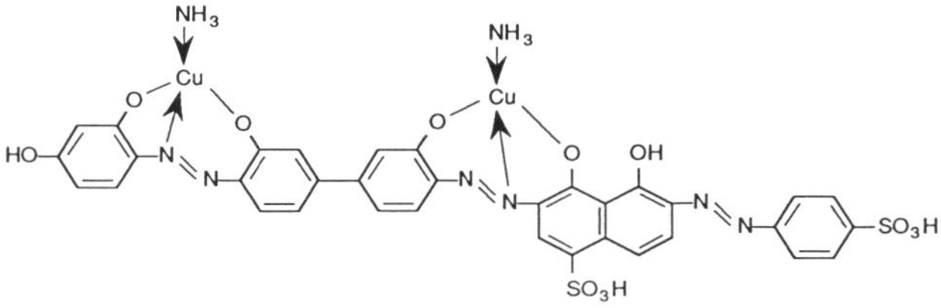
NH_3
Cu
O
O
HO
NH_3
Cu
O
O
OH
SO_3H
SO_3H

59

metabolism (reductive-cleavage), by replacing terminal *meta*-phenylene-diamine with 1,3,5-trihydroxybenzene to prevent the metabolic generation of 1,2,4-triaminobenzene, and by using intermediates capable of metal-complexation following their incorporation into dye structures. No doubt the third of these variations shields the azo linkages against reduction processes. It is worth noting, however, that the aquatic toxicity of copper ions and the uncertainty of its bioavailability even in bound form may make this approach the least attractive of the three in the current regulatory climate. It is also interesting to note that these dyes contain sulfonated and carboxylated derivatives of diamines (*cf*. **14** and **15**) mentioned in section 7.3.1.

Gregory[59] formulated structure–carcinogenicity relationships based on carcinogenicity data that had been reviewed by Longstaff[60] for a group of 97 azo dyes. The dyes were found to fall into six general categories

(1) carcinogenic 4-aminoazobenzene solvent dyes such as C.I. Solvent Yellow 2 (**26**; R = Me) and C.I. Solvent Yellow 3 (**35**);
(2) non-carcinogenic 2-aminoazonaphthalene dyes such as **60**. In this case, it was suggested that a nitrenium ion formed from **60** would be ideally situated for benzotriazole formation (*cf*. **61–62**), and that such a transformation could explain the absence of genotoxicity in this type of dye;
(3) non-carcinogenic arylazophenols existing predominantly in the azo form;
(4) carcinogenic arylazonaphthols which generated a genotoxic aromatic amine upon reductive-cleavage of the nitrogen–nitrogen

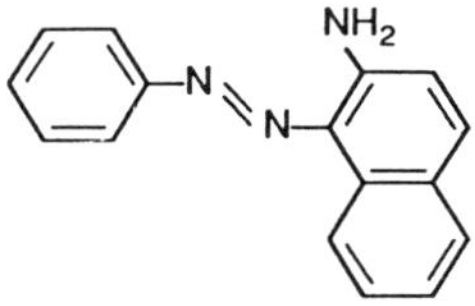

60

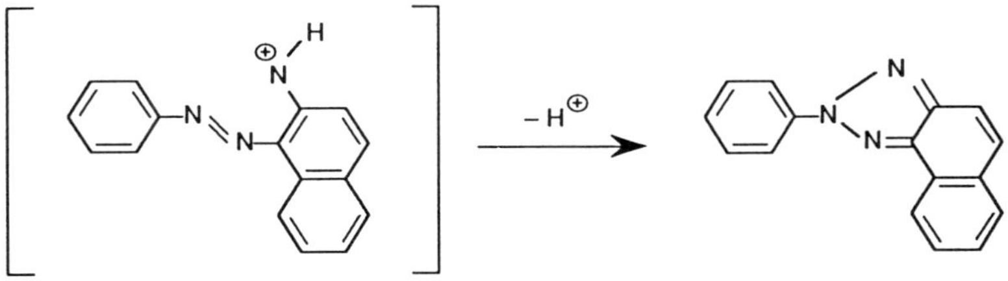

61   **62**

bond. It is known that such dyes exist predominantly in the hydrazone form (e.g. **63**);

(5) non-carcinogenic arylazonaphthols which generate a non-genotoxic aromatic amine upon cleavage of the nitrogen–nitrogen bond (e.g. **64**);

(6) non-carcinogenic highly insoluble diarylide pigments such as C.I. Pigment Yellow 12 (**49**).

In other studies, Garner and Nutman[61] examined a group of ten monoazo dyes in *Salmonella* TA 1538, nine of which are arylazo naphthols, and found only two to be mutagenic in the presence or absence of rat liver enzymes (S9). Among the dyes found non-mutagenic in the study were 1-phenylazo-2-naphthol (**38**) and **45**. A group of 21 aniline derivatives and five naphthylamines were also examined in the *Salmonella* mutagenicity assay and the results showed sulfonated amines to be non-mutagenic. In the case of non-sulfonated aromatic amines **65**, where $R_1/R_2/R_3$ include a variety of combinations of alkyl, amino, halo, and nitro groups, it was reported that compounds containing NH_2 groups as the only ring substituents require metabolism in the liver to produce a

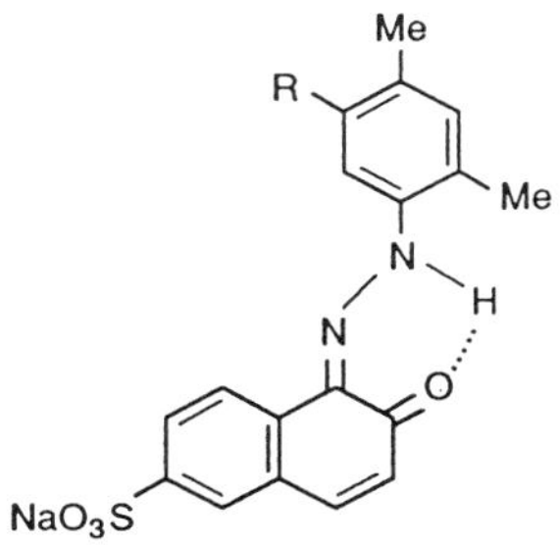

63

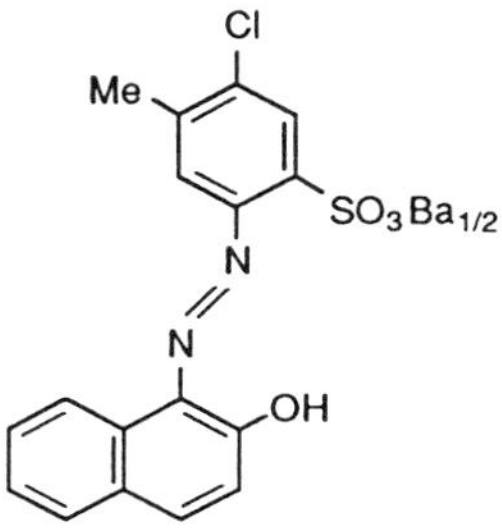

64 D & C Red 9

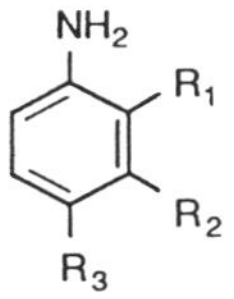

280   PHYSICO-CHEMICAL PRINCIPLES OF COLOR CHEMISTRY

65 (R = amino, alkyl, halo, nitro)

genotoxic response. On the other hand, nitro-substituted anilines, especially nitrated aminophenols, were found to be mutagenic without prior liver enzyme activation.

In work reported from our laboratory[25] involving aminoazobenzene dyes **36** and their reductive-cleavage products, it was shown that the mutagenicity of dye **39** could be removed by the concomitant incorporation of a bulky alkoxy group *ortho* to the amino group and an NEt$_2$ group in the 4'-position. Interestingly, dye **37** was one of the more mutagenic dyes in the study, despite the fact that the reductive-cleavage products of this hydrophilic monoazo dye are non-mutagenic. This work was extended[62] to the synthesis of analogs of Congo Red in which Pr and OBu groups were placed *ortho* to the amino groups. The resulting dyes (**66**) were non-mutagenic in the standard *Salmonella* assay, but were weakly mutagenic in the Prival modification.

In studies involving some relatively simple dye structures, Sanhu and Chipman[63] evaluated the five monoazo dyes identified as components in Gurr Chrysoidine Y (*cf*. Figure 7.6) and found that those components having a methyl group between the two NH$_2$ groups were especially mutagenic (*cf*. dyes c). The authors also reported results from the use of a human liver hymogenate to metabolize the dyes studied. It was found that the behavior of the human liver enzyme protocol corresponded closest to that observed in uninduced rat S9.

7.5.2 Computer-assisted

Rosenkranz and Klopman[64] developed a computer-assisted method for predicting azo dye genotoxicity. Their system (CASE; Computer-*Auto*-

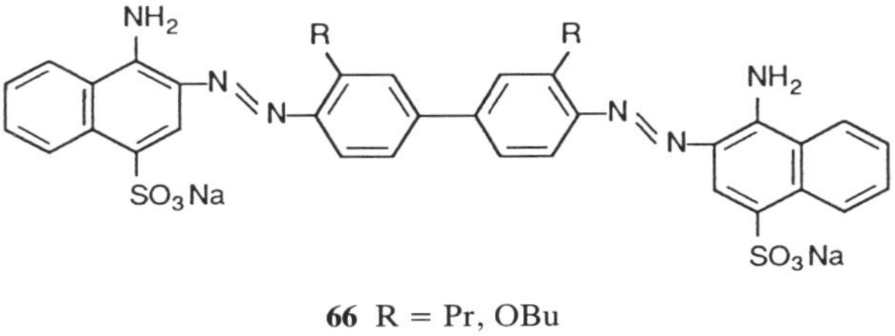

66 R = Pr, OBu

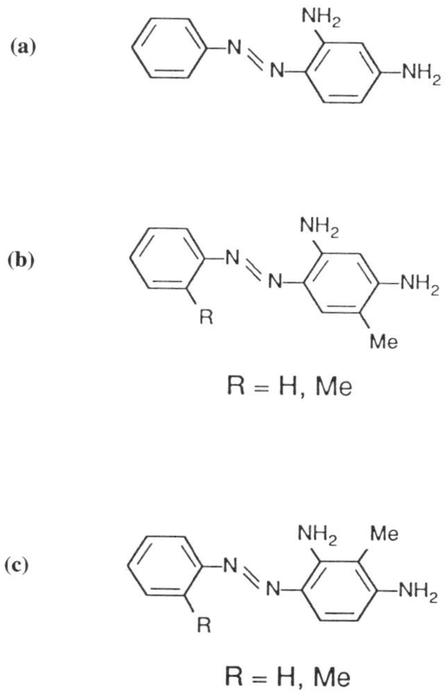

Figure 7.6 Gurr Chrysoidine Y components.

mated *S*tructure *E*valuation) appears to have been designed for hydrophobic monoazo dyes such as C.I. Solvent Yellow 2 (**26**, R = Me) whose genotoxicity is dependent upon an intact azo bond. The ultimate purpose of the system is the identification of potential human carcinogens. While their method of predicting genotoxicity, which is based on connectivity rules, does not take into account the contribution of biochemical interactions, it does represent an interesting approach. For their study, the authors chose to employ a database which included non-azo dye structures, rather than limiting their comparisons to azo dyes alone. This is an interesting idea because it requires us to consider the possibility that organic compounds of diverse molecular structure may actually interact with common or very similar binding sites on DNA *en route* to eliciting a genotoxic (particularly mutagenic) response. It is envisioned that this method will serve as a plausible forerunner to *in vivo* testing by detecting non-genotoxic carcinogens.

In a follow-up paper, Rosenkranz and Klopman[65] used the CASE system to predict the mutagenicity of azo dyes containing the 1-amino-2-naphthol residue. The authors concluded that this residue is inherently mutagenic and that the incorporation of sulfonic acid groups to remove mutagenicity must occur at specific sites. Moreover, CASE was used to identify these required positions. Specifically, sulfonation at C-3, C-4, or C-6 results in the destruction of the biophores (structural units giving rise to mutagenic compounds) which characterize the parent compound and affords a predicted lowering of mutagenicity. The database employed contains data from testing more than 800 organic compounds from which CASE generated 29 biophores and three biophobes (structural units inhibiting mutagenic activity) having a 95% probability of being associated with mutagenicity.

In related studies, Enslein and Borgstedt[66] employed a database comprising a group of 185 structurally heterogeneous compounds to develop a model for predicting the carcinogenic potential of compounds not found in the database. Their studies included the evaluation of azo dyes such as C.I. Disperse Yellow 3 (**67**), D&C Red 9 (**64**), FD&C Yellow 6 (**68**, R = SO_3Na), C.I. Solvent Yellow 14 (**38**) and azobenzene itself, and led to a structure–activity relationship equation for predicting the carcinogenicity potential of an azo dye. Calculation of genotoxicity probability is based on molecular connectivity indices and the cumulative affects of groups having a positive or negative impact on carcinogenicity. As with prior computer models, validation of their model gave rise to some false positives, and the authors cited limitations resulting from minor structural variations which have a major impact on the genotoxicity observed and the need to incorporate parameters into the model to reflect biotransformations and pharmacokinetics associated with such compounds as possible reasons. Examples illustrating how the model works are given.

Claxton et al.[67] developed an SAR analysis system known as ADAPT (*A*utomated *D*ata *A*nalysis by *P*attern recognition *T*echniques) to

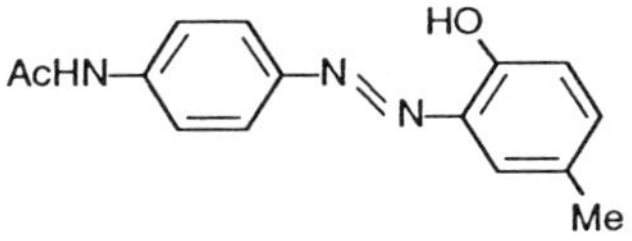

67 C.I. Disperse Yellow 3

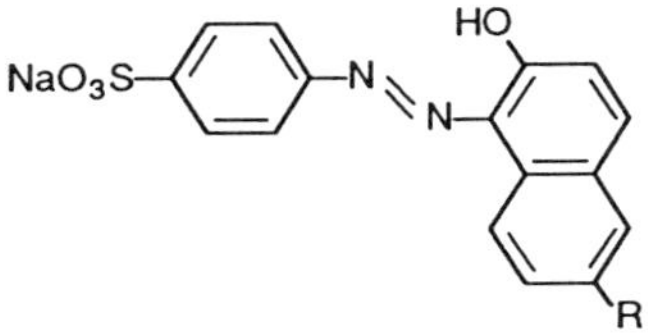

68 R = H, SO_3Na

generate structure–mutagenicity relationships for hydrophilic and hydrophobic azo dyes and their reductive-cleavage products. This system employs a force-field molecular modeling program to optimize the geometry of the dye structure undergoing evaluation for mutagenic potential, and generates a set of molecular descriptors for the parent dye and its reductive-cleavage products. ADAPT then uses pattern recognition techniques to classify the dye as mutagenic or non-mutagenic. In this study, the mutagenicity of 44 azo dyes was predicted using five basic descriptors. ADAPT correctly predicted 34 mutagenic and two non-mutagenic dyes, and incorrectly predicted seven non-mutagenic dyes. There were no mutagenic dyes incorrectly predicted. The overall predictability of 84% using ADAPT compared favorably with a 55% chance-alone predictability. This study employed databases arising from a compilation of mutagenicity data in the peer-reviewed literature on 108 compounds.

7.6 Environmental

7.6.1 General

Few would argue against the proposition that concerns in this area have spawned much of the dyestuffs research taking place today. Consequently, we would be remiss if we were to end this chapter without commenting on at least a few of the published studies.

Without a doubt, the preponderance of the research activities in this important area have been devoted to addressing ecological problems arising from the presence of color, salt, and metals in industrial wastewater. Reife[68] has published a summary of the variety of chemical, biological, and physical methods for removing synthetic dyes from wastewater effluents, and the details of that review will not be repeated here. In addition, Smith[69] has outlined recent developments aimed at enhancing the level of fixation of fiber reactive dyes on cotton. With regard to the latter point, there remains the sobering thought that even with modern technology many dyeings are carried out at 85–90% exhaustion, and in the case of fiber reactive dyes for cellulosic fibers, between 50 and 90% fixation[70] is generally achieved, 40 years after the dyes were first commercialized.

Due to the aquatic toxicity that occurs when effluents high in salt concentration are released from industrial plants, there has been a flurry of activity associated with the development of low salt direct and reactive dyes, since the application of these dyes utilizes the highest levels of electrolytes. Examples of papers published on this subject include that of Herlant[71] and a more recent paper.[72]

The last few years has seen increased concern over the environmental fate of disperse azo dyes in aquatic systems. Since these dyes are hydrophobic with no fully ionic groups present on the molecule, they are likely to migrate to bottom sediments during waste treatment. Here, anaerobic degradation is possible, and reductive-cleavage to potentially genotoxic aromatic amines can occur. Indeed, Weber[73] has demonstrated the anaerobic reduction of the high volume disperse dye C.I. Disperse Blue 79 (**69**). Within 30 min of exposure, multiple degradation products were identified, indicating that rapid degradation in the anaerobic bottom sediment had occured.

Although it has been demonstrated that synthetic dyes can be decolorized using a variety of agents, there is a significant void of published information pertaining to the toxicological properties of the products of chemical and biological degradation processes. In an effort to generate a colorless effluent following the manufacture and application of dyes, it is important to ensure that compounds more toxic than the parent dyes are not released into the environment. Extensive research is needed in this area.

7.6.2 Metallized dyes

The environmental impact of metals in wastewater effluents is probably the most important issue facing dye manufacturing and applications industries today. Certainly it can be said that the US Environmental Protection Agency (EPA) has taken a hard look at this matter in recent years. This is an important point because a significant number of dyestuffs are metallized dyes. Commonly employed metals are chromium, cobalt, nickel and copper, all of which are designated as priority pollutants by the US EPA.

Arguably, the most important metal for dye complexation is chromium, since it provides bright dyes having high fastness properties. Although chromium is an essential trace element for humans, with a required daily intake of between 50 and 200 μg,[74] it is well known that hexavalent chromium (Cr^{6+}) is highly toxic and carcinogenic, and since practically all sources of naturally occuring chromium exist in the relatively non-genotoxic trivalent state (i.e. Cr^{3+}), Cr^{6+} is essentially a man-made product. Trivalent chromium can not easily penetrate cells, but penetration by

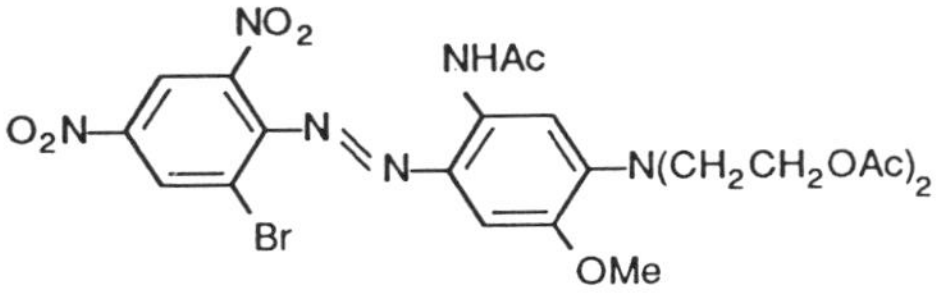

69 C.I. Disperse Blue 79

Cr^{6+} is a facile process. Within the cell, Cr^{6+} is metabolized and it is believed that its high genotoxicity results from a propensity to cause DNA lesions (strand breaks, interstrand cross-links, and cross-links with proteins). Trivalent chromium is known to produce DNA/protein cross-links but has not been shown unequivocably to be genotoxic, presumably because it cannot cross cellular membranes.

As there are still a number of commercially used Cr^{6+}-complexed dyes, particularly among the mordant dyes, there is the lingering question of the fate of such dyes in the environment. This is an important issue because the aquatic toxicity of hexavalent Cr has been shown to be most potent in a solubilized form.[75] If the metallized dye were to be absorbed onto sediment, for example, the toxic effect should be reduced; however, there remains the possibility that chromium ions could still leach into free solution.

Recently, Shukla et al.[76] investigated the toxicity of a chrome dye towards *diazotrophic cyanobacterium* bacteria. This organism is important for the oxygenation of aquatic ecosystems, and has been used as a biofertilizer to enhance productivity of soil and aquatic based products such as rice.[76] A concentration of $12-16$ mg dm^{-3} of Metomega Chrome Orange GL has been found in rivers downstream from wool and carpet textile plants utilizing this colorant. This concentration increases the biological oxygen demand (BOD) and reduces the reoxygenation of the receiving river. Thus, one would anticipate that the ability of the receiving water to sustain aquatic life is adversely affected by solutes which hamper the growth of bacteria such as the aforementioned one. Interestingly, the authors attributed the toxicity of this chrome dye to its anionic charge rather than to the possible presence of free (unbound) Cr^{6+}. It was hypothesized that the ability of the sulfonic acid groups on the dye to form complexes with metal ions such as Mg^{2+} and Ca^{2+} ions, which are essential for the growth of the bacterial cells, was the reason for the observed toxicity. Unfortunately, data pertaining to the concentration of uncomplexed Cr^{6+} were not reported.

In another interesting paper, Tratnyek et al.[77] recently investigated the ability of textile dye wastewaters to contribute to light-induced sensitization of singlet oxygen. This type of process could produce undesirable effects in receiving rivers, such as inhibition of microbial growth and oxidation of pollutants and natural products. Indeed these effects were demonstrated when a sample of dyebath wastewater from an industrial plant inhibited the growth of *Escherichia coli* bacteria and oxidized phenolic compounds in the presence of sunlight.[77] In fact, the use of some sensitizing dyes as solar disinfectants in sewage waters has been proposed.[78]

Relevant to a discussion of the environmental impact of metals in textile dye effluents is the recent renewed interest in natural dyes.[79-82] In addition to the potential environmental impact associated with generating a significant volume of vegetable waste matter following extraction of minute quantities of dye, natural dyes require high concentrations of mordanting agents to provide satisfactory fastness properties. These agents often contain priority pollutant metals and the concentration of these metals in the ensuing effluent would be considerably higher than that currently allowed by the various regulatory bodies in the western world. For example, it has been estimated[80] that concentrations of 220 mg/l of Cr^{6+} would be released when potassium dichromate is used as a mordant for natural dyes, while current regulations mandate less than 0.5 mg/l in effluent.[79,80,83] Hence, from an environmental standpoint, natural dyes do not offer a viable ecological alternative to synthetic colorants.

Considerable effort is being devoted to optimizing as much as possible the fixation of metal complexed dyes to the substrate, due to the ever increasing cost of pollution control of this class of colorants.[83] However, in line with an EPA directive calling for the minimization of toxic waste at its source, work is under way in our laboratories aimed at developing environmentally friendly alternatives to metallized dyes based on Cr and Cu. One approach has been to investigate the synthesis and properties of iron-complexed dyes.[84,85] Iron (the second most abundant metal) is significantly less toxic than Cr and Cu towards humans and aquatic life, and is an essential component in a large number of biological processes. Its role in the transport, storage and activation of molecular oxygen, in addition to a number of other processes, means that it is required for the growth of almost all living organisms.[86] In fact, physicians often prescribe iron in large amounts as a dietary supplement.

Although the use of iron in metal complex dyes is not new, few commercial Fe-complexed dyes are available. It is possible the wide belief that Fe complexes are appreciably inferior to those derived from Cr, for example, in fastness, brightness, and range of possible shades accounts for the fact that they have not heretofore been considered as viable alternatives to presently used high volume Cr complexes. Commercially important Fe-complexed dyes are generally used as colorants for leather. However, the development of black 1:2 Fe complexed azo dyes of type **70**, which are suitable for application to wool, nylon and leather and which exhibit high all around fastness,[87] will no doubt stimulate others to take another look at the commercial potential of iron complexes for substrates other than leather. In addition, mutagenicity studies showed these potentials alternatives to the popular Cr-based dye C.I. Acid Black 52 (**71**) to be non-mutagenic.[85]

Although a significant level of effort has been expended to reduce the level of toxic metals released in dye effluents, few studies have focused on the bioavailability of metals in textile wastewater. Hill et al.[88] discussed

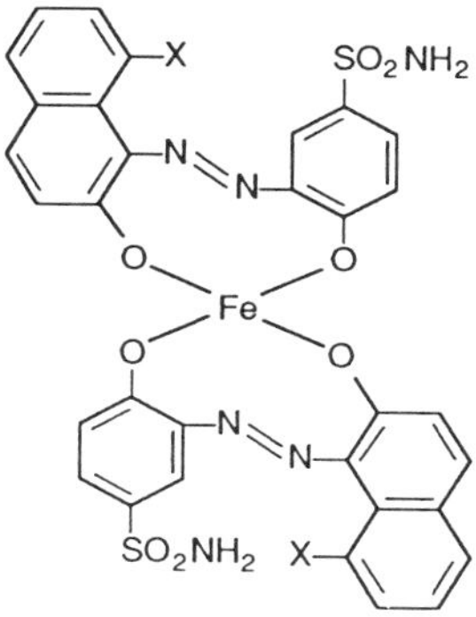

70 R = H, NHAc

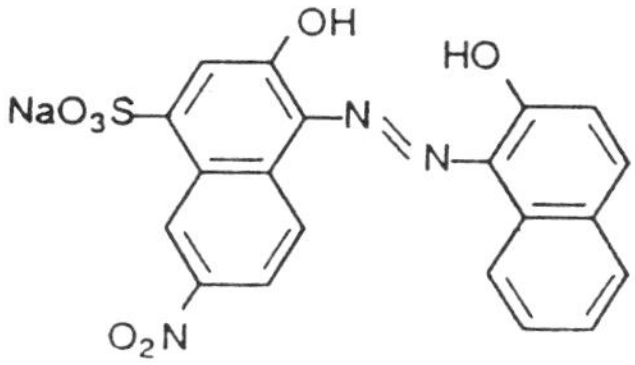

71 C.I. Acid Black 52
(a 2:3 Cr complex)

the need for a good test method to quantify the amount of unbound metal in solution relative to the amount that is present in a stable (bound) complex. It is known that metals held in a stable complex are generally less toxic to the environment than when present in weakly bound or unbound form. In their study, these workers examined three metallized reactive dyes and three metallized acid dyes. After determining the total metal (Cu, Ni, or Cr) content present using atomic absorption spectroscopy, the amount of free metal present was determined by mixing solutions of the dye with a chelating resin. The amount of metal absorbed by the resin indicated the amount of uncomplexed or weakly complexed copper. The reactive dyes appeared to form the most stable complexes, with less than 10% total metal being removed by the chelating resin in each case, while up to 67% was removed from the acid dyes. Thus, the strength of the complex may be an important factor in determining the bioavailability of the metal. Interestingly, dye decolorization experiments involving ozonation released more than 70% of the metal from the reactive dyes, and this was removed from the solution using the chelating resin.

Wastewater from a textile plant known to contain Cu-complexed dyes was also tested by Hill et al.[88] and was found to contain 30–40% of the total copper present in free solution or weakly complexed. Following ozonation, practically all of the metal existed in free form. Consequently, decolorization processes (e.g. ozonation) used to decolorize wastewater containing metal complex dyes will require a treatment step to ensure the release of effluents having little or no aquatic toxicity.

In addition to the use of chelating resins, one plausible way in which metals can be removed from wastewater is by precipitation. Such a method for reducing the concentration of chromium in wastewater was described by Gorzka et al.[89] In this case, the focus was on metal ion containing wastewater arising from the preparation of 1:1 and 1:2 Cr-complexed dyes, which commonly involves using an excess of a chrome-containing intermediate (e.g. sodium chromosalicylate) to drive the metallization step to completion. Therefore, the resulting wastewater can contain as much as 150–300 mg dm^{-3} metal ions,[89] in addition to chromium in bound form. These workers found that the treatment of the wastewater with alkaline calcium chloride decomposed the complexes and precipitated chromium as calcium chromite. As much as 98.5% of the total available chromium was removed using this procedure. Furthermore, the toxicity of the purified water was found to be 30 times lower than the corresponding untreated wastewater.

Aside from complexation and precipitation, a number of alternative methods have been investigated to reduce the concentration of toxic metals in dye wastewater. In view of the fact that commercial chelating resins are relatively expensive, Shukla and Sakhardande[90,91] investigated the use of inexpensive cellulose-based materials as possible adsorbents for heavy metals such as Cu^{2+}, Pb^{2+}, Hg^{2+}, Fe^{2+}, Fe^{3+}, Zn^{2+} and Ni^{2+}. Common waste materials such as bamboo pulp and sawdust were shown to adsorb the above cations. Interestingly, when these adsorbents were dyed with certain reactive dyes, the efficiency of adsorption of the cations increased significantly. Adsorption of 40–90% was achieved when a cation was mixed with the dyed substrate using a liquor/adsorbent ratio of 50:1. The reason for the increased adsorption was suggested to be chelation between the dye and the metal. Importantly, these workers found that almost complete removal of adsorbed substances could be achieved by simple treatment with a 1 M solution of a mineral acid, thereby facilitating reuse of the adsorbent.

In this section we have chosen for discussion some of the more interesting and recent research studies pertinent to the reduction of the concentration of certain toxic metals from industrial waste streams. Although this discussion was in no way an attempt to provide a complete review of all the relevant literature, and should not be viewed as such, it is clear that the scientific community is taking seriously the necessity for improving the toxicity of waste streams from dye manufacturing facilities and textile plants. However, significantly more research is required to determine the fate of dyes and their degradation products in aquatic environments. Currently, the driving force for the reduction of potentially toxic wastewater is the steady flow of federal regulations aimed at protecting human health and our environment. It is likely that the flow of such measures will continue in the near term.

7.7 Conclusions

Historically, the primary motivation for the design of novel, marketable dyes and pigments has been the need for colorants having improved technical performance. This perspective has led to significant improvements in brightness, tinctorial strength, wash- and lightfastness, and chemical stability. In more recent years, however, it has become clear that the toxicological properties of dyes and their precursors must also be factored into the dye design equation. This means that the development of environmentally friendly colorants must embrace all aspects of the life cycle of synthetic dyes, from manufacture through dye application and handling of residual dyebath color. In this regard, it is also clear that dye chemists must continue to work closely with genetic toxicologists and environmental toxicologists to assure the viability of our industry without compromising human health or the environment.

As for the specific role of dye chemists, it is imperative for us to work especially closely with genetic toxicologists to enhance our understanding of the nature of molecular interactions between azo dye molecules and DNA. Of particular interest is the nature of the binding site(s) with which azo dyes interact *en route* to eliciting a genotoxic response. The ability to define these sites and develop working models would shed light on the reason closely related dye structures can differ significantly in genotoxicity. Models of protein binding sites have been formulated and used in the past by medicinal chemists to design pharmaceuticals having improved efficacy, and are derived from MO-based molecular modeling techniques. Of course, the real challenge to the dye chemist is to design non-genotoxic dyes while preserving the desirable technical properties of the dyes targeted for replacement.

We also believe that efforts aimed at using computer models in developing structure–activity relationships should not be discounted solely on the basis that they lack features which consider pharmacodynamics and related phenomena. Particularly interesting and informative are approaches that utilize diverse molecular structures in predicting genotoxicity. Here, we believe computer-based molecular modeling and studies directed towards characterizing binding sites, when used in tandem, will be especially beneficial in enhancing our understanding of the basis for azo dye genotoxicity.

As for a future logical contribution of genetic toxicologists in this area, it would be especially good to see the perfecting of methods for conducting DNA binding studies without the need for radiolabeled compounds. Recent, unpublished, developments at the US EPA suggest that certain post-labeling techniques circumvent the need for radiolabeled compounds. We hope that such techniques will prove effective and be made available to the scientific community at large and that studies involving azo dyes will ensue. Similarly, the development of short-term, *in vitro*, assays for azo dyes which afford mutagenicity data correlating better with carcinogenicity than current *Salmonella* tests would advance progress in this area.

In the area of environmental toxicology, there is a need for studies which focus on identifying the products arising from chemical decolorization of wastewater and a correlation of the results with genotoxicity. While much has been published from studies directed towards decolorizing dyebath wastewater, there is a dearth of published work in which the focus is the characterization of the products formed by the various chemical and biological degradation methods. This is an important matter because it is quite likely that in some cases the products formed by such decolorization methods pose greater hazards than the intact dyes themselves.

We believe this chapter demonstrates that a significant body of information on the subject of azo dye genotoxicity has been published from which useful design approaches can be gleaned as short-term solutions. However, the joint US EPA/ETAD initiative of waste minimization through source reduction/elimination is the most ecologically sound long-term approach to solving dye-related environmental problems. It is our view that the catalyst for the latter lies not within the dyehouse, or even within the dye manufacturing plant, but rather is the dye chemist.

While the unraveling of the basis for genotoxicity in azo dyes to generate a long-term solution presents a challenging problem, it is by no means an insoluble one. Interestingly, the necessary first step is the characterization of the biological target(s) we are trying to miss!

References

1. A. R. Gregory, J. Elliott, and P. Kluge (1981) *J. Appl. Toxicol.*, **1**(6), 308.
2. B. N. Ames, J. McCann and E. Yamasaki (1975) *Mutat. Res.*, **31**, 347.
3. D. M. Maron and B. N. Ames (1983) *Mutat. Res.*, **113**, 173.
4. T. Yahagi, M. Degawa, Y. Seino, T. Matsushima, M. Nagao, T. Sugimura and Y. Hashimoto (1975) *Cancer Lett.*, **1**(2), 91.

5. M. J. Prival and V. D. Mitchell (1982) *Mutat. Res.*, **97(2)**, 103.
6. R. Gingell and R. Walker (1971) *Xenobiotica*, **1**, 231.
7. R. R. Scheline (1973) *Pharmacol. Rev.*, **25**, 451.
8. R. Walker (1970) *Food Cosmet. Toxicol.*, **8**, 659.
9. H. Schroder and D. E. S. Campbell (1972) *Clin. Pharmacol. Ther.*, **13**, 539.
10. T. Watabe, N. Ozawa, F. Kobayashi and H. Kuruta, (1980) *Food Cosmet. Toxicol.*, **18**, 349.
11. J. A. Robertson, W. J. Harris and D. B. McGregor (1982) *Carcinogenesis*, **3(1)**, 21.
12. T. M. Reid, K. C. Morton, C. Y. Wang and C. M. King (1984) *Environ. Mutagen.*, **6**, 705.
13. T. R. Reid, K. C. Morton and C. M. King (1983) *Mutat. Res.*, **117**, 105.
14. J. P. Brown, G. W. Roehm and R. J. Brown (1978) *Mutat. Res.*, **56**, 249.
15. C. P. Hartman, A. W. Andrews and K. T. Chung (1979) *Infect. Immunol.*, **23**, 686.
16. C. P. Hartman, G. E. Fulk and A. W. Andrews (1978) *Mutat. Res.*, **58**, 125.
17. D. Dillon, R. Combes and Z. Erroll (1994) *Mutagenesis*, **9(4)**, 295.
18. D. J. Kornbrust and T. R. Barfknecht (1984) *Mutat. Res.*, **136**, 255.
19. E. C. Miller, J. A. Miller and M. Enomoto (1964) *Cancer. Res.*, **24**, 2018.
20. J. K. Selkirk (1980) *Carcinogenesis*, **5**, 1.
21. *IARC Monographs on the Evaluation of the Carcinogenic Risk of Chemicals to Man* (1975) **8**, IARC, Lyon.
22. *IARC Monographs on the Evaluation of the Carcinogenic Risk of Chemicals to Man* (1978) **16**, IARC, Lyon.
23. *IARC Monographs on the Evaluation of the Carcinogenic Risk of Chemicals to Man* (1982) **27**, IARC, Lyon.
24. R. H. Horning (1981) *AATCC Environ. Symp. Papers*, **28**, 46.
25. H. S. Freeman, M. Esancy, J. Esancy and L. D. Claxton (1991) *Chemtech*, **21**, 438–4.
26. W. Bauer and J. Ritter (1993) in *Chemistry of Functional Dyes*, Z. Yoshida and Y. Shirota, eds, Vol. **2**, Mita Press, Tokyo, p. 649.
27. V. R. Holland and B. C. Saunders (1974) *Tetrahedron*, **30**, 3299.
28. R. Anliker (1977) *Rev. Prog. Coloration*, **8**, 60.
29. G. Della Porta and T. A. Dragani (1981) *Cancer Lett.*, **14**, 329.
30. H. S. Freeman, J. F. Esancy, M. K. Esancy, K. P. Mills, W. M. Whaley and B. J. Dabney (1987) *Dyes Pigm.*, **8**, 417.
31. K.-T. Chung, G. E. Fulk and A. W. Andrews (1978) *Mutat. Res.*, **58**, 375.
32. M. M. Shahin, A. Bugaut and G. Kalopisis (1980) *Mutat. Res.*, **78**, 25.
33. M. M. Shahin, D. Rovers, A. Bugaut and G. Kalopisis (1980) *Mutat. Res.*, **79**, 289.
34. J. F. Esancy, H. S. Freeman and L. D. Claxton (1990) *Mutat. Res.*, **238**, 1.
35. J. F. Esancy, H. S. Freeman and L. D. Claxton (1990) in *Colour Chemistry—The Design and Synthesis of Organic Dyes and Pigments*, A. T. Peters and H. S. Freeman, eds, Elsevier Applied Science, London, p. 85.
36. M. A. Brown and S. C. Devito (1993) *Crit. Rev. Environ. Sci. Technol.*, **23**, 249–324.
37. T. Kimura, M. Kodama and C. Nagata (1982) *Carcinogenesis*, **3**, 1393.
38. T. Kimura, M. Kodama and C. Nagata (1979) *Biochem. Pharmacol.*, **28**, 557.
39. J. P. Brown and P. S. Dietrich (1983) *Mutat. Res.*, **116**, 305.
40. M. Degawa, Y. Shoji, K. Masuko and Y. Hashimoto (1979) *Cancer Lett.*, **8**, 71.
41. F. Joachim, A. Burrell, and J. Anderson (1985) *Mutat. Res.*, **156**, 131.
42. H. S. Freeman and J. P. Clemmons, unpublished work in progress.
43. D. J. Winton, B. Flaks and A. Flaks (1988) *Carcinogenesis*, **9**, 987.
44. M. Stiborova, B. Asfaw, P. Anzenbacher and P. Hodek (1988) *Cancer Lett.*, **40**, 327.
45. M. Kojima, M. Degawa, Y. Hashimoto and M. Tada (1991) *Biochem. Biophys. Res. Commun.*, **179**, 817.
46. R. H. Dashwood, R. D. Combes and J. Ashby (1986) *Food Chem. Toxicol.*, **24**, 708.
47. Y. Mori, T. Niwa and K. Toyashi (1983) *Carcinogenesis*, **4**, 1487.
48. K. T. Chung (1983) *Mutat. Res.*, **114**, 269.
49. M. Stiborova, M. Matrka and J. Hradec (1980) *Biochem. Pharmacol.*, **29**, 2301.
50. M. J. Prival, S. J. Bell, V. D. Mitchell, M. D. Peiper and V. C. Vaughan (1984) *Mutat. Res.*, **136**, 33.
51. H. S. Freeman, J. F. Esancy, L. D. Claxton and M. K. Esancy (1990) *Book Pap.—AATCC Int. Conf. Exhib.*, 188.

52. J. C. Kennelly, P. J. Hertzog and C. N. Martin (1982) *Carcinogenesis*, **3**, 947.
53. C. E. Cerniglia, J. P. Freeman, W. Franklin and L. D. Pack (1982) *Carcinogenesis*, **3**, 1255.
54. E. J. Lazear, J. G. Shaddock, P. R. Barren, and S. C. Louie (1979) *Toxicol. Lett.*, **4**, 519.
55. H. S. Freeman, J. F. Esancy and L. D. Claxton (1990) *Dyes Pigm.*, **13**, 55.
56. K. T. Chung and C. E. Cerniglia (1992) *Mutat. Res.*, **277**, 201.
57. H. S. Freeman and L. G. Cleveland, unpublished work in progress.
58. G. H. Y. Lin and W. E. Solodar (1988) *Mutagenesis*, **3**, 311.
59. P. Gregory (1986) *Dyes Pigm.*, **7**, 45.
60. E. Longstaff (1983) *Dyes Pigm.*, **4**, 243.
61. R. C. Garner and C. A. Nutman (1977) *Mutat. Res.*, **44**, 9.
62. H. S. Freeman, M. K. Esancy and L. D. Claxton (1993) in *Chemistry of Functional Dyes*, Z. Yoshida and Y. Shirota, eds, Vol. **2**, Mita Press, Tokyo, p. 10.
63. P. Sandhu and J. K. Chipman (1990) *Mutat. Res.*, **240**, 227.
64. H. S. Rosenkranz and G. Klopman (1989) *Mutat. Res.*, **221**, 217.
65. H. S. Rosenkranz and G. Klopman (1990) *Mutagenesis*, **5**, 137.
66. K. Enslein and H. H. Borgstedt (1989) *Toxicol. Lett.*, **49**, 107.
67. L. D. Claxton, D. B. Walsh, J. F. Esancy and H. S. Freeman (1990) *Prog. Clin. Biol. Res.*, **340**, *Mutat. Environ*, Part B, 11.
68. A. Reife (1993) *Kirk-Othmer Encyclopedia of Chemical Technology*, fourth ed, Vol. **8**, John Wiley, 753.
69. R. Smith (1993) *Kirk-Othmer Encyclopedia of Chemical Technology*, fourth ed, Vol. **8**, John Wiley, 809.
70. U. Sewekow, (1993) *Book Pap.—AATCC Int. Conf. Exhib.*, Montreal, Canada, 235.
71. M. A. Herlant, (1991) *Book Pap.—AATCC Int. Conf. Exhib.*, Charlotte, North Carolina, USA, 287.
72. Anon (1994) *International Dyer*, April ed., 6.
73. E. J. Weber, (1988) *Book Pap.—ACS National Meeting*, Los Angeles, California, USA, September 25–30, 1988, **28(2)**, 177.
74. J. Alexander, (1993) *Scand. J. Environ. Health*, **19(1)**, 126.
75. T. C. Gendusa, T. L. Beitlinger and J. H. Rodgers (1993) *Bull. Environ. Contam. Toxicol.*, **50**, 144.
76. S. P. Shukla, A. Kumar, D. N. Tiwara, B. P. Mishra and G. S. Gupta (1994) *Environ. Pollution*, **84**, 23.
77. P. G. Tratnyek, M. S. Elovitz and P. Colverson (1994) *Environ. Toxicol. Chem.*, **13**, 27.
78. A. J. Acher and I. Rosenthal (1977) *Water Res.*, **11**, 557.
79. B. Glover (1993) *Book Pap.—AATCC Int. Conf. Exhib.*, Montreal, Canada, 229.
80. S. K. Bhardwaj (1994) *Book Pap.—AATCC Int. Conf. Exhib.*, Charlotte, North Carolina, USA, 472.
81. S. I. Ali (1993) *J. Soc. Dyers Colour.*, **109**, 13.
82. G. Dalby (1993) *J. Soc. Dyers Colour.*, **109**, 8.
83. H. Thomas, R. Kaufmann, R. Peters, H. Hocker, M. Lipp, J. Goschnick and H-J. Ache (1992) *J. Soc. Dyers Colour.*, **108**, 186.
84. J. Sokolowska-Gajda, H. S. Freeman and A. Reife (1994) *Text. Res. J.*, **64**, 388.
85. H. S. Freeman, J. Sokolowska-Gajda, and A. Reife, L. D. Claxton and V. S. Houk (1995) *Text. Chem. Color.*, **27**, 13.
86. M. Fontecave and J. L. Pierre (1993) *Biochimie*, **75**, 767.
87. H. S. Freeman, J. Sokolowska-Gajda and A. Reife (North Carolina State University) (1994) US 5, 376, 151.
88. W. E. Hill, W. S. Perkins and G. S. Sandlin, (1993) *Text. Chem. Color.*, **25(11)**, 26.
89. Z. Gorzka, J. Kraska and H. Lawniczak (1984) *Dyes Pigm.*, **5**, 263.
90. S. R. Shukla and V. D. Sakhardande (1991) *Dyes Pigm.*, **17**, 11.
91. S. R. Shukla and V. D. Sakhardande (1991) *Dyes Pigm.*, **17**, 101.

Index

acid dyes *see also* dye solubility values;
  lyotropic liquid crystals;
  photodegradation of dyes
 C.I. acid orange 7 89
 C.I. acid red 1 83
 C.I. acid red 151 95
 C.I. acid red 266 99
  deuterium spectra 102
  optical micrograph 100
  x-ray diffraction patterns 103
 C.I. food yellow 3
  deuterium spectra 101
  effect of urea on aggregation 104
  optical micrograph 100
  order parameters 102
  phase diagram 101
  x-ray diffraction patterns 102
 edicol sunset yellow 68, 69
 general description 83
 orange 2 (C.I. acid orange 10) 68, 69
acidities
 low acidity
  defined 8
 moderate acidity
  definition 11
 high acidity
  definition 13
Ames test 255
anionic azo dyes
 aggregates 88–89
 C.I. acid orange 7
  interplanar separation 88
 important classes 83–84
azo coupling
 highly acidic aromatic diazo
  components 5
 limitation involving 2-amino-4-
  chloro-5-formylthiazole 31

benzidines
 dyes derived from
  mechanistic studies 265, 273
  mutagenicity testing 258
 genotoxicity
  N-acetylated derivatives 273, 275
  3,3'-dichlorobenzidine 275
  3,3'-dihydroxybenzidine 260
  removal of 260–262
  ortho-tolidine 260
Bessel functions
 disperse dyes on polyester 46

 reactive dyes on cotton 67
Bragg angle 188

chromotropic acid 264
cotton
 pore size and distribution
  methods for characterization 72–73
critical micelle concentration 84
crystal form and habit
 control of
  during pigment synthesis 137
  following pigment synthesis 137–141
 definition 107
crystal formation
 growth 138
 single crystals
  quinacridones 114
crystal habit
 agglomerates 126
 crystal shape 132–136
  effect on rheological properties 134
 crystal size and state of
  aggregation 126–131
  effect on color strength 126
  light scattering theory 127
 flocculates 126
 morphology 125–126
 packing 129
 primary particles 126
crystal lattice properties
 crystallinity 125
  techniques for assessment 125
 dpp pigment molecules 119
 isomorphism 123–125
  C.I. pigment blue 15 124
  C.I. pigment yellow 3 analogs 124
  diarylides 124
 molecule versus crystal 112–121
 polymorphism 121–123
  C.I. pigment red 3 123
  C.I. pigment yellow 17 123
  copper phthalocyanine 123
  quinacridone 123
  solid solutions 124–125
  x-ray diffraction analysis 122

4,4'-diaminoazobenzene 262
4,4'-diaminobenzanilide 262
diazotization 1
 2-amino-4-chloro-5-formylthiazole
  in highly acid media 31

diazotization *contd*
  one-stage process 40–41
  protonation steps 35
2-amino-4-chloro-5-formyl-3-thiophene-carbonitrile
  competitive reactions 34, 36
  maximum yield 37
2-aminothiazole
  byproducts of 34
  characterization of diazonium ion 29
  nmr data 33
  protonation 29
  rate and selectivity 29–30
analytical methods for monitoring 24–25
and azo coupling
  one-stage process 40–41
  two-stage process 35–40
contact zone 41
2-bromo-6-cyano-4-nitroaniline
  competitive acid-catalyzed hydrolysis 27–29
  maximum yield of diazonium ion 31
2-bromo-4,6-dinitroaniline
  half-life time 27
  kinetics 26–29
  nmr data 28–29
  uv–visible data 28
2-cyano-4-nitroaniline
  kinetics 35–40
  maximum yield 39
cyano-substitued aromatic amines 21–23, 35
  competitive acid-catalyzed hydrolysis 21–23, 35–40
  kinetics 21–22
  rate constants 22
  reactions in highly acid media 22–23
2,6-dichloro-4-nitroaniline
  kinetics 25–27
  nmr data 25, 27
  uv–visible data 26
heteroarmatic amines
  4-aminopyridine 17–19
  5-amino-3-R-1,2,4-thiadiazole 19
  2-aminothiazole 20
in highly acidic media 13, 22
kinetics and mechanisms 1
  dimensionless quantities 38–39
mechanistic pathways 12, 19
multistage 8
rate constants 7, 21–22, 36–38
rate curves 26–29
rate equations 9–13, 15–16, 18
limiting step 10
  proton transfer 15
reversibility
  4-chloro-3-cyano-5-formyl-2-thiophenediazonium ion 19

5-phenyl-1,3,4-thiadiazole-2-diazonium ion 19
under industrial reaction conditions 23–41
  influence of reactor 40
  methods for optimizing procedures 34–41
weakly basic aromatic amines 4
  decomposition of diazotized form 6–7
diazotization rates
  carbocyclic aromatic amines 8–17
    high acidities 13–17
    low acidities 8–11
    moderate acidities 11–13
  dependence on acidity 7–23
  heterocyclic aromatic amines 17–20
    4-aminopyridine 17–19
    5-amino-3-R-1,2,4-thiadiazole 19
    difference from carbocyclic aromatic amines 19
  4-nitroaniline and derivatives 17
  rate-acidity profile 16
  acidity constants 16
diffusion coefficient 47, 49
  for C.I. disperse violet 1 46–47
  for reactive dyes on cotton 69
diffusion layer theory 5 48
diffusional boundary layer 47, 54
  diffusion through 53
direct dyes
  C.I. direct brown 202 96
  C.I. direct red 2 90
  C.I. Direct Red 28 (congo red) 82
  general description 83
  non-genotoxic, for ink-jet printing 262
disperse dye particle size
  influence on thermodynamic solubility 47
  rate of dissolution 47–49
    factors affecting 49–50
  saturation solubility 47
disperse dyes *see also* dye solubility values; dyeing with; photodegradation of dyes
  blue monoazo
    examples 3
    synthetic route 2
  C.I. disperse blue 56 48
  C.I. disperse orange 3
    diffusion coefficient 57
  C.I. disperse red 15
    activation energy for diffusion 56–57
  C.I. disperse red 60 48
  diffusion coefficients
    effects of free volume 58
    relationship to molecular volume 58
    typical values 56–57
  factors affecting dissolution 49–50

INDEX

crystal polymorphism 49
thermodynamic parameters 50
dyeing cellulose with fiber reactive dyes
  activation energy
    for fiber diffusion 76
    for fixation 76
  adaptation of Fick's second law 63
  associated boundary
    conditions 63–64
  competitive hydrolysis reaction
    associated rate equations 64
    schematic representation 62
  fiber diffusion
    activation energy 76
    key factors 70–73
    utility of available models 73
    water swollen pore 71
    with simultaneous chemical
      reaction 73–79
  mechanism
    adsorption immobilization 78
    cellophane as a model 74–75
    ionization of cellulose 61
    nucleophilic substitution 76
  transitional kinetics
    dependence of half-time of
      dyeing 55
  uptake of dye from bulk solution
    acid dyes as models 68
    microscopic diffusional boundary
      layer 65
    interfacial flux 68
    internal fiber interface 65
    into surface pore structure 67
    rate equation 69
    representative reactive dyes 67
    surface induced association 70
dyeing processes
  mathematical treatments
    associated equations 46
    for a finite dyebath 46
    for an infinite dyebath 46
    origin 46
dyeing with disperse dyes
  batchwise dyeing of polyester 48
  carriers
    mode of action 58
    diffusion within the fibre 56–59
      Arrhenius equation 56
      descriptive models 58
      Fick's first law 56
      volume fraction 57
    finite *vs* infinite kinetics 50–52
      determining factors 51
      general description 51
      transitional kinetic regime 52
  heterogeneous kinetics
    dissolution process 47
    dyeing polyester 44–61

flow cell used 66
leveling
  description of process 59
  driving force 59
  relation to partition coefficient 61
  schematic representation 60
  transport to fiber surface 52–56
    rate equation 52
    equilibrium partition coefficient 53
dye solubility
  computational methods 184–185
    activity coefficients of mixtures 184
    utility and limitations 185
  empirical/estimation
    experimental methods 192–193
  influence of particle size and form 148
  measured solubilities 150–184
    in organic solvents 180–184
    in water 151, 179–180
    influence of non-surface-active
      solutes 179
    miscellaneous contributing
      factors 180
  measurement methods 149–150
    choice of filter materials 149
    C.I. disperse blue 79 150
    C.I. disperse red 121 149
    Reactive Bright Red 6S 150
    Reactive Violet 4K 150
  regression analysis 190–192
  solubility parameters 187–189
    applications 189–190
    cohesive energy 187
    free energy of mixing 188
    heat of mixing 187
  solubility values 152–163
    acid dyes 182, 183
    basic dyes 154, 155, 161
    disperse dyes 152–153, 156
    miscellaneous
      anthraquinones 152–153
    miscellaneous azo dyes 156–160
    reactive dyes 182
    solvent dyes 156, 160
    vat dyes 162
dyestuff aggregation
  acid dyes *see also* acid dyes, C.I. food
    yellow 3
    low molecular weight dyes 88
    sulfonic acid groups 88
  aggregate stacks
    intermolecular spacing 85
  direct dyes
    aggregation numbers 90
  governing factors 84–86
    electrostatic repulsive forces 86
    hydrogen bonding 85
    pi–pi interactions 85
    Van der Waals forces 85–86

dyestuff aggregation *contd*
  hydrophobic effects
    alkyl chains in anionic dyes   89
    micelle formation   84
  reactive dyes
    effect on properties   90
    x-ray analysis   90
  role in textile dyeing   103-104
  techniques for measuring
    associated difficulties   86-88
    diffusion methods   86-87
    nmr   87
    small angle x-ray scattering (saxs)   87
    visible spectrophotometry   87
    wide angle x-ray scattering
      (waxs)   88

enthalpy changes   85
environmental issues
  fate of azo disperse dyes
    C.I. disperse blue 79   284
    metallized dyes   284-288
      free *vs* bound metal   286-287
    natural dyes   286
  removal of dyes from wastewater
    physical methods   284
  removal of free metals from
    wastewater   288
  sediments
    reducing property   284
  singlet oxygen sensitization   285
  toxicity of metal ions   284-286

Fick's second law of diffusion   46, 63
flavin mononucleotide (FMN)   257, 258
functional dyes
  photolytic behavior   237-239, 242
  properties and uses   137-138
  structures   240-241

genotoxicity   *see also* benzidines
  aromatic amines
    anilines and related amines   264
    benzidines and related diamines   260
    4,4'-diaminoazobenzene   262
    4,4'-diaminobenzanilide   262
    N,N-dimethyl-1,4-
      dimethylaminobenzene   264
  assessment by instrumental methods
    electron microscopy   270
    esr   267
  biophores
    defined   282
    C.I. acid black 52   264
    C.I. acid yellow 151   276
    C.I. d&c red 9   282
    C.I. d&c yellow 6   282
    C.I. direct black 17   275
    C.I. direct black 19   276

C.I. direct black 38   273, 276
C.I. direct blue 15   273, 276
C.I. direct blue 218   276
C.I. direct orange 6   265
C.I. direct red 2   273
C.I. disperse yellow 3   282
C.I. pigment yellow 12   273, 279
C.I. solvent yellow 2   267, 273, 278
C.I. solvent yellow 3   269
C.I. solvent yellow 14   282
congo red   273, 282
defined   254
gurr chrysoidine y   281
in vitro   255
in vivo   255
mechanisms and metabolism
  DNA interactions   270-272
  disazo dyes   273-275
  endoplasmic reticulum (ER)   270
    disaggregation
  flavin-induced activation   269
  N-hydroxylation   267-270
  metabolic activation   265, 268, 271
  monoazo dyes   267-273
  nitroxide radical formation   268
  reductive-cleavage   269, 272, 273-274
  t-RNA interactions   272-273
methyl orange   264
mutagenicity
  pathways leading to   256
  schematic diagram of the assay   256
phenylenediamines
  removal of   265
structure-activity-relationships
  computer-assisted   280-283
  empirical   275-280
test methods for assessing
  Fusobacterium sp. 2   258-259
  hamster liver hepatocytes   259
  intestinal anaerobes   258-259
  Prival modification   257-258
  rat cecal flora   258
  Salmonella mutagenicity
    assay   255-257
glucopyranose structure   62
glucose 6-phosphate dehydrogenase
  (NADH)   257

h-acid   264
Hammett acidity function   8
heterogeneous dyeing process
  schematic representation   45
hydrogen bonding
  organic pigments   114, 121

interplanar spacing
  dpp molecules   116

Jablonski diagram 198

lightfastness
　physical factors influencing 202–203
lyotropic liquid crystal formation
　definition 91
　measurement techniques
　　nmr spectroscopy 92–93
　　optical microscopy 91–92, 100
　　x-ray diffraction 93–94
　principal categories 91
　stable mesophases
　　description 91–92
　techniques for 91
lyotropic liquid crystals
　chromonic 94
　　C.I. acid red 266 98
　　C.I. food yellow 3 98
　nematic and middle mesophases
　　C.I. food yellow 3 99
　　herringbone array 96, 97
　p-type systems
　　C.I. acid red 266 99–100
　　C.I. food yellow 3 100
　　C.I. reactive red 3:1 98
　　orthorhombic symmetry 98
　quasi-crystalline form 97
　semetic-like mesophases
　　acid dyes 95
　　direct dyes 95–96

metabolic activation 255, 265, 268, 272
micelle formation
　disperse dye solubility 48–49
　impact of disperse dye uptake 50
mutagenic activity
　quantitative measure 255
mutagens
　direct acting 256
　promutagens 255, 271
mutations
　reverse 255
　types 255

nitrosating electrophiles
　diazoanhydrides 4, 6
　dinitrogen trioxide
　　mechanism 9
　k_{exp}-expressions 10–11
　nitrosonium ion 1
　　mechanism 10
　^{15}N nmr signal 14
nitrosyl sulphuric acid
　decomposition of 5
　^{15}N nmr signal 14–15
　reactions of amides in 23
nuclear magnetic resonance (nmr)
　analysis of liquid crystals 93, 101–102
　carbon (^{13}C) spectra

　　diazotized amines 13
　Fermi-contact shift 15
　fundamental principles
　　dipole–dipole interactions 92
　　order parameter 93
　　quadrapole interactions 93
　　Zeeman interaction 92
　^{23}Na spectra
　　C.I. food yellow 3 102
　nitrogen (^{15}N) spectra
　　nitrosyl sulphuric acid 13–14
　　nitrous acid–nitrosonium ion 15
　proton (^{1}H) spectra
　　diazotized amines 13, 25, 27
Nusselt number 49

organic pigments
　aggregates 112
　anthraquinoid
　　C.I. pigment blue 60 110
　　C.I. pigment red 177 110
　azo
　　C.I. pigment orange 36 110
　　C.I. pigment red 144 110
　　C.I. pigment yellow 183 110
　azomethine
　　C.I. pigment yellow 110 110
　　C.I. pigment yellow 139 110
　characteristics 108
　C.I. pigment blue 15 123–124
　C.I. pigment blue 15:3 132
　　TEM 133
　C.I. pigment blue 60 110
　C.I. pigment orange 36 110
　　TEM 133
　C.I. pigment orange 43 134
　C.I. pigment orange 65
　　TEM 142
　C.I. pigment red 1 132
　C.I. pigment red 3 123, 132
　C.I. pigment red 31 125
　C.I. pigment red 57:1 132
　　TEM 133
　C.I. pigment red 112 132
　C.I. pigment red 144 110, 124
　　TEM 133
　C.I. pigment red 177 110
　C.I. pigment red 179 110
　C.I. pigment red 214 110
　C.I. pigment red 224 110
　C.I. pigment red 254 110
　　TEM 133
　C.I. pigment violet 19 110
　C.I. pigment violet 23 110
　C.I. pigment yellow 1 132
　C.I. pigment yellow 12 124
　C.I. pigment yellow 13 127, 132
　C.I. pigment yellow 14 124
　C.I. pigment yellow 17 123

organic pigments *contd*
  C.I. pigment yellow 24   130–131
  C.I. pigment yellow 63   124
  C.I. pigment yellow 74   132
  C.I. pigment yellow 110   110
  C.I. pigment yellow 139   110
  C.I. pigment yellow 183   110
  classical
    definition and examples   109–111
    utility   110
  copper phthalocyanine
    beta-modification   114
    crystal mixtures   130
    pleochroic behavior   120
    scanning electron micrographs   113, 115
  crystal formation
    driving forces   114
  crystal shape   132–136
  diketopyrrolopyrrole
    C.I. pigment red 254   111
    polarized reflection spectra   121
    visible absorption spectra   120
    SEM photographs   117
    TEM photographs   139
    TGA measurements   122
    x-ray crystal structure   118, 119
    x-ray diffraction diagrams   116
  dioxazine
    C.I. pigment violet 23   110
  dispersibility   129
  high performance
    definition and examples   110, 111
    utility   110
  molecular structure
    influence on crystal properties   110–111
  naphthol   119
  performance and selection criteria   106–107
  perylene
    C.I. pigment red 179   110
    C.I. pigment red 224   110
  quinacridone
    C.I. pigment violet 19   110
    structural features   108
    utility   107, 141–142
Ostwald ripening
  copper phthalocyanine pigment   130
  dpp pigment   138
Ostwald's law   114
oxygen molecule
  electronic states   202

photochemistry
  basic principles   197–202
    electron transitions   197–198
    potential energy diagram   199

  fluorescence and phosphorescence   199
  photo–redox reactions   200
  photosensitization   199–200
  singlet oxygen involvement   201–202, 205–208, 210, 215–218, 220–224, 229–230, 236, 239
photodegradation of dyes
  arylazonaphthols   207–208
  arlyazopyrazolones   208–209
  C.I. acid blue 40   233
  C.I. acid blue 277   214
  C.I. acid green 25   215, 226, 234
  C.I. acid orange 60   205, 234
  C.I. acid orange 156   234
  C.I. acid red 143   226
  C.I. acid red 361   234
  C.I. acid violet 47   226
  C.I. acid yellow 29   226
  C.I. acid yellow 127   226
  C.I. basic violet 3 (crystal violet)   217–218, 220, 239
  C.I. disperse blue 14   227
  C.I. disperse blue 56   226
  C.I. disperse red 15   227
  C.I. disperse yellow 50   226
  C.I. vat green 1   19
  C.I. vat orange 9   216, 230
  C.I. vat yellow 2   230
  C.I. vat yellow 26   230
  catalytic fading
    acid dyes   226
    disperse dyes   226–228
    quinophthalone   226
  malachite green   218
  mechanisms
    anthraquinone dyes   212–217
    azo dyes   204–212
    azomethine dyes   225
    fluorescent brightening agents   223–224
    indigoid dyes   220–221
    quinophthalone dyes   221–222
    rhodamine dyes   224
    triphenylmethane dyes   217–220
  mono-substituted azobenzenes
    cis–trans isomerization   211–212
    ortho-nitrosubstituted azo dyes   209, 211
photodegradation of dyed polymers
  natural fibers
    phototendering of cotton   228–232
    yellowing of wool   229
  polyamide fibers   232–234
  polyester fibers   235
  polypropylene fibers   235–236
  titanium oxide-sensitized   236
photostabilization of dye polymers   232–233, 236–237

pi–pi interactions 108, 114, 118
prototropic equilibrium 20

reactive dyes
  C.I. reactive blue 4 67, 79
  C.I. reactive blue 19
    x-ray scattering data 90
  C.I. reactive orange 4 67, 79
  fixation
    interfacial kinetics 79
    fixation to cellulose 84
    general description 84
    hydrolysis 104
reductive-cleavage
  chemically-induced 258
regular solution theory
  basic premise 187
revertant colonies 255

Salmonella typhimurium
  bacterial strains 255
saturation solubility 48
Scherrer–Bragg approximation 88

solvent mapping 140–141
solubility parameters
  for commmon solvents 141
spontaneous revertants 256
specific free energy 49
Stokes–Einstein equation 87

TEM (transmission electron micrograph)
  pictures
    organic pigments 133, 135, 139, 142
tortuosity 58

uv/visible spectroscopy
  charge-transfer bands 13
  diazotized amines 26
  nitrosyl sulfuric acid solution 13

x-ray diffraction *see also* acid dyes, C.I. acid red 266 and C. I. food yellow 3
  diffraction patterns
    for lyomesophase structures 93–94
    of a middle phase 94
  utility 129